$32.00

(Amazon.com price is $88.00)

Stereolithography and other RP&M Technologies

Stereolithography and other RP&M Technologies

from Rapid Prototyping to Rapid Tooling

Paul F. Jacobs, PhD

Society of Manufacturing Engineers
in cooperation with the
Rapid Prototyping Association of SME
Dearborn, Michigan

ASME Press
American Society of Mechanical Engineers
New York, New York

987654321

Library of Congress Catalog Card Number: 95-070369
International Standard Book Number: 0-87263-467-1

Additional copies may be obtained by contacting:

Society of Manufacturing Engineers
Customer Service
One SME Drive
Dearborn, MI 48121
1-800-733-4763

American Society of Mechanical Engineers
22 Law Drive
Fairfield, NJ 07007
1-800-THE-ASME

SME staff who participated in producing this book:

Larry Binstock, Senior Editor
Rosemary Csizmadia, Operations Administrator
Dorothy Wylo, Production Assistant
Sandra Suggs, Editorial Assistant
Sonya Cashner, Cover Design

Printed in the United States of America

This book is dedicated to

STARLA KAY JACOBS
My Wife, and Companion for Life

Through beauty, elegance, grace, wit, humor, and semi-infinite patience, she has made this book possible. Furthermore, while I was attempting to coordinate the efforts of 33 authors residing in five different countries, she would often remind me, with a twinkle in her eye, that whether the task is writing a book, doing research in the lab, or painting the house, "if it's not one thing, it's another."

Paul Francis Jacobs
La Crescenta, California

Table of Contents

Preface

When *Rapid Prototyping & Manufacturing: Fundamentals of StereoLithography* was first published in July 1992, there were roughly 300 RP&M systems installed worldwide. As this volume goes to press, there are about 1,000 RP&M systems operating in over 40 countries on six continents. This corresponds to a compound annual growth rate of almost 50% during this period.

Recognizing that the RP&M field is moving ahead at a remarkable pace, SME believed that a new book was appropriate. This second volume is intended to detail the advances that have occurred during the past three years, as well as the expanding range of practical applications enabled by these emerging technologies. The initial volume quite naturally focused on the fundamental science underlying the predominant method in use at that time. This volume, however, explores a number of important and very pragmatic advances and applications involving the use of RP&M-generated patterns in shell investment casting, sand casting, soft tooling, and hard tooling. Simply stated, the goal of the first book was to help readers understand how the technology worked, while this book presents an update on the state-of-the-technology, and what these systems can do for the user in terms of reducing cost, speeding time to market, and improving product quality.

In the spring of 1994, as the original outline of the book was taking shape, the intent was to have key personnel at each North American supplier of commercially available RP&M systems write at least one chapter about their particular technology. Those who provided information have been included. Significantly, Helisys, Inc. contributed an excellent chapter on the recent advances and applications of laminated object manufacturing.

The reader will note a wide array of important topics, ranging from hardware, software, materials, and process advances during the past three years, to the development of entirely new methods of generating patterns for direct shell investment casting. Other chapters describe foundry perspectives on recent RP&M developments, the use of RP&M at Texas Instruments, Inc., Sandia National Laboratories, Ford Motor Company, and some novel applications involving flow visualization, photoelastic stress testing, dynamic vibration analysis, and surgical preplanning in the medical field. Furthermore, detailed step-by-step procedures are included to assist users in generating soft and hard tooling. Finally, for those readers interested in exploring the benefits of RP&M, but unable for whatever reason to purchase a system, there is a chapter on RP&M service bureaus, including an extensive directory of those companies worldwide that currently provide CAD modeling, rapid prototyping, and/or various secondary processes on either a part-by-part or project-by-project basis.

While the name of the principal author/technical editor may appear on the cover, it is important to state that this book is the product of the experience, technical knowledge, enthusiasm, dedication, and hard work of 33 authors. It has been a remarkable experience to work with them over the past year. Collectively, we all hope the reader will find the contents informative and valuable.

Paul F. Jacobs

CHAPTER 1

Introduction

To see the world in a grain of sand
and heaven in a wild flower,
Hold infinity in the palm of your hand
and eternity in an hour.

William Blake

1.1 Physical Prototypes

"Let's get physical," sang Olivia Newton-John, never realizing the importance of her message for today's competitive product arena. In the words of Michael Schrage of the MIT Sloan School of Management, "For companies that genuinely care about incremental and breakthrough innovations, organizational redesign and core process re-engineering are not enough. *Companies that want to build better products must learn how to build better physical prototypes.*"[1]

Physical models and prototypes are fundamental to superior product development and production, despite the rising interest in computer simulations of 3-dimensional objects, and virtual reality. In a recent address at the IMS International Conference on Rapid Product Development, Daimler-Benz's Dr. Werner Pollmann said, "Purchase of a car depends strongly on subjective impressions. Next to technical properties like horsepower or security equipment, properties like noise, handling, or styling are key factors for a purchase decision. But these properties can only be evaluated by physical prototypes. For that reason, availability of high quality functional prototypes will remain an important element of product development and cannot be substituted by digital models and analysis."[2]

1.2 Adoption, Diffusion, and Infusion of RP&M

The benefits of physical prototypes probably explain the 35% compound annual growth in the number of installed rapid prototyping and manufacturing

---

By Richard P. Fedchenko, VP, Strategy & Market Development, and Paul F. Jacobs, Ph.D., Director of Research & Development, 3D Systems Corporation, Valencia, California.

(RP&M) systems since the publication of the previous volume on rapid prototyping and manufacturing in July 1992.[3] From a total of roughly 300 systems at that time, the world population of RP&M machines grew to about 750 by June 1994, as shown in Table 1-1.[4] An RP&M system is installed around the world almost every business day. By the time this book is published, the total number of operational RP&M systems will probably exceed 1,000.

Table 1-1. RP&M System Installations as of June 1994

RP&M Technology	Vendor	Installations
1. Stereolithography	3D Systems	469
	CMET/Mitsubishi	43
	DMEC/Sony	28
	EOS GmbH	19
	Teijin-Seiki	3
	Mitsui Engineering	1
	Subtotal:	563
2. Laminated Object Manufacturing	Helisys	52
	Sparx	15
	Kira	10
	Subtotal:	77
3. Selective Laser Sintering	DTM	43
	EOS GmbH	4
	Subtotal:	47
4. Fused Deposition Modeling	Stratasys	44
	Sanders	1
	Subtotal:	45
5. Solid Ground Curing	Cubital	18
	Subtotal:	18
	Total:	**750**

There are a number of techniques currently under development, such as Ballistic Particle Deposition (BPM Corporation and Incre, Inc.), Shape Melting (Babcock & Wilcox Co.), MD or selective spray metal deposition (Carnegie-Mellon University), and Three-Dimensional Printing (MIT/Soligen). These approaches have been described in the literature,[5-7] but are either in beta test or are not yet commercially available.

Growing User List. New entries are being added to the list of adopting industries at an accelerating rate. Some of the latest organizations to join the ranks of RP&M users are as diverse as a leader in the toy industry, U.S. Army-Watervliet Arsenal, a gardening equipment firm, the FBI, a supplier of medical diagnostic systems, an investment casting foundry, NASA Manned Spacecraft Center, a major producer of plastic injection molding dies, and a manufacturer of vacuum cleaners. The spread of a successful new technology into the broader marketplace as the technology becomes more routine is called *diffusion*, and diffusion of RP&M is occurring at an increasing rate.

For example, RP&M is not only being used by virtually all the world's leading automobile manufacturers, but now even the emerging free market automotive industry spawned by the breakup of the former Soviet Union has moved into this technology through acquisition of several stereolithography apparatus (SLAs) in both Russia and Byelorussia. Diffusion has also been accelerating in such industries as aerospace, medical devices, and consumer products.

An encouraging sign regarding the health of the RP&M industry is the growing trend toward multiple purchases, single companies acquiring two or more machines at the same time. Although not yet the rule, it is no longer rare to find such multiple purchases. This represents a type of internal diffusion as firms establish a growing utilization of RP&M. In some cases, this has resulted in more than a dozen machines in the same company.

Using the Range of Benefits. The real payoff occurs when firms utilize the full range of RP&M benefits, taking advantage of all applications that can contribute to competitiveness and profitability. The process of developing applications of a new technology to their maximum potential is called *infusion*, and signs of high RP&M infusion are appearing with increasing regularity. For example, at Ford Motor Company, RP&M is being used to build models for a multiplicity of purposes. First, designers and engineers can hold examples of their concepts for early *visualization*, *verification*, *iteration*, and *optimization*. Second, the models can serve as a communication tool for simultaneous engineering. Third, they are valuable for form and fit tests of components. Fourth, they can serve as test samples for marketing studies of consumer design preferences. Fifth, RP&M parts can help with production planning to determine the need for tools, jigs, and fixtures. Sixth, they complement bid packages for tooling quotations. Seventh, they are useful in dunnage design, to hold component parts shipped to another site for assembly. Eighth, metal prototypes are being made with QuickCast™, Laminated Object Manufacturing™ (LOM), Selective Laser Sintering (SLS™), and Fused Deposition Modeling™ (FDM) technologies, to be used for functional testing. And, ninth, in a very exciting recent development, Ford engineers are using QuickCast tooling to fabricate core and cavity tooling pairs directly from SLA masters, using QuickCast technology and subsequent shell investment casting with tool steels.[8] The resulting tooling can then be used to injection mold numerous prototypes in the desired end-use materials. Tools of this type may ultimately be used for the production of injection molded components.

These efforts represent infusion of RP&M to the highest degree. Ford is squeezing every ounce of benefit possible from their investment in RP&M. Furthermore, additional applications in design, development, engineering, tooling, and manufacturing are still evolving. Within a few years, it is quite likely that continued infusion will result in a whole range of RP&M applications not even mentioned in this book.

1.3 RP&M Systems

The basic principles of operation for each of the leading commercially available RP&M systems as of July 1992 were described in some detail in the previous RP&M book.[3] However, in the nearly three-year period between the publication of that book and this one, considerable progress has been made by each system vendor. Therefore, in the interest of both timeliness and completeness, the essential characteristics of each RP&M technology listed in Table 1-1 will be described in the order listed.

Stereolithography

In 1987 an entirely new application of computer aided design (CAD) was developed. Stereolithography, or 3-dimensional printing, was the first commercially available layer-additive process to enable the rapid generation of physical objects directly from a CAD data base.[9] The stereolithography (SL) process begins with the generation of a CAD model of the desired object. Solid CAD models make the process easier, but surface CAD models that are "watertight" have also been used successfully.

Next, the boundary surfaces of the CAD description are tessellated—formed as a connected array of triangles. This step is performed using an interface specification developed by 3D Systems Corporation, Valencia, California, the company that pioneered the RP&M industry.

The triangles may be as large or as small as desired. Smaller triangles result in finer resolution of curved surfaces and improved part accuracy through reduced chordal deviations, while larger triangles minimize the system storage requirements, at the expense of accuracy. As work stations attain higher processing speeds with greater memory and data storage capacity, file size limitations have become less problematic.

The STL File. The software generates a tessellated object description known as an STL or **ST**ereo**L**ithography file. This file basically consists of the X, Y, and Z coordinates of the three vertices of each surface triangle, as well as an index that describes the orientation of the surface normal. The latter feature is necessary to ensure that a clear distinction is made between inner and outer surfaces. Since its inception, the STL format has become the *de facto* standard of the RP&M

field, and is now supported by over 40 CAD system suppliers, including every major CAD vendor. Furthermore, it has been adopted by all the various RP&M system suppliers as the primary, if not the sole, basis of their CAD interface.

Although other software approaches have been developed, including various contour methods, the STL format continues to be the most widely used. It is conceptually simple, topologically robust and, when used with sufficient resolution, is capable of high accuracy. It is important to note that errors introduced through the use of the STL format can be made arbitrarily small relative to other system-related sources of part error.

Slicing the File. Once the STL file has been generated from the original CAD description of an object, the next step involves slicing that file. Here, a series of closely spaced horizontal planes, like the floors of a tall building, are mathematically passed through the tessellated object file. The result is an SLI or **SLI**ce file, which represents a series of closely spaced 2-dimensional cross-sections of the 3-dimensional object, each at a slightly different Z coordinate value.

In the early days of SL, the most common slice thickness was 0.020 inch (0.5 mm). As SL achieved finer precision and improved accuracy, slice thickness values have decreased to the point where 0.004-inch (100-micron) layers (approximately equal to the thickness of a dollar bill) are now commonplace. Values as low as 0.002 inch (50 microns) (roughly the diameter of a human hair) have recently been demonstrated in the laboratory. This is important because finite layer thickness causes stair-stepping errors in the Z axis, as shown in Figure 1-1. *All current RP&M systems exhibit this characteristic.* As layer thickness is decreased, the object generated can more faithfully approach the design intent, and surface quality is correspondingly improved.

The Final Build File. After the STL files have been sliced to create SLI files, multiple SLI files are merged into a final build file. One can now begin the process of generating an actual physical object. The build process starts by fabricating a series of supports. These are necessary for a number of reasons. Supports act like fixtures in conventional machining. Simply stated, they hold the object in place during the build process. They also provide a means of securing isolated segments or islands that would otherwise float away. Supports are also used to restrain certain geometries (specifically free cantilevers) that might undergo distortion as the photopolymer resin is laser cured.

This distortion, known as curl, was formerly a significant source of SL part error. With the recent development of very low shrinkage epoxy resins by Ciba-Geigy and vinylether resins by Allied-Signal, curl restraint is now much less problematic. Nonetheless, manual support generation is time consuming and support removal can be tedious, especially on complex geometries. Fortunately, automatic support generation software is now available. Two such programs that have met with widespread user approval are Bridgeworks from Solid Concepts, Valencia, California, and MAGICS from Materialise, Heverlee, Belgium.

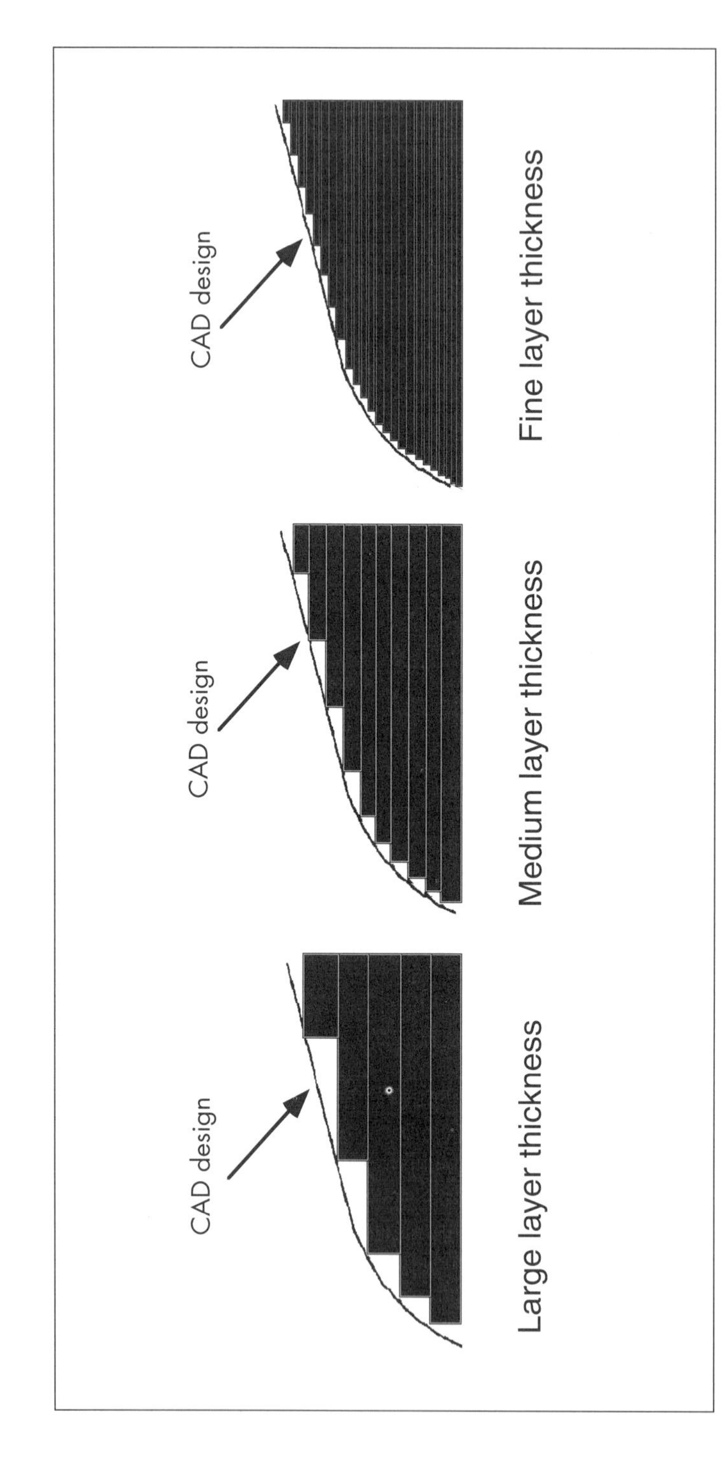

Figure 1-1. Effects of stair-stepping as a function of layer thickness.

The automated laser drawing process, which is fundamental to both support and part building, involves a series of commands from a control computer reading the build file. This information is sent to a pair of orthogonal, servo-controlled, galvanometer-driven mirrors. These mirrors deflect a focused ultraviolet laser beam downward onto the surface of a vat containing liquid photopolymer.

At present, 3D Systems produces four different classes of SLA. The smallest unit is the SLA-190, often used as a university research tool. The most popular system in the RP&M field by a wide margin is the SLA-250. Both these systems utilize helium-cadmium lasers emitting at 325 nm wavelength. The larger SLA-400, as well as the top of the line SLA-500, use higher power argon-ion lasers emitting at 351 and 364 nm. Detailed specifications for all these machines are available from 3D Systems Corporation.

All SLA units direct the laser beam downward, at near normal incidence, onto the free surface of the liquid photopolymer. Provided the resin surface receives sufficient actinic laser exposure, the resin will undergo a transformation from the liquid phase to the solid phase. Basically, the resin becomes locally solidified. The details of the photopolymerization process are described in numerous texts on polymer chemistry, with Reference 10 being directly relevant to the SL process.

The focused laser spot, with a diameter of about 0.008 inch (200 microns), is positioned by the scanning mirrors. First, the laser beam traces the boundaries of the particular slice cross-section being drawn. Once the borders have been completed, the laser then accomplishes the solidification of the appropriate cross-sections. If the photopolymer underwent the liquid-to-solid phase change with zero shrinkage, this step, known as hatching, would be straightforward. Unfortunately, this phase transition does involve finite volumetric, and hence linear, shrinkage. Nonetheless, if the shrinkage were perfectly uniform and no distortion took place, excellent part accuracy could still be achieved through an appropriate scaling factor when generating the build file. However, in certain geometries involving intersecting thick and thin sections, *nonuniform resin shrinkage becomes the engine of part distortion*.

Consequences of Hatching Styles. Extensive research and development over the past five years has shown that the specific hatching algorithms employed during the part building process have a significant effect on overall part accuracy.[3] Recent work is described in detail in Chapter 5, and is summarized in Figure 1-2. From this graphical perspective, it is clear that the advances in SL part accuracy over the past five years have been truly remarkable. One can see how the development and implementation of various hatching styles such as WEAVE™, STAR-WEAVE™, QuickCast, and ACES™, along with the release of Ciba-Geigy epoxy resins Cibatool SL 5170 and SL 5180, have dramatically reduced user-part RMS error.

Although the data strictly applies only to the user-part, many SL operators have indicated similar accuracy improvements for the typical objects they regu-

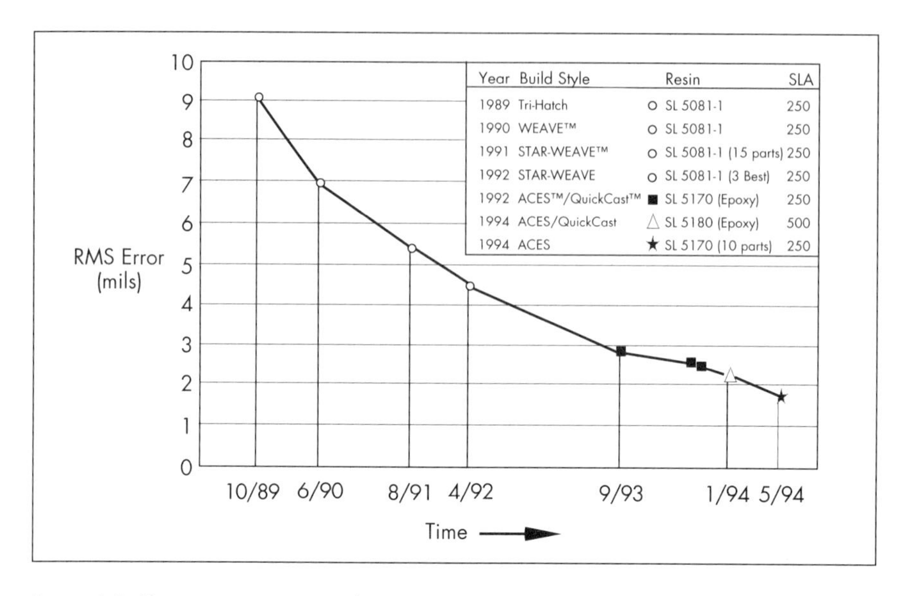

Figure 1-2. User-part accuracy timeline.

larly build.[11,12] While SL accuracy was initially poor relative to computer numerically controlled (CNC) machining, recent progress in software, hardware, build styles, and photopolymer resins have significantly changed the picture. As this book goes to press, SL is approaching CNC machining accuracy levels for complex parts.

Positioning the Platform. After the first layer is completed, the platform holding the object is lowered. The platform motion is directed by the control computer and is driven by a precision stepper motor. Liquid resin now flows over the recently polymerized layer. Excess resin is removed with a recoater blade. The platform is then positioned so that the liquid resin level is one layer thickness above the previously exposed layer. The procedure is now repeated for the next layer. Data from the build file in the computer forwards commands to the servo system, actuating the scanning mirrors. The system automatically provides the correct exposure so that all the border and hatch vectors achieve the correct cure depth to assure that the second cross-section of the object will adhere to the first. The process continues, with no human intervention required, on an additive layer-by-layer basis until the entire physical object has been generated, from the bottom to the top.

When the final layer has been completed, the platform is elevated and the solidified object literally emerges from the vat of liquid resin. After excess resin drains back into the vat, the operator takes the platform, supports and part from the SLA. Uncured residual liquid resin is cleaned off the part. Next, the supports are removed. This must be done carefully to avoid damaging the downfacing surfaces. Finally, the object is placed in a postcure apparatus (PCA), where it is flooded with ultraviolet radiation of an appropriate wavelength to effect a thorough postcure. This step is necessary to achieve full resin strength and to complete the photopolymerization or crosslinking process. After about an hour, postcuring is complete and the part may undergo a variety of finishing steps such as glass bead blasting, sanding, milling, drilling, tapping, polishing, painting, electroplating, etc.

Laminated Object Manufacturing

The Laminated Object Manufacturing (LOM) process was initially developed by Helisys, Inc. (formerly Hydronetics, Inc.), Torrance, California. This process also builds objects in thin layers, but as the name implies, LOM parts are fabricated using laminated sheet material. Consecutive layers are joined using an adhesive that is both temperature and pressure sensitive. The individual cross-sections are cut using a 25- or 50-watt carbon dioxide (CO_2) laser, emitting in the infrared, at a wavelength of 10.6 microns.

Adhesive Coated Paper. The LOM operational sequence proceeds essentially in the following manner.[5,13,14] As with SL, the first step involves generating a CAD description of the object. Next, an STL file is created. Subsequent to slicing at a thickness identical to that of the sheet material being used, the computer contains a file of all the boundary information for each lamination. The material (typically adhesive coated paper) is then introduced onto the working area in a continuous manner from a feed roller. Unused material is spooled onto a take-up roller at the other end of the sheet. A computer-driven platform, capable of being lowered as each layer is formed, acts as the working substrate. The CO_2 laser beam is then delivered to the desired location by means of an X-Y positioning system that also contains the final focusing optics for the 10.6 micron radiation. A heated roller is also located above the laminate. This roller activates the adhesive required to bond adjacent layers.

After the sheet is registered on the platform, the LOM computer then actuates the X-Y positioning unit to move the focused laser spot. The high-power infrared beam cuts the boundary of the first cross-section of the part by burning a thin strip of the working material. The depth and width of the cut depend on the laser power, scan speed, and physical properties of the material being laminated. Control of these parameters impacts overall part accuracy. To date, LOM systems do not apply a correction for the finite width of the laser cut, which results in a small but systematic dimensional error on each object cross-section.

Block of Laminated Material. The entire surface of the laminate is coated with adhesive. Thus, every portion of a given layer adheres to the material below. A beneficial aspect of this approach is that the surrounding material forms a natural support for cantilever sections or isolated islands. A negative aspect of this technique is that the surrounding volume must be diced to enable it to be broken away from the part upon completion. At the end of the LOM process, an entire block of laminated material is produced, and the user must then remove all the unwanted surrounding material to obtain the final part. For large solid parts, this is generally not a problem. However, removing laminated material from complex interior passages or finely detailed geometries can be difficult.

A major advantage of the LOM technique is that only the boundaries of the object cross-section need to be drawn (other than dicing regions exterior to the part, as discussed earlier). Thus, no interior hatching is required, which definitely speeds the build process. Although the CO_2 laser can cut vertically *through* sheet material, it cannot cut horizontally *along* the junction *between* layers. As a result, all intended upfacing and downfacing surfaces on a part will adhere to the block of surrounding material. This adhesion can be minimized by cutting the mating portion of the unwanted laminate with a closely spaced cross-hatch pattern. This process locally reduces the strength of the adjacent layer, enabling easier separation of the part from the block.

Looks Like Wood. LOM parts made from paper resemble wood. They even look, feel, and smell like wood. Thus, it is perhaps natural that a primary LOM application involves patterns for sand casting, as many sand casting patterns have traditionally been made of wood. Since sand casting tolerances are relatively loose, pattern accuracy is not especially critical. Furthermore, sand castings rarely involve delicate features, extremely thin sections, or very fine detail, while models containing these characteristics can be problematic for the LOM method.

The materials used in the LOM technique are low cost (adhesive coated paper) and do not require special handling. Also scrap material is easily discarded. However, a focused, 50-watt CO_2 laser is invisible to the human eye and capable of serious tissue damage. Consequently, good laser safety practice must be carefully followed. Also, when the laser is incident on a sheet of paper laminated to numerous other layers of paper, the large thermal mass is capable of dissipating the absorbed laser energy. However, if the beam is incident on a single sheet, the energy deposition can be sufficient to locally burn the paper beyond the desired part cross-section.

Finally, despite a search of the relevant literature, including the proceedings of numerous RP&M conferences held during the past four years, no statistically significant accuracy results have been published with respect to the LOM process. Therefore, reliable statements regarding LOM part accuracy and repeatability are not possible at this time.

Selective Laser Sintering

The Selective Laser Sintering (SLS) process, originally developed at the University of Texas in Austin, is now sold by DTM Corporation, also of Austin. The SLS technique is based on the selective fusing or sintering of small particles by means of a high power, 50-watt CO_2 laser.[5,15,16] As with stereolithography, SLS is also a layer-additive procedure for generating physical objects from a CAD data base. The Sinterstation 2000 accepts STL files, which are sliced, generating the usual series of cross-sections at closely spaced *Z* coordinate values. A quantity of powdered material is moved upward by a piston within a feed cylinder, and is spread over the working area by means of a counter-rotating roller. The result is a thin layer of fine powder deposited over the working area. Excess powder is moved onto a second feed cylinder and is later used to form the next layer.

A 50-watt CO_2 laser beam, emitting infrared radiation at 10.6 microns, is then directed by a pair of orthogonal mirrors, and focused with appropriate optics, onto the working surface. Depending on the specific powder material and its infrared spectral absorption characteristics, some portion of the directed laser energy is absorbed and transformed into heat. To minimize the required laser output, the powder is maintained at an elevated temperature, just below its fusing point.[5] Further, to avoid oxygen contamination of the bonding surfaces, as well as a potential combustion/explosion hazard involving numerous small powder particles with large surface-to-volume ratios at elevated temperatures, the process chamber of the machine is operated in an inert nitrogen environment. The oxygen concentration in the chamber is best maintained below 2%.

A Quasi-solid Material. The pattern for a given cross-section is drawn by the scanned laser beam. For amorphous materials, such as polyvinyl chloride (PVC) or polycarbonate, the absorbed laser energy causes the powder particles to soften and bind to one another at their points of contact, forming a porous quasi-solid material. As an example, SLS parts built from PVC powder typically have about 60% of the density of molded PVC. For polycarbonate powder, the corresponding value is roughly 85%. On the other hand, crystalline materials, such as nylon or wax, are locally melted in the SLS process and can produce essentially fully dense parts.

Unfortunately, the volumetric shrinkage for crystalline materials is quite large, leading to the potential for significant distortion in complex geometries. This leads to a trade-off between diminished mechanical properties resulting from part porosity, and increased dimensional errors due to excessive shrinkage.

The cross-section of an SLS part is scanned in a raster fashion. While all RP&M systems are subject to stair-stepping errors, raster scanning also produces similar X-Y errors (also known as horizontal aliasing). Thus, a diagonal line will actually be drawn in a zig-zag fashion typical of dot matrix printers. This characteristic contributes to the rough surface finish of SLS parts.

Another source of surface roughness is the intrinsically discontinuous nature of the powder working material. The particles used in the SLS process are approximately spherical and range in diameter from about 0.002 to 0.005 inch (50 to 125 microns). The amorphous materials, as noted earlier, tend to fuse at their respective contact points, leaving significant interstices, which become the dominant contributor to surface roughness.

Wide Range of Materials. An advantage of the SLS process is the ability to sinter a wide range of materials. To date, PVC, polycarbonate, nylon, and wax powders have been used. In principle, this would afford the designer or engineer an opportunity to generate prototypes in the desired end-use material. Unfortunately, from Reference 17, the mechanical properties of sintered powders typically scale as the fourth power of the ratio of their density relative to the density of the same material in its fully solid form. A material that is 60% dense will have mechanical properties about $(0.6)^4 \approx 13\%$ of those for the solid material. Thus, SLS sintered PVC will exhibit about one eighth the tensile strength of conventional PVC, and sintered polycarbonate parts about half the strength of molded polycarbonate parts.

SLS parts emerge contained within a large powder cake. A potential advantage of this approach is that the powder surrounding the object can serve as a support for isolated islands and cantilevers. Nonetheless, some geometries still require direct supports, or "anchors." However, the part must be extracted from the cake in a breakout station. The cake also requires slow cooling for a few hours to avoid part distortion. Next, the bulk of the surrounding powder is manually cut away using spatulas. Finally, the residual powder is removed with various brushes, dental tools, and low-pressure air.

The University of Texas and DTM are working on the sintering of binder-encased metal powders. Here the laser is used to fuse the binder material, and, somewhat loosely, the metal powder as well. After the green part has been built, it is placed in an oven where the binder material is burned away, and the metal particles are fused together. Subsequently, a lower melting point metal, such as molten copper, can be infiltrated to fill the interparticle voids. Unfortunately, shrinkage is quite large while the binder is burned away, and the resulting part distortion errors have not yet been established.

The Sandia Experience. The only statistically significant accuracy data regarding SLS parts comes not from the University of Texas, or DTM, but rather from Sandia National Laboratories.[11] Here, a test part somewhat smaller than the user-part was the basis for the measurements. Both the SLS and SL processes were used to build multiple versions of this part. A total of 21 diameters, 21 X and Y measurements, 30 radius points, and 38 Z measurements were taken for each part. The results from Reference 11 indicate that for the SLS part, built in polycarbonate, the overall RMS error was 0.0064 inch (0.16 mm). For the corresponding SL part built in ACES using epoxy resin SL 5170, the overall RMS error was 0.0016 inch (0.04 mm). The significance of these relative accuracy levels can be seen by inspection of Figure 1-2.

Fused Deposition Modeling

Fused Deposition Modeling (FDM), developed by Stratasys, Inc., of Eden Prairie, Minnesota, uses spools of thermoplastic filament as the basic material for part fabrication.[5,18,19] The material is heated just beyond the melting point in a delivery head. The molten thermoplastic is then extruded through a nozzle in the form of a thin ribbon and deposited in computer controlled locations appropriate for the object geometry. The FDM system builds parts in multiple thin layers, as is the case with all current RP&M methods.

The Thin Ribbon. The FDM technique is based upon the controlled extrusion and rapid subsequent cooling of a thin ribbon created from a number of possible thermoplastic materials. Essentially, the spooled filament material, having a diameter of 0.070 inch (1.78 mm), is fed into the delivery head and heated to a precisely controlled temperature just above the melting point of the thermoplastic being used. On demand, the working material is extruded through a nozzle, and onto the previous layer.

The feed rate of the filament is used to monitor and control the amount of material extruded at any given time. The previously formed layer, which is now the substrate for the next ribbon, must be maintained at a temperature just below the solidification point of the thermoplastic material.[5] This is important in order to assure good interlayer adhesion. The model is built on a computer controlled piston that is lowered as each layer is completed. This motion enables the delivery head to remain a fixed distance above the working surface, which is essential for deposition repeatability.

Layer Thickness. The thickness of each layer is determined by the material's physical properties, delivery headspeed, extrusion pressure, and dimensions of the nozzle exit. The surface finish depends on the temperature of both the liquefier and the substrate.[19] If the substrate temperature is too high, the surface may exhibit rippling, while, if too low, delamination is a problem. Conversely, if the liquefier temperature is too low, nozzle plugging will occur and, if too high, poor part detail can result, as the extruded thermoplastic may remelt the substrate.

Layer thicknesses are typically in the range from 0.004 to 0.020 inch (0.1 to 0.5 mm), with 0.040 inch (1.0 mm) occasionally still being used. The layer thickness is varied by changing the speed of the delivery head, with a maximum speed of $\approx$15 inches/sec (38 cm/sec). The ribbon width can range from about 0.01 to 0.24 inch (0.25 to 6 mm). Delivery head precision is listed as $\pm$0.001 inch ($\pm$25 microns). However, supporting test results covering the extremes of head position, scan velocity, or scan acceleration have not been presented in the literature. The head of the FDM system must be kept in continuous motion. If the nozzle pauses, material melts near the tip and forms small bumps that are visible on the surface.[5] Also, in any geometry requiring very small horizontal holes in a vertical wall, it has not been demonstrated that the extruded thermoplastic material can be interrupted rapidly enough to reproduce the necessary fine detail. As an example, the reliable generation of a series of 0.020-inch (0.5 mm) square holes, passing horizontally through a vertical wall, has not been established for this technique.

Using a Wide Array of Material. An advantage of the FDM approach is that a wide array of thermoplastic materials may be used to build models. Among those already used are an investment casting grade wax, ICW04; a machineable wax, MW01; a polyolefin, P200; and a polyamide, P300. The investment casting wax enables FDM models to serve as patterns for the shell investment casting process. The P200 material allows the operator to generate a range of concept models that possess sufficient strength to be passed around a room for evaluation, without concern for breakage during normal handling. From Reference 19, the following are midrange test values for material P200: tensile strength 1.3 kpsi (9 MPa), tensile modulus 90 kpsi (620 MPa), flexural strength 1.5 kpsi (10 MPa), flexural modulus 90 kpsi (620 MPa), and elongation-to-break 4.7%. By comparison, from Reference 20, the corresponding values for the epoxy resin SL 5170 are: tensile strength 8.7 kpsi (60 MPa), tensile modulus 570 kpsi (3,930 MPa), flexural strength 15.5 kpsi (107 MPa), flexural modulus 425 kpsi (2,930 MPa), and elongation-to-break 16%.

Used in Varied Environments. Another advantage of the FDM system is that it can be operated not only in a laboratory or shop floor environment but, in principle, in an office environment as well. No high-powered lasers are used, so laser related safety issues are not relevant. The materials are supplied in spool format, and present neither special handling nor environmental concerns. While the delivery head is heated to elevated temperatures (210 to 225° F [99 to 107° C]), this is not a safety hazard in normal operation.

Once again, despite a search of the relevant RP&M literature, no statistically significant FDM part accuracy data has been published either by Stratasys or any of the FDM users. Therefore, reliable statements regarding FDM part accuracy and repeatability are not possible at this time.

Solid Ground Curing

The Solid Ground Curing (SGC) process was originally developed in 1988 by Cubital, Ltd., an Israeli company. This technique involves the use of photopolymer resins but, unlike SL, SGC does not utilize lasers. Rather, each layer is generated through a multi-step procedure.[5,21] As with all RP&M systems, the method begins with a CAD model of the desired object. Cubital's software runs on a work station known as the "Data Front End," or DFE. The CAD data file is read into the DFE. The Cubital system accepts a number of possible formats including STL files, VDA-FS files commonly used in the European automotive industry, as well as the Cubital Facet List (CFL) format. All acceptable files are automatically converted to the CFL format.

CFL File. The CFL file generates the necessary cross-section information for each layer. Data from the CFL file is used to produce a mask. Ionography, similar to xerography, is used to create a charge distribution on a glass plate, corresponding to the cross-section for a given layer. Black toner powder is then spread over the image, with the toner powder selectively adhering at charged locations. Excess toner is then removed.

Optical Mask. At this point, the black toner powder has formed an optical mask. A thin layer of resin is then prepared on a support carriage/substrate, which is moved to the exposure station. The glass plate, bearing the mask for that cross-section, is registered above the substrate, and then exposed to flood ultraviolet radiation from a high power UV emitting lamp. The liquid photopolymer resin is then selectively cured wherever the mask is transparent. Conversely, the resin is not cured in those locations where black toner powder has adhered to the glass plate with sufficient optical density to block the ultraviolet radiation. An air knife subsequently removes residual uncured liquid resin.

This is followed by a wax application system that coats the surface with a layer of molten wax over both the solidified resin and the surrounding region. A cooling plate is then applied to rapidly solidify the wax. Next, a mechanical fly-cutter mills the combined polymer/wax structure to the correct thickness, providing a flat working surface for the next layer. Finally, a vacuum system removes the wax/resin chips just after they are produced. Throughout this operation, the carriage shuttles the developing structure from station to station, while also moving downward to accommodate new layers.

Predicting Build Time. An advantage of the SGC process is that the time to build every layer is the same, independent of either part geometry or the number of parts being built on the carriage. Thus, the build time can be predicted quite accurately by simply multiplying the time per layer by the number of layers. The latter is readily determined by dividing the height of the tallest object by the layer thickness.

Also, since the surrounding material acts as a block support, parts can be built in any nested arrangement within the working volume. This allows the operator to build parts on top of one another. However, all material in a volume equal to the working area times maximum build height is consumed in each build. Thus, a large amount of resin may be wasted. Further, when the parts are completed, a large block emerges from the machine. Removing the surrounding wax, necessary to free the desired parts, involves the use of a high temperature water/citric acid wash. Elimination of all the unwanted wax is not trivial.

Full-time Monitoring. As a direct result of the numerous subsystems, position changes and moving parts, the SGC system requires a full-time operator in order to maintain proper system function. In fact, SGC is the only current RP&M method that requires full-time monitoring. This is an important aspect of total system cost, especially in building a large part over a weekend or a holiday, where three shifts of personnel are needed to support the process. Also, resin mixed with wax particles cannot be used for additional part building. It must either be discarded or recycled, resulting in further operating expense.

Finally, the relevant RP&M proceedings show no record of statistically significant accuracy measurements for SGC parts. Therefore, reliable statements regarding SGC final part accuracy and repeatability are not possible at this time. However, the basic SGC image mask is generated at 300 dots per inch. This corresponds to a current system resolution limit of 0.0033 inch (0.084 mm) per spot. Other sources of error resulting from mechanical design and resin shrinkage will add to this value.

1.4 Recent Advances in RP&M

Continuous improvement has marked RP&M since its commercial inception in 1988. An aid to understanding this step-by-step advancement is shown in Figure 1-3, which we might refer to as "The Wheel of Progress." The hill to be climbed is the increasing performance of RP&M systems. The wheel that moves up this hill, if sufficient driving force is applied, has segments representing the hardware, software, building process, and materials used to make parts.

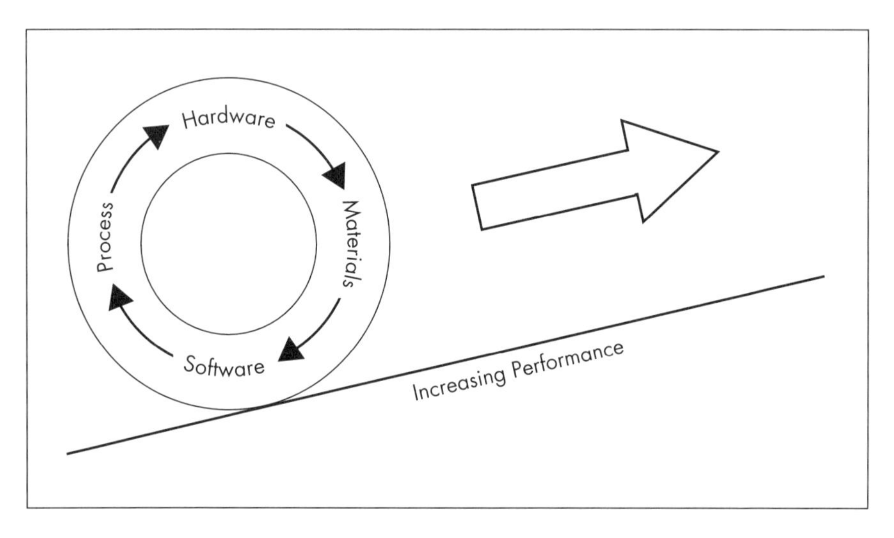

Figure 1-3. The Stereolithography Wheel of Progress.

Steady Progress. Using 3D Systems' experience as the model, once a technique is configured with all four components in place, it soon becomes clear that one of the components is a limiting factor, impeding progress up the hill. Research, development, and engineering attention to the appropriate topic eventually yields improved capability of that component, and allows the wheel to travel uphill a little further. But no sooner has that limitation been nullified than another takes its place. And so it goes. Improved hardware requires enhanced software, enabling better processing techniques, after which the materials will probably require upgrading. And the cycle begins again because, at least at this point, stereolithography is not facing any near-term fundamental limits that represent barriers to further progress.

Chapters 2 through 5 review the most recent advances in these four segments of system performance, and are written by specialists closely involved with those advances. Included are experimental test results supporting the release of epoxy resin photopolymers from Ciba-Geigy, called by some RP&M practitioners the second most significant development in the history of RP&M (the first being the invention of stereolithography by Charles Hull in 1986). The epoxy resins SL 5170 and SL 5180 deliver unparalleled accuracy with mechanical properties approximating medium impact polystyrene.

Improvements to SLA hardware are also chronicled. For the SLA-250/40, the list includes diode laser leveling, a subsystem that allows more accurate and repeatable generation of 0.004 inch (0.1 mm) layers, while also speeding build time significantly. Further enhancements are a bridge recoater system and conveniently interchangeable vats.

Today's SLA-500 is not the same machine either. The enhanced SLA-500/30 system now uses the Orion Imaging Technology™, which provides a scan velocity as high as 200 inches (≈5 m) per second while maintaining an RMS drawing error of 0.0004 inch (10 microns), as well as a solid-state Acousto-Optic Modulator (AOM), capable of interrupting the laser beam in less than one microsecond. Diode laser leveling has also been added to the SLA-500, which further improves leveling accuracy and speed. With these hardware upgrades, the SLA-500 recently held the world's record for RP&M accuracy, and it can accomplish this outstanding result at twice its previous drawing speed. Faster drawing and more accurate at the same time . . . true advancement in RP&M.

Easier to Use. Advances have also been made in SLA software. Equipment is now easier to use because there are fewer decisions required of the operator. This "de-skilling" allows outstanding parts to be built without the need for extensive experience. User interfaces are simpler, more intuitive and graphically oriented, speeding and simplifying build preparation. Processing speed for such operations as slicing a part has been greatly enhanced. Laser beam positioning is now more accurate, even at higher speeds, as a result of the recent implementation of feed forward digital signal processing logic. These and many other similar advances in software have made today's SLA systems significantly easier to use, with higher speed and improved performance.

Process developments have greatly increased the scope of applications for stereolithography. Specifically, the release of the quasi-hollow QuickCast build style has enabled the generation of accurate patterns for direct shell investment casting. As a result, a user can now proceed from a CAD model, to an accurate QuickCast pattern, to a high quality functional metal prototype, ready for testing, without the need for any tooling whatsoever. The savings have been documented by numerous users.

The ACES™ build style also allows users to build solid parts of remarkable accuracy and clarity. These parts have already seen application as: (1) flow test models for automotive cooling studies,[22] (2) parts for photoelastic stress analysis of complex structures,[23] and (3) models for dynamic testing of vibrational resonance characteristics.[24] Additional specialized applications involving ACES negatives as the basis for wax tooling are also being investigated.[25]

1.5 The Expanding World of RP&M Applications

Like the universe, the world of RP&M is rapidly expanding. Within just the past five years, a number of annual RP&M conferences have already been established, with attendance levels ranging from about 100 to as many as 900. Some of these conferences specialize in RP&M applications, while others are of a more academic nature. All of the following provide a forum for the dissemination of new ideas in this field.

- The North American Stereolithography Users Group Meeting, held in March, at various sites within North America.
- The SME RP&M Conference, held in late April or early May in Dearborn, Michigan.
- The University of Dayton-sponsored International Conference on Rapid Prototyping, held in June in Dayton, Ohio.
- The University of Nottingham-sponsored European Conference on RP&M held in early July in Nottingham, England, or other European sites.
- The University of Texas-sponsored Solid Freeform Fabrication Symposium, held in early August in Austin, Texas.
- The European Stereolithography Users Group Meeting, held in late October or early November, at various European sites.

Attendance at one of these conferences or symposiums will quickly open the eyes of the most casual observer to the wide variety of RP&M applications available today. From exciting but humble beginnings primarily as a visualization tool, RP&M is now being used throughout many companies that have come to rely upon these techniques for more sophisticated applications. Two major contributing factors have helped catapult this technology into such prominence: the performance advances discussed earlier and detailed in Chapters 2 through 6, as well as the dedicated vision and expertise of practitioners.

Together, these factors have driven the movement from simple applications, requiring only approximate conformance to the CAD design, to highly demanding uses such as making tooling core and cavity masters that require outstanding accuracy, fine detail, and exceptional surface finish. There is also limited production of plastic parts, requiring not only accuracy and surface finish, but material strength, tensile modulus, and toughness as well.

Providing Many Models. In Reference 3, a case was made for considering the initial applications to be *visualization*, *verification*, *iteration*, *optimization*, and *fabrication*. It is now clear that an even more fundamental application has become quite prominent, one in which designers, engineers, or developers can give free rein to their imagination. Instead of making only one prototype of one design concept, RP&M can provide many models of designs that are only slight variations of one another, or where each design is radically different. The time and cost of obtaining the models are so reasonable as to encourage experimentation. Let us call this precursor activity *innovation*. It gives full rise to the creative spirit and leads the user to a design or family of designs that demands more detailed investigation. A prominent practitioner of RP&M has stated that, "The SLA is used as a 3-dimensional sketch pad by our designers. They explore many design ideas before committing to a final idea." The result is a more adventurous design process and ultimately better products for us all.

The Apple Experience. An outstanding example of this innovative spirit became evident in the development of Apple Computer's new concept in multimedia computing. Combining a high resolution video monitor and a superb stereo sound system into a package that would probably be placed on top of a computer became a highly challenging engineering problem. Further complicated by the need to respond to consumer preferences, the design was subjected to extensive testing via SL models. Beginning with fairly simple sound chamber concepts, the designers were able to advance to much more sophisticated and successful approaches because of freedom to explore results of their creative ideas. They were able to build a wide variety of sound chamber designs based on extremely different concepts. In a short time, the team was able to identify designs that provided the required sound quality without negatively impacting the video quality or affecting the operation of a hard disk drive placed directly below. The design group was able to explore almost without compromise, and the final result was considered by Apple to be a superior product.

In fact, according to Michael Schrage, "Effective prototyping may be the most valuable core competence an innovative organization can hope to have . . . one around which an organization could build a successful culture."[1] That certainly applies to Apple Computer, Ford Motor Company, Texas Instruments, United Technologies, Pitney-Bowes, and scores of other users who apply the power of RP&M to provide "more value than our competitors" (Apple), and "the highest value to the customer" (Ford).

1.6 RP&M as a Communication Tool

The complexity of modern business and technology demands specialization. In one form or another, we are all experts in some limited field or fields. For our specialization to become truly effective, organizational management must create an environment for coordination, cooperation, and collaboration. "As companies grow, however, the systemic nature of the organization gets hidden. Distances increase as functions focus on their own needs, support activities multiply, specialists are hired, and reports replace face-to-face conversations. Before long the clear visibility of the product and the essential elements of the delivery process are lost or, at least, blurred. Instead of operating as a smoothly linked system, the company becomes a tangle of conflicting constituencies whose own demands and disagreements frustrate the customer."[26]

In an effort to combat this self-defeating situation, management frequently forms teams of one type or another. Despite the benefits of the team-based approach, simultaneous development teams often face communication problems—

a Tower of Babel syndrome. Engineering, marketing, production, and other disciplines have specialized languages . . . people in these disciplines have different points of view, which affects their ability to make the trade-offs needed for products to be marketable, manufacturable, and profitable.[27]

The Prototype as Common Denominator. So what's a manager to do? According to many practitioners and observers of RP&M, the answer lies directly in the physical models we produce. Models represent the emerging product in a reasonably neutral language and provide visibly accessible symbols of the finished product, helping to unify the design team.[27] In some innovative cultures, prototypes effectively become the *media franca* of the organization—the essential medium—for information, interaction, integration, and collaboration.[1]

Why is the use of models so helpful in solving this communication dilemma? Partly because not many of us can visualize in three dimensions without error. Even after studying many 2-dimensional drawings of the various views of an object, we are often surprised when we see the actual, physical object. According to Chrysler's Vern Schmidt, "The average life of an SL model is 10 minutes. After a few Oh no's, it's back to the CAD tube." This from experienced designers.

The power of a 3-dimensional model to clarify reality is extraordinary indeed. Can we expect real understanding of a design presented in two dimensions from individuals not experienced in converting them into 3-dimensional conceptualizations? Very few company presidents, CEOs and CFOs are truly comfortable with blueprints. However, the physical model that is readily generated by any of the various RP&M techniques is what Sean O'Reilly of Ford Motor Company has called, *"an incredibly wide bandwidth medium of communications."*

Daimler-Benz's Werner Pollmann states that the lower the abstraction level of information, the easier it is to exchange that information between people with different backgrounds and interests. To say nothing of different agendas. Dan Droz observes that, "Models keep everyone focused on the possibilities of the final product, not their own point of view." Many users of RP&M have experienced this communication phenomenon. Even after three, four, or five years of rapid prototyping, and having already achieved high infusion, as much as 40% of an organization's models may be used for communication purposes alone.

One SL veteran was heard to comment, "Before the SLA, concurrent engineering was only a theoretical concept. Now that we use SL models to communicate across functional boundaries, concurrent engineering is a reality." Seen as a communication tool, RP&M serves two vital roles—giving disparate disciplines a common language for discussion, and helping to better integrate and link activities that bring a new concept to market. It can minimize errors and misunderstandings. It not only serves as an internal communication tool, but also provides an opportunity to interface with potential customers at a very early stage in the development cycle.

1.7 Automating the Manufacturing Middle

In an address to the annual meeting of the North American Stereolithography Users Group in 1993, 3D Systems CEO Arthur B. Sims presented the concept diagrammed in Figure 1-4, which can be applied to the entire field of RP&M. Shown schematically is the process of conceiving, designing, and manufacturing a new product. At left is the front-end concept and design portion of the cycle. This sequence has been automated to a high degree through the use of CAD and CAE. The benefits to the design community have been well documented.

Similarly, the right side of the diagram shows the production end of the cycle. This sequence has been automated via CAM, CNC, robotics, and similar manufacturing advances. Again, the benefits of such technologies are widely documented and are presently being realized at many thousands of organizations.

Serving the Manufacturing Middle. RP&M now serves the central portion of the product design, development, and production cycle. This domain has been labeled the *manufacturing middle*. Previously the generation of models and prototypes had eluded the computerization so often applied to both the design and manufacturing areas. RP&M brings urgently needed solutions to disciplines characterized by high labor costs, long lead times, and a shrinking pool of aging artisans and craftsmen.

Minimizing these constraints unleashes creativity. While innovation was always latent in the talented ranks of designers, model makers, and engineers, it had often been restricted by the dual constraints of tight budgets and critical deadlines. Simply stated, making models and prototypes accurately, rapidly, and economically is very important to a wide range of organizations.

1.8 RP&M Benefits

Ask any user of RP&M what the key benefits are and you'll hear much the same answer. *"It's faster, costs less, and quality is improved."* These are not empty claims. The literature is replete with success stories giving specific comparative examples of time and cost savings relative to the more traditional ways of completing the same steps in the product development and commercialization cycle. Typically, these savings vary between 40 and 80% for both cost and time, and in some cases go even higher. In most companies a 10% cost or time saving on a major development program would be considered excellent. A 20% time or cost saving is considered outstanding. A 30% cost or time saving would be considered phenomenal. But a 50% time or cost saving is a paradigm shift that can overwhelm competition.

Quality's Role. Quality is a more elusive component to measure. Everyone is convinced that quality is improved through the use of RP&M, but facts and figures are hard to gather. There are certainly a number of reasons to explain why quality should advance. First, *more time is available to make and test prototypes.*

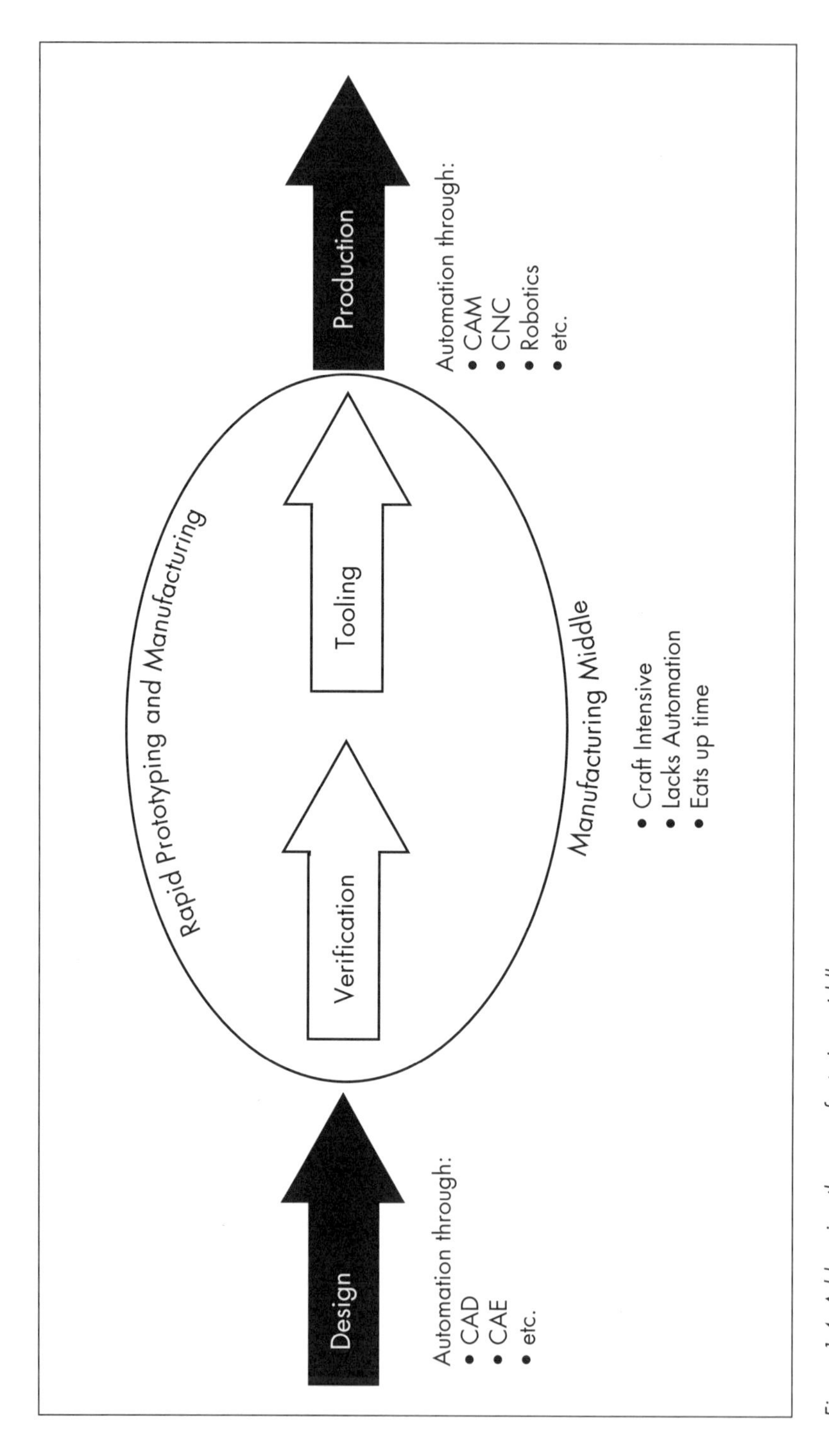

Figure 1-4. Addressing the manufacturing middle.

This allows the designers and engineers to find weaknesses and failure modes early enough to address them in the final design. Second, *more time can be spent in exploring alternate designs* that improve reliability, longevity, ease of operation, and overall customer satisfaction. Third, *designs can be finalized much later in the process*, ensuring that the latest advances in technology can be applied to the product. In the case of an automobile, where cycles have been reduced from 60 months to under 36, this can be a tremendous contributor to quality and competitiveness. And fourth, designers and engineers enjoy *a continuity of process, including rapid feedback on design concepts*. This enhances their ability to dwell on identifying problems and finding solutions, while still remaining on schedule and within budget.

Recently, a hot topic in many business publications has been *rapid time to market*. It has been determined that being six months late to market relative to your competitor can sacrifice as much as 30% of the available profit from a given product life-cycle. By speeding up the product development and commercialization process, RP&M contributes to faster product launch and potentially higher profits. An allied benefit deals with the number of new product launches.

In certain markets, competition is based, at least in part, on the sheer number of new models that can be introduced. It is no secret that product life cycles have shrunken drastically. For some computer products, the average life expectancy is now under nine months. All this demands the ability to introduce more products at a faster rate. The benefits of RP&M in this area alone can be dramatic.

Launching products that fail is something few companies can afford. RP&M helps in this respect because products can be conceived, designed, and manufactured very close to the time of perceived market demand. Also, the products being developed can be tested by potential customers at an early stage of product evolution. It is a tremendous advantage to be assured that the product being introduced truly meets customer needs, preferences, and expectations.

Product Cost. In these days of severe competition, product costs are paramount in importance. RP&M contributes to reduced manufacturing costs. Design for assembly and design for manufacturability are two key ways in which RP&M can help control expenses. By addressing these issues in the earliest stages of product development, designers, engineers, production supervisors, and others are given a basis for reaching agreement on initial design criteria. The final design can then reflect the highest level of thinking with respect to ease of assembly, reducing part count, and minimizing vendor quotes for components and subassemblies.

With the impending advent of rapid tooling capability as an outgrowth of RP&M, the cost of tooling will come down, and tooling decisions could be made later in the process, saving expensive tooling rework due to either mistakes or design evolution. Because tooling will become significantly less expensive and more rapidly available, design changes intended to improve customer acceptance or to improve product quality, which might otherwise have been postponed due to the high cost of new tooling or tooling rework, will be completed early enough to impact market share.

Many hundreds of companies around the globe are now experiencing these benefits on a daily basis. Documented studies show hundreds of cases where economic savings and other advantages of RP&M are real and often very substantial. As Peter Sferro of the Alpha Manufacturing Center at Ford Motor Company puts it, "It is no longer a question of whether you will experience time and cost savings with RP&M, it's a question of how fast can I get it, and how fast can I start using it?"

References

1. Schrage, M., "The Culture(s) of Prototyping," *Design Management Journal*, Winter 1993, Vol. 4, No. 1, pp. 55-65.
2. Pollmann, W., "Prototyping at Daimler-Benz: State of the Art and Future Requirements," IMS International Conference on Rapid Product Development, Stuttgart, Germany, January 31-February 2, 1994.
3. Jacobs, P.F., *Rapid Prototyping & Manufacturing: Fundamentals of StereoLithography*, SME, Dearborn, Michigan, 1992.
4. Wohlers, T. "Future of Rapid Prototyping & Manufacturing Around the World," Proceedings of the Third European Conference on RP&M, University of Nottingham, Nottingham, England, July 6-7, 1994, pp. 1-11.
5. Lightman, A. and Cohen, A., "Alternate Approaches to RP&M," Chapter 16, Jacobs, op. cit., pp. 397-423.
6. Burns, M., *Automated Fabrication: Improving Productivity in Manufacturing*, Prentice-Hall, Englewood Cliffs, New Jersey.
7. Rapid Prototyping Report, various issues including: June 1991, August 1991, December 1991, January 1992, March 1992, December 1993 and May 1994.
8. Denton, K. and Jacobs, P., "QuickCast & Rapid Tooling: A Case History at Ford Motor Company," Proceedings of the Third European Conference on Rapid Prototyping & Manufacturing, University of Nottingham, Nottingham, England, July 6-7, 1994, pp. 53-72.
9. Hull, C., "Apparatus for Production of Three-Dimensional Objects by Stereolithography," U.S. Patent 4,575,330, March 11, 1986.
10. Hunziker, M. and Leyden, R., "Basic Polymer Chemistry," Chapter 2, Jacobs, op. cit., pp. 25-58.
11. Atwood, C., McCarty, G., Pardo, B. and Bryce, E., "Advances in Rapid Prototyping," Proceedings of the Rapid Prototyping & Manufacturing '94 Conference, SME, Dearborn, Michigan, April 24-26, 1994.
12. Blake, P., Baumgardner, O., Haburay, L. and Jacobs, P., "Creating Complex Precision Metal Parts using QuickCast™," Proceedings of the Rapid Prototyping & Manufacturing '94 Conference, SME, Dearborn, Michigan, April 26-28, 1994.

13. Feygin, M., "LOM System goes into Production," Proceedings of the Second International Conference on Rapid Prototyping, University of Dayton, Dayton, Ohio, June 23-26, 1991, pp. 351-357.
14. Warner, M., "From Rapid Prototyping to Functional Metal and Plastic Parts," Proceedings of the Fourth International Conference on Rapid Prototyping, University of Dayton, Dayton, Ohio, June 14-17, 1993, pp. 325-332.
15. Marcus, H., Beaman, J., Barlow, J. and Bourell, D., "Solid Freeform Fabrication: Powder Processing," *Ceramics Bulletin*, Vol. 69, No. 6, p. 1030, 1990.
16. Barlow, J., Sun, M. and Beaman, J., "Analysis of Selective Laser Sintering," Proceedings of the Second International Conference on Rapid Prototyping," University of Dayton, Dayton, Ohio, June 23-26, 1991, pp. 1-5.
17. Knudsen, F., "Dependence of the Mechanical Strength of Polycrystalline Specimens on Porosity and Grain Size," *Journal of the American Ceramics Society*, Volume 42, No. 8, p. 376, 1959.
18. Crump, S., "The Extrusion Process of Fused Deposition Modeling," Proceedings of the Third International Conference on Rapid Prototyping, University of Dayton, Dayton, Ohio, June 7-10, 1992, pp. 91-100.
19. Comb, J. and Priedeman, W., "Control Parameters and Material Selection Criteria for Fused Deposition Modeling," Proceedings of the Solid Freeform Fabrication Symposium, University of Texas, Austin, Texas, September 1993, pp. 86-101.
20. Pang, T., "Stereolithography Epoxy Resins SL 5170 and SL 5180: Accuracy, Dimensional Stability, and Mechanical Properties," Proceedings of the Solid Freeform Fabrication Symposium, University of Texas, Austin, Texas, August 1994.
21. Herskowits, V. and Meller, A., "In situ Thermal Monitoring of the Solid Ground Curing Process," Proceedings of the Fourth International Conference on Rapid Prototyping, University of Dayton, Dayton, Ohio, June 14-17, 1993, pp. 175-184.
22. Steinhauser, D., "Flow Analysis of Complex Configurations: Another Successful Application of RP&M," 1994 European Stereolithography Users Group Meeting, Strasbourg, France, November 7-8, 1994.
23. Steinchen, W., Kramer, B. and Kupfer, G., "Photoelasticity Cuts Part Development Costs," *Photonics Spectra*, May 1994, pp. 157-162.
24. Dornfeld, W., "Direct Dynamic Testing of Stereolithography Models," ASME International Gas Turbine and Aeroengine Congress, The Hague, Netherlands, June 13-16, 1994, pp. 1-6.
25. Mieritz, B., "Making Lost Wax Models from SLA-Molds," Report of the Danish Technological Institute-EEC Brite/Euram Instantcam Project, Proceedings of the European Stereolithography User Group Meeting, Bürgenstock, Switzerland, October 18-19, 1993.

26. Bower J., and Hout, T., "Fast-Cycle Capability for Competitive Power," *Harvard Business Review*, Nov.-Dec. 1988, pp. 110-118.
27. Droz, D., "Prototyping: A Key to Managing Product Development," *Journal of Business Strategy*, Vol. 13, May/June 1992, pp. 34-38.

CHAPTER 2

Advances in Stereolithography Photopolymer Systems

Double, double, toil and trouble; fire burn and cauldron bubble.
Filet of a fenny snake in the cauldron boil and bake.
Eye of newt, and toe of frog, wool of bat, and tongue of dog,
Adders fork, and blind worms sting, lizards leg, and howlets wing.
O, well done! I commend your pains. And every one shall share the gains.

Act IV Scene I MacBeth—William Shakespeare

2.1 Background

The number of RP&M machines continues to increase worldwide. As more and more people learn how RP&M technologies can be utilized in their own applications, they become very creative. The properties of the RP&M materials, especially in the case of SL photopolymers, have been instrumental in expanding the number and variety of new applications that could not possibly have been achieved with earlier materials. The resulting applications, which are discussed in subsequent chapters, have largely been made possible due to advances in the dimensional accuracy and mechanical properties of parts built from recently developed SL photopolymer systems. New applications will continue to arise as an even greater variety of materials, possessing special properties, becomes available.

What are the properties that determine a good SL photopolymer? This question sounds simple. However, SL technology, just like other emerging technologies, requires a new perspective based on new information.

Much description of materials for SL technology should also be applicable to other RP&M technologies, albeit with some differences. A key aspect of the selection process is the need for users to establish a new set of standards. This is so

By Thomas H. Pang, Ph.D., Technical Manager, Stereolithography Resins, Ciba-Geigy Corporation, Los Angeles, California.

because SL results frequently prove highly nonintuitive, and simple linear extrapolations of properties derived from conventional materials often do not adequately describe the behavior of RP&M materials.

First, there is a significant difference between choosing materials for RP&M machines versus selecting conventional materials. Consider the basic criteria used in selecting engineering plastics for injection molding. Here, as shown in Figure 2-1, one might begin by looking at the end-use mechanical properties. When a material is chosen, the suitability of the plastic for the intended injection conditions such as pressure, temperature, and releasability from the mold may be determined. There need not be much concern about the accuracy of the part at this point.

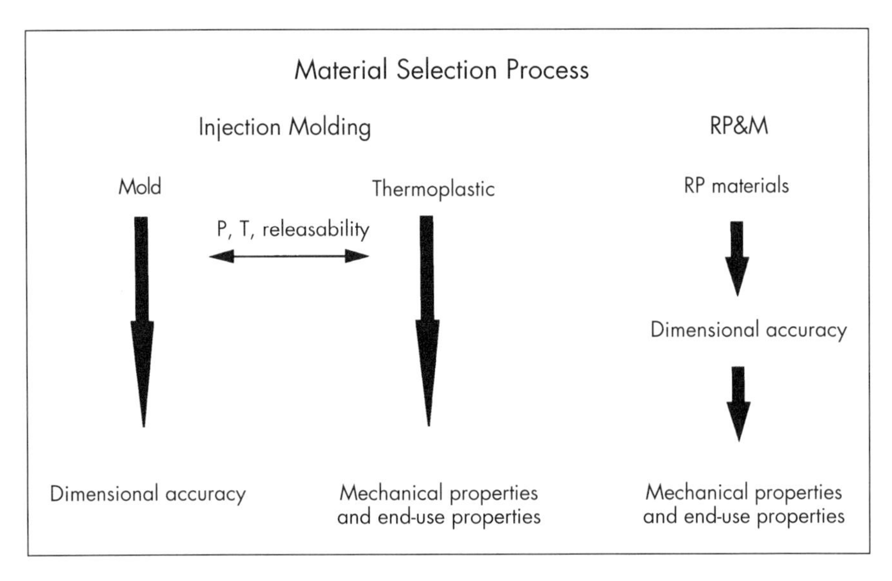

Figure 2-1. Dimensional accuracy of the RP&M material must be considered simultaneously with the end-use properties. In conventional shape-forming techniques the two can be essentially considered separately.

In the case of injection molding, and other common molding applications, correlation between materials and part accuracy is relatively weak. Admittedly, some fraction of the achievable accuracy is a function of the injection molded material. However, the final part accuracy is determined primarily by the design and accuracy of the mold itself.

Such is not the case for SL, or for other RP&M processes. The difference lies in the interrelationship between build materials and the level of achievable part accuracy. Each RP&M material has an inherent dimensional accuracy that depends on the chemical or physical transformation associated with generating and adhering the build layers. RP&M part accuracy depends strongly on the material. The RP&M interdependence of materials and accuracy is usually quite nonintuitive, and makes material selection for RP&M systems difficult.

Interdependence of Machine and Material

RP&M system accuracy is only one factor in building accurate parts. In SL, for example, no matter how accurately the laser beam may be positioned on the liquid resin surface, if the photopolymers distort during or after laser photo curing, the inherent machine pointing accuracy is compromised. SL photopolymers and SLA systems are mutually dependent, and the performance limitations of either component can result in inferior parts. This interdependence of machine and materials does not apply only to SL, but also to other RP&M systems. Each type of RP&M system has its own set of strengths and weaknesses. Some of these may be obvious, while others may not be apparent to the users. This makes it relatively difficult to judge the quality of the parts that could be generated from a particular combination of RP&M machine and its materials, simply by inspecting published machine specifications such as the positioning accuracy of laser beams, elevator translation, extruding heads, X-Y plotters, or jetting heads.

Since SL resins need to achieve high dimensional accuracy, photopolymer development requires an SLA. The accuracy achievable today with Ciba-Geigy SL resins is sufficiently high that machine errors associated with non-SLA scanning systems are intolerable. The resin behavior must be determined on a system capable of outstanding laser pointing accuracy. For these reasons, almost all SL photopolymer development, carried out jointly at 3D Systems, Valencia, California, and the Ciba-Geigy Research Center in Marly, Switzerland, is performed on SLA systems.

Impact of Polymers

The performance of the objects generated from the SL process depends on the availability of high-quality photopolymer systems. The quality is characterized by the resin dimensional properties, namely accuracy and dimensional stability, as well as its physical and mechanical properties. These photopolymer properties directly impact the user's applications. It is imperative that current and future users of SL and other RP&M technologies thoroughly inspect the dimensional accuracy and mechanical properties actually achieved by the materials, in addition to the machine specifications, and compare these with the required tolerances and material property specifications appropriate for their needs.

The applications of SL parts depend strongly on the performance achieved by the photopolymer systems. For example, with limited accuracy RP&M parts may only be useful as concept models and for design visualization purposes. Interference checks will not be appropriate where part accuracy is poor. With inferior mechanical properties, direct functional testing using RP&M parts is impossible. Dimensional accuracy and mechanical properties cannot be separated; rather, they must be considered simultaneously for RP&M applications.

2.2 Photopolymers for SL and Radiation Curable Coatings

Various types of liquid photopolymers can be solidified by exposure to electromagnetic radiation over a wide range of wavelengths including gamma rays and x-rays, ultraviolet (UV), and visible.[1] Electron-beam (EB) and UV curing of polymers are clearly the most established and widespread commercial processes.[2] The majority of radiation curing technology of commercial importance today uses either EB or UV radiation.

UV Curable Photopolymers

The first UV curable photopolymers were developed initially during the late 1960s,[3] as a way to reduce air pollution from solvent-based coatings. These were based on acrylates, a class of chemicals known to undergo rapid photopolymerization when exposed to appropriate actinic radiation in the presence of free-radical photoinitiators. Today, acrylates continue to be the most popular and widely used class of photopolymerizable resin systems in the UV coating and printing industry. Many types of acrylate monomers that give rise to various final cured properties are commercially available.

Today's SLAs utilize lasers that emit in the UV,[4] but the use of lasers emitting at other wavelengths is not ruled out. The ability to undergo photopolymerization by exposure to UV radiation is simply a prerequisite for current SL photopolymers. The laser wavelengths used are 325 nm, generated from a helium-cadmium (HeCd) laser, or 351 and 364 nm, emitted by an argon ion laser, with the latter operating in either monochromatic or dual wavelength mode. This is quite different from the radiation sources employed in the conventional photocurable coating industry,[5] which are typically UV lamps with a wide distribution of actinic output wavelengths ranging from 200~400 nm, as well as some wavelengths in the visible and IR.

SL is currently one of the most exciting applications of photopolymer chemistry. In the UV curable coating and printing industry, where the majority of commercial photopolymers are used, the applications are limited to the geometrical manipulation of essentially two dimensions. Consequently, the relevant physical properties of films and coatings are essentially limited to surface properties such as adhesion, wetting characteristics, and scratch resistance.

On the other hand, in SL the photopolymer is grown, layer by layer, into a physical object having true 3-dimensional structure. The object height is limited only by the system-build envelope. These parts may be inches to tens of inches tall, generated by building thousands of layers using the SL process. The third dimension allows you to test the mechanical properties of the cured photopolymer. Provided the measurements are made under the same protocol, the properties of the cured photopolymer systems can now be compared directly with various engineering plastics used in the market today. Measuring mechanical properties is not very new for thermoplastics or thermosets. However, these measurements are relatively new for objects generated from multiple layers via RP&M processes.

Many properties of multi-layered cross-linked polymeric materials, especially the various distortion mechanisms, are not fully understood. The interdisciplinary nature of the field is another characteristic of the SL photopolymer technology. Specialty areas in the fields of surface, organic, photopolymer, and physical chemistry must be pulled together to decipher the dynamic mechanisms underlying SL photopolymer behavior. These are some of the reasons why SL photopolymer development, initiated only about seven years ago, is a field breaking new ground.

2.3 The Chemistry of SL Photopolymers

SL photopolymers for UV curing are generally formulated from photoinitiators and reactive liquid monomers. Some specialty photopolymers may also contain fillers and other chemical modifiers. The sequence of events for free-radical photopolymerization are schematically shown in Figure 2-2. The photoinitiator molecules, P, mixed with the monomers, M, are exposed to a UV source of actinic photons designated by $h\nu$. The photoinitiators absorb some of the photons and yield an excited photoinitiator species, P*. A small fraction of the P* is converted into reactive initiator molecules, P•, after undergoing various complex chemical energy transformation steps. These molecules then react with a monomer molecule to form a polymerization initiating species, PM•. This is known as the chain initiation step.

Once activated, additional monomers continue to react during the chain propagation step forming PMMMMM• until a chain inhibition process terminates the polymerization reaction. If the reaction is sustained for a sufficient period, a high molecular weight can be achieved. With difunctional monomers, having two reactive sites per molecule, the polymerized material becomes highly viscous, and eventually becomes a super viscous solid. However, if the monomer molecules have three or more reactive chemical groups, then the resulting polymer will be cross-linked. Cross-linking generates an insoluble continuous network of molecules.

Cross-linking. For SL, it is important for the polymers to be sufficiently cross-linked that the polymerized molecules do not redissolve into the constituent liquid monomers. The photopolymerized species must also possess sufficient green strength to remain structurally intact, while the laser cured resin is exposed to various viscous forces during the deep-dipping and the recoating processes of SL.

As the laser cured SL photopolymer is postcured with additional UV radiation, the reaction proceeds even further, increasing the degree of polymerization and the cross-link density. The cross-linking is formed by strong covalent bonds, which connect the polymer chains to each other.

The chemical bonds must first be broken before molecular flow can be realized. Therefore, solidified SL photopolymers do NOT melt upon heating as a

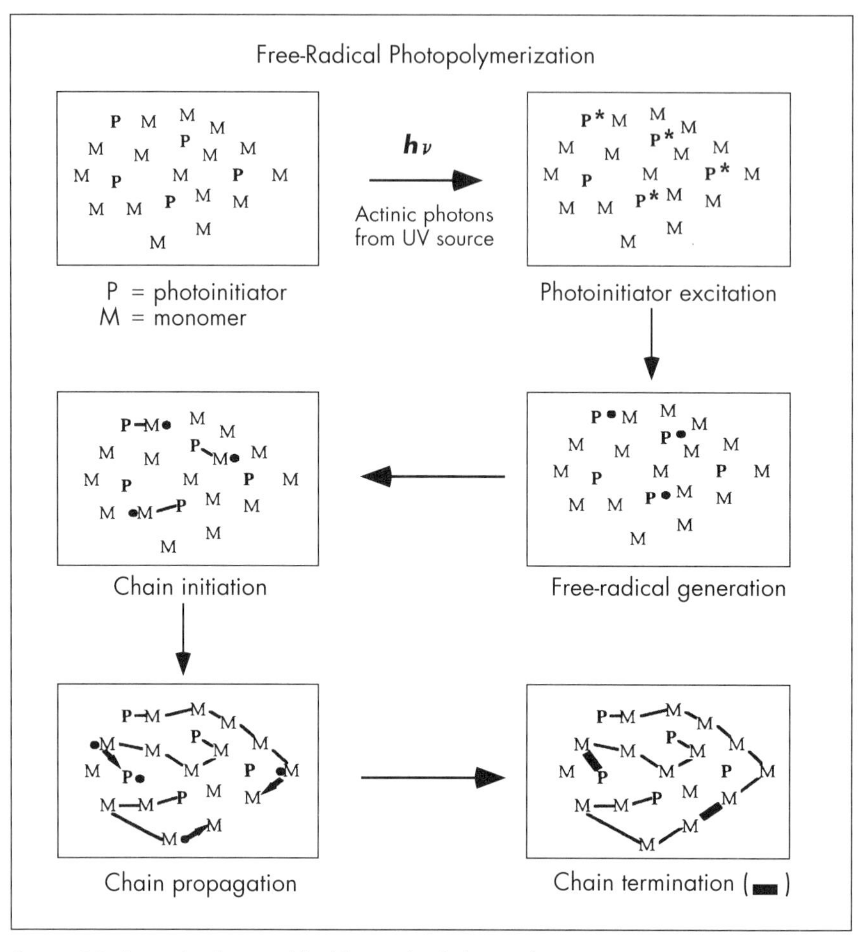

Figure 2-2. Example of a simplified free-radical photopolymerization sequence.

result of their covalent chemical cross-linking. They can, however, soften significantly upon heating, and eventually will thermally decompose at a very high temperature. Of course, the thermal behavior can be altered tremendously by changing the monomers used in the formulation.

Acrylate Chemistry. The first photopolymer system used for SL was based on acrylate chemistry,[6] which polymerizes via a free-radical mechanism. A generalized acrylate structure is seen in Figure 2-3. The photospeed of acrylate-based SL photopolymers is generally quite high. However, one of the drawbacks in terms of the chemistry of free-radical polymerization is oxygen inhibition.[7] This may result in surfaces that are either not fully cured or are not quite "tack-free" in the event that the UV irradiance and exposure are insufficient.

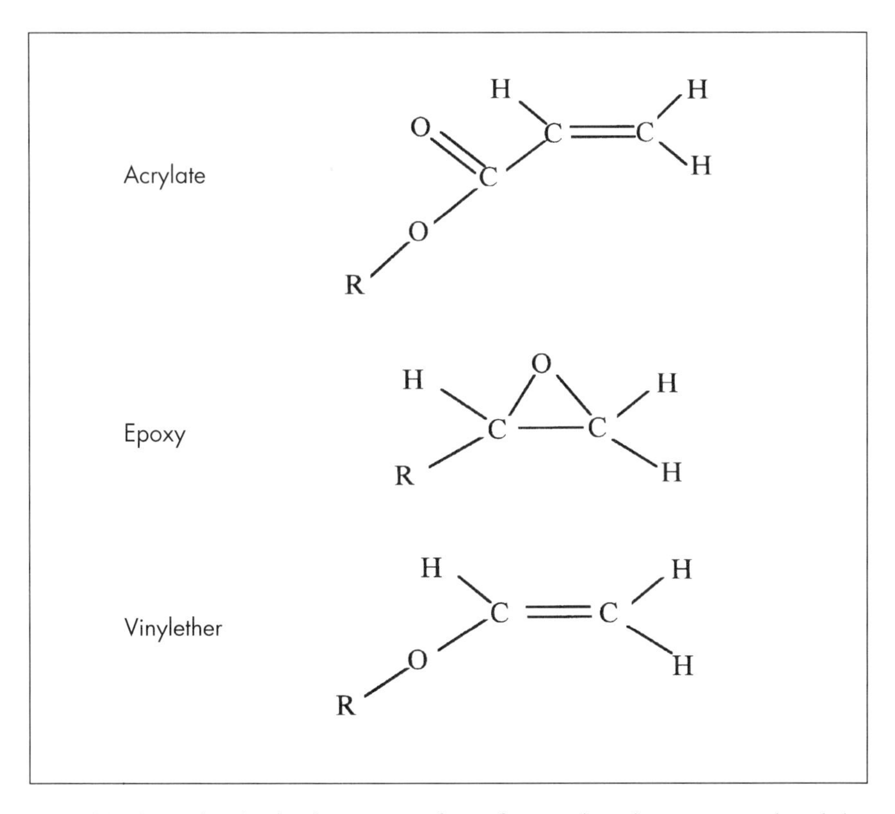

Figure 2-3. Generalized molecular structure of monofunctional acrylate, epoxy, and vinylether structures.

Cationic Photopolymerization. The second type of chemistry used more recently in SL is *cationic* photopolymerization. Some of the available cationic photoinitiators and monomers are discussed elsewhere.[8,9] The very first photoinitiated polymerization of epoxy resins was reported in a U.S. Patent in 1965.[10] J. V. Crivello was instrumental in carrying out experiments leading to the commercialization of practical cationic photoinitiator systems.[11,12,13] Two classes of cationically photopolymerizable monomers are used in SLAs today.

The first class is epoxies. As shown in Figure 2-4, epoxy monomers form polymers by undergoing ring-opening reactions in the presence of cationic photoinitiators.[14] Ring-opening is known to impart minimal volume change on reaction, because the number and types of chemical bonds are essentially identical before and after the reaction. This is unlike acrylates, or vinylethers, which convert one double bond into two single bonds. Epoxy resins are well known, in consumer products such as two-part epoxy adhesives, and for industrial uses such as castable epoxy for molding, as well as adhesives for automotive and aerospace applica-

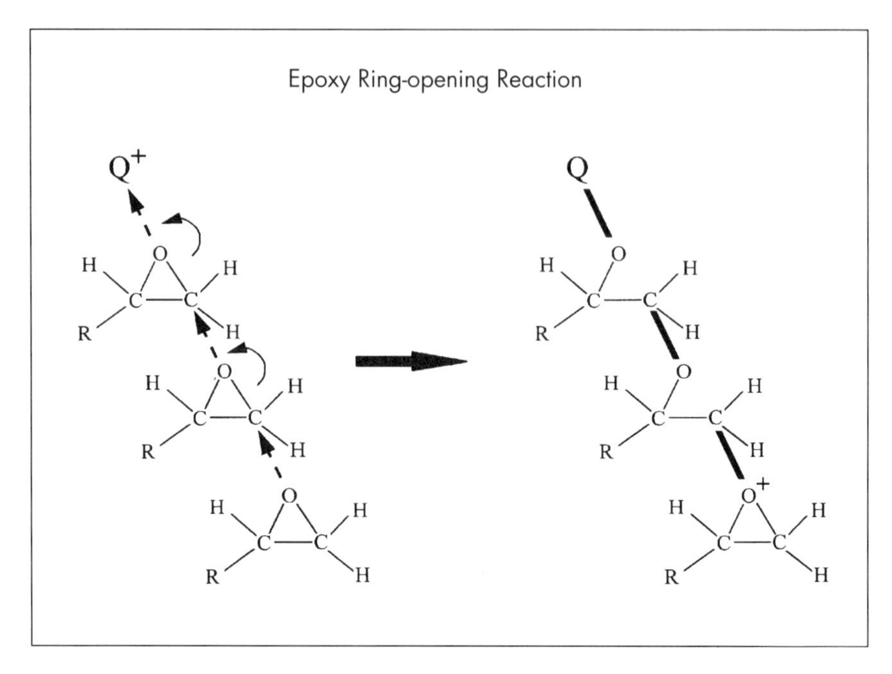

Figure 2-4. Simplified ring-opening homopolymerization of epoxy monomers. Q+ is an activated cationic photoinitiator, Lewis acid, or a proton.

tions. One difference, however, is that many of the commonly known epoxy systems are NOT cationically photoinitiated, but rather thermally cured in the presence of amines or other types of strong bases or acid anhydrides.

Another class of cationically photopolymerizable monomers is vinylether systems. Vinylethers do not have a ring structure as do epoxies. Rather, they possess double bond structures that can be reacted cationically, due to the strongly electron donating nature of the adjacent oxygen atom. Vinylethers react efficiently via cationic photoinitiation.[15] When combined with electron deficient molecules containing allylic functional groups (characterized by a carbon atom adjacent to a carbon carbon double bond), a substantial fraction of their reaction mechanism is believed to occur via a free-radical pathway.[16]

Cationically initiated SL photopolymers possess a number of chemical advantages compared to photopolymers that are free-radically initiated. First, photopolymerization based on cationic reactions is not inhibited by ambient oxygen, unlike acrylate-based systems. Second, cationic photopolymers manifest a continued thermal curing following photocure. In acrylate-based resins, the green strength does not substantially change once the part has been laser cured. Increase of green strength due to continued thermal reaction, even at room temperature, may be substantial for cationic reactions, and particularly beneficial for SL.

For example, the green flexural modulus of epoxy-based Cibatool SL 5180 increases from 88,000 to 157,000 psi (607 to 1,082 MPa) as the part is allowed to age at room temperature from 10 minutes to 1 hour.[17] After laser-cured resin has been UV-postcured, the cationic reaction continues, and the mechanical strength of the postcured parts increases asymptotically up to a limiting level characteristic of that resin. For SL 5180, the postcured flexural modulus reaches 366 kpsi (2,520 MPa). This behavior is normally not observed for acrylate-based photopolymers. Also, cationically cured layers are generally known to have better adhesion, due to lack of surface oxygen inhibition, and continued thermal reaction.

Commercial versions of cationically curable liquid SL photopolymers were successfully formulated with lower viscosity than the earlier acrylate-based formulations, while maintaining or even exceeding their cured mechanical properties. The mechanical properties of the epoxy resins will be discussed in detail later in this chapter. The viscosity of the earlier acrylate SL resins ranged from about 800 to 3,000 centipoise at 86° F (30° C). With the advent of the vinylether and epoxy resin systems, much lower viscosity values became achievable for SL. Now, for example, the Cibatool epoxy resins SL 5170 and SL 5180 have viscosities less than 200 centipoise at 86° F (30° C). Thus, these epoxy resins exhibit about an order of magnitude lower viscosity than those of the earlier acrylate resins.

However, cationic photopolymer resins also have some disadvantages. The polymerization of cationic systems is often inhibited by bases and water, including humidity. The inhibition effect depends on the monomers and photoinitiator systems that constitute the photopolymer formulation. Another disadvantage of the initial cationically curable SL resins was their relatively slow photospeed. The slow photospeed is generally believed to be due to the relatively high activation energies of monomer propagation for cationic reactions as opposed to the low values for free-radical reactions.[18] This is why the rate of free-radical reaction is generally quite insensitive to temperature change.

On the other hand, the high activation energy means that cationic SL resin reaction rates can theoretically be accelerated if the temperature is raised. Nevertheless, cationically cured SL photopolymers, having photospeeds comparable to the conventional acrylate-based systems at normal SLA operating temperatures, are feasible through appropriate chemical modifications.

2.4 Epoxy, Vinylether, Acrylate Systems

Today's SL photopolymers are often classified into epoxy, vinylether, or acrylate systems. However, the respective advantages and disadvantages associated with each class of photopolymers may not always be true for all photopolymer formulations. Associating performance of an SL photopolymer simply with the class to which it belongs is not recommended. It is important to remember that the classification of SL photopolymer systems into epoxy, vinylether, or acrylate is based on the type of chemical reaction responsible for enabling cross-linking.

This means that chemical links holding the monomers together are formed by the reaction of epoxy, vinylether, or acrylate functional groups.

By analogy, if a person was a monomer capable of reacting with another monomer (person), as in holding hands to form cross-linking, the epoxy, the vinylether, or the acrylate portion is analogous to the hands of a person. Hands are only a small portion of the whole body, and do not represent the rest of the body. In other words, virtually an infinite number of formulations of epoxy, vinylether, or acrylate formulations can be made, all with different properties.

Therefore, while some of the characteristics of each class of photopolymer may certainly be due to the formation of a particular type of chemical linkage, it is good practice not to assume that all epoxy, vinylether, or acrylate resins have the good or the bad attributes often quoted for each class of photopolymer system. Always examine the dimensional accuracy and mechanical properties of each individual photopolymer formulation to determine suitability for your application.

Vinylether-based Stereolithography Photopolymers

The first cationic SL photopolymer introduced was a vinylether-based system exclusively for the SLA-190 and SLA-250 using HeCd lasers operating at 325 nm. In March 1992, Allied Signal commercialized EXactomer™ 2201 resin,[19] and in March 1994, EXactomer 2202 SF.[20] Both vinylether formulations were noted for their low viscosity, resulting in SL parts that were easy to clean relative to the earlier moderately high viscosity acrylate-based photopolymers. The primary disadvantage of the original EXactomer 2201 vinylether resin was its slow photospeed. Nevertheless, it generated parts with relatively low curl. The curl factor for 2201 was reported to be less than 1%, and that for 2202 SF was less than 3%.

EXactomer 2202 SF is a vinylether system that has improved photospeed compared to EXactomer 2201. The relative photospeed at a cure depth C_d = 12 mils (0.3 mm) is given in Table 2-1, along with curl factor, photospeed parameters (i.e., penetration depth D_p,and critical exposure E_c), as well as resin viscosity

Table 2-1. Data on EXactomer 2201 and EXactomer 2202 SF

Photopolymer	Relative Photospeed @ C_d = 12 mils	Curl Factor (%)	D_p (mil)	E_c (mJ/cm^2)	Viscosity (cps @ 30° C)
EXactomer 2201	1.0	<1	7.0	27	205
EXactomer 2202 SF	2.9	<3	6.6	8.5	230

at 30° C. EXactomer 2202 SF is almost three times faster than 2201 at this cure depth.

Epoxy-based Stereolithography Photopolymers

In July 1993, 3D Systems Corporation announced the availability of the first epoxy-based SL resin, Cibatool SL 5170, developed jointly with the Ciba-Geigy Research Center, Marly, Switzerland. This epoxy-based photopolymer, manufactured by Ciba-Geigy in Los Angeles, is suitable for part building on an SLA-190 or SLA-250 using an HeCd laser emitting at 325 nm. In March 1994, an epoxy resin for the SLA-500 was also released. SL 5180 resin was developed specifically for use with argon ion laser-based systems operating at 351 nm.

These epoxy resin formulations have demonstrated many significant advantages over the earlier acrylate SL resins.[21] Parts built in either SL 5170 or SL 5180 resin show greatly reduced effective linear shrinkage, negligible curl, minimal creep distortion in the green state, improved flatness, and almost zero swelling. These properties allow the generation of exceptionally accurate SL parts. Also, parts built in these resins exhibit substantial improvements in their mechanical properties, relative to earlier SL acrylate photopolymer systems. These resins were essential for QuickCast applications that enabled direct shell investment casting of functional metal prototypes from SL patterns.

2.5 Dimensional Accuracy and Stability

Various measurements are used to demonstrate the accuracy achieved by SL 5170 and SL 5180 epoxy resins, relative to the earlier acrylate-based systems. Among numerous diagnostic tests available to characterize SL resins and build processes, the following methods were selected. Curl distortion is tested by the twin cantilever curl test.[22] The in-plane dimensional stability of the laser-cured ("green") photopolymer is demonstrated by the green creep distortion test.[21,23] Part flatness is determined by the Slab 6×6 diagnostic test.[22] Finally, overall part accuracy, especially in the X-Y SL build plane, is established by the statistically significant user-part analysis.[24,25,26]

Photopolymer Shrinkage and SL Part Accuracy

SL photopolymers must be capable of generating parts with excellent dimensional accuracy to satisfy the tolerance requirements for various applications. Otherwise, the parts would be almost useless, unless they were intended solely for visualization purposes. Accuracy generally requires photopolymers with minimal shrinkage during building, and also during the postcuring process. However, the most relevant shrinkage parameter for SL parts is NOT the total volumetric shrinkage of the liquid SL resin upon photocuring, but rather the phenomenologically measured effective linear shrinkage determined, for example, from a CHRISTMAS-TREE™ diagnostic part[22] built on an SLA.

Curl Distortion. Curl distortion is quantitatively described by curl factors.[6,22] Curl factors are basically rise over run for a cantilevered section, expressed in percent. This description is similar to a commonly used measure of the steepness of roads. The twin cantilever diagnostic test, shown in Figure 2-5, indicates how much the extended cantilever sections of a part will curl if they are not properly supported.

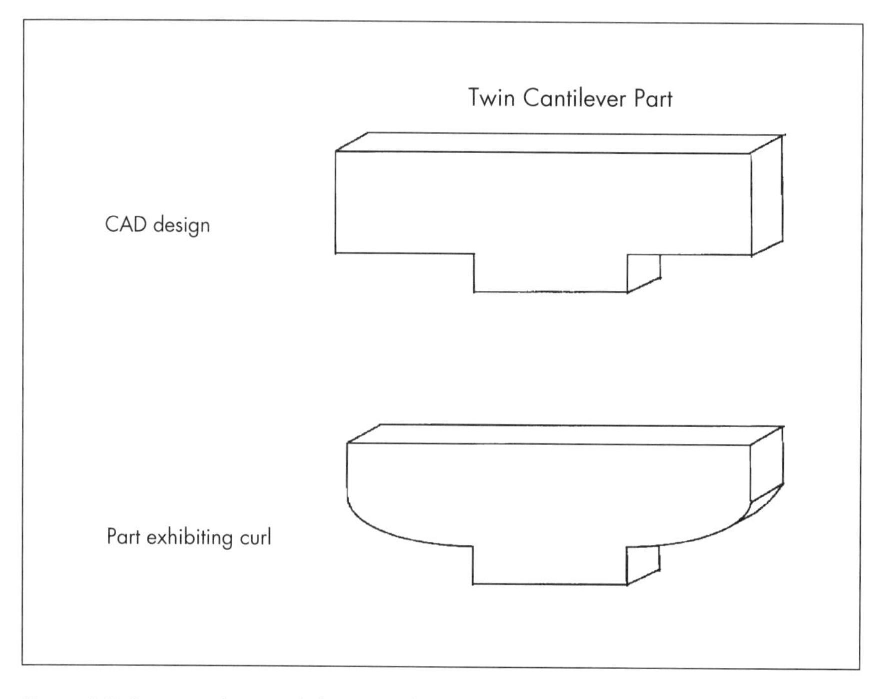

Figure 2-5. Twin cantilever curl distortion diagnostic test part.

Furthermore, curl distortion is a phenomenon that is characteristic of objects built in multiple layers. If the laminates undergo shrinkage (or expansion) when an additional layer is applied over a cured layer, the object will distort. This distortion may be instantaneous or delayed. The susceptibility to curl is present not only for the SL process, but for all layer-additive RP&M techniques, including SLS, FDM, and SGC, that involve chemical or physical changes at the layer adhesion interface. Even the adhesion process in LOM leads to residual stresses in the final part.

Volumetric Shrinkage. Intuitively, volumetric shrinkage and curl distortion appear to be directly related. If resins did not undergo any shrinkage, there would be no residual stresses generated within the multilayered SL part. However, curl distortion has been shown NOT to correlate directly with the total resin volumetric shrinkage. This was clearly evident in a study performed on two model acrylate and methacrylate SL resins systems.[6] In this study, the acrylate-based system resulted in much greater curl distortion than the methacrylate system, even though the total volumetric shrinkage values for the two different resin systems were essentially identical.

The Ciba-Geigy epoxy-based resins SL 5170 and SL 5180, which have been shown to produce parts with excellent accuracy, is another set of examples that corroborate this observation. Both SL 5170 and SL 5180 shrink about 6-7% by volume when the liquid resin is completely postcured to the solid state. The densities of representative acrylate and epoxy resins in the liquid and fully solidified states are presented in Table 2-2.

Volumetric shrinkage of ~6% corresponds to approximately 2% linear shrinkage for both acrylate-based SL 5149 and SL 5081-1 and the epoxy-based SL 5170 and SL 5180. If curl distortion was proportional to bulk volumetric shrinkage, one would expect that all these resins would curl by ~10%, similar to the acrylates. However, the curl distortion for the epoxy-based resin SL 5170 is about an order of magnitude less than that for acrylate-based SL resins.

The curl factors for the earlier acrylate resins manufactured by Ciba-Geigy are typically between 8~13%. There are other commercially available SL resins that have curl factors substantially greater than 10%.

This result indicates that the epoxy resins have an almost negligible tendency to undergo curl distortion during the building process. Resins that undergo high curl distortion are not ideally suited for SL. They reduce productivity by increasing the incidence of blade collisions. Also, the final part will not be geometrically accurate due to warping. Worst of all, flat parts often cannot be obtained from these resins.

Cantilever Curl. Cantilever curl is indeed one measure of the inherent tendency for the resin to undergo warpage during the building process, when the part is not properly supported. This was a useful measure in the past when feasibility studies were performed to see whether or not parts could be built with minimal supports, or to prevent catastrophic curl.[6] Today, however, cantilever sections of real SL parts are built such that negligible unsupported cantilever sections occur. When cantilever sections are properly supported, curl distortion can be minimized or even effectively eliminated altogether.

Therefore, while the twin cantilever diagnostic part is still very useful for comparing different resin systems, it has limited applicability to real SL parts built with proper supports. The diagnostic test discussed in the following section addresses the tendency for supported parts to undergo delayed or latent curl distortion. This phenomenon, known as creep distortion, poses a potential problem after the supports have been removed.

Table 2-2. Densities of Liquid and Solid Stereolithography Resins at 25° C

Resin	Liquid	Solid*	Bulk Volume Shrinkage (%)	Effective Linear Shrinkage** (%)		Curl Factor (%)	
SL 5149	1.12	1.20	6.1	0.7 [WEAVE]		10	
SL5081-1	1.14	1.21	7.1	0.9 [WEAVE]		10	
SL 5170	1.14	1.22	7.0	0.05 [QC]	0.05 [ACES]	1 [QC]	2 [ACES]
SL 5180	1.15	1.22	6.1	−0.2 [QC]	0.2 [ACES]	6 [QC]	3.5 [ACES]

* Postcured in PCA for one hour.

** CHRISTMAS-TREE built on an SLA and postcured in a PCA for one hour. Dimensions were measured, and average linear shrinkage was determined. Values correspond to recommended SLA linear shrinkage compensation.

QC = QuickCast™, quasi-hollow buildstyle.

WEAVE™ = solid build style for acrylate resins. ACES™ = solid build style for epoxy resins.

Green Creep Distortion

SL parts are designed to be postcured as soon as they are built. However, some SL parts are so complex that cleaning and support removal may take many hours before postcure can occur. Also, a part may be unintentionally left sitting in the green state, simply because it finished building at midnight or during the weekend, and no one was available to carry out the cleaning work. These are both very realistic issues that raise questions about dimensional stability in the intermediate stage, that is, in the green state. How would an SL part behave if left in the green state for an extended period?

In addition to building accurately in an SLA, SL parts must also remain dimensionally stable over a reasonable period. Most SL applications, including visualization, verification, iteration, and design optimization, as well as functional testing or investment casting, require postprocesses that take some time. For example, green parts must be cleaned and postcured. Postcured parts are usually sanded, finished, or painted before they are used. QuickCast parts, for example, are investment cast into metal.

In this case, depending upon part complexity, the finishing and part preparation job may take a few hours, or may even take as long as a few days. In the typical shell investment casting process, it takes about 1 to 2 weeks for the SL QuickCast pattern to be dimensionally stabilized by the applied ceramic shell, including the time that it may sit around due to scheduling delays. If the pattern deforms over this period, it may no longer accurately represent the original CAD data, even if the part was accurate immediately after being built in the SLA.

So what types of accuracy problems do the parts typically exhibit? Experience with the earlier acrylate resins showed that when SL parts are left in the green state for a period of days without postcuring, their dimensional errors increase. Furthermore, the longer these parts remain in the green state, the more inaccurate they become.

For visualization purposes, the dimensional change with time may be relatively small and often causes no problems. However, dimensional instability and its time dependence must be identified and understood in order to build highly accurate, dimensionally stable parts that meet the requirements for direct shell investment casting applications. For rapid tooling applications where accuracy is critical, resins and build processes must be found that can generate parts with superior fidelity to the CAD model.

Draining the Resin. New dimensional stability requirements for SL resins in the green state became apparent. With QuickCast, SL parts are built in a quasi-hollow structure. When the QuickCast part rises out of the SLA vat, the part is initially filled with liquid resin, trapped in cells between the thin, outer boundaries of cured resin. Next, a set of vent and drain holes are generated at appropriate locations, either manually or using the latest QuickCast 1.1 software during SL file preparation. These holes provide a means to drain uncured liquid resin from the part. In a QuickCast pattern, the internal volume is topologically simply connected, allowing liquid resin to flow from one internal section of the part to another. This enables drainage of the liquid resin before a QuickCast pattern is further processed for shell investment casting.

Due to the finite viscosity and surface tension of the liquid resin, drainage does not take place instantaneously. The epoxy-based resins have viscosities of about 200 cps at 86° F (30° C), which are about an order of magnitude lower than the viscosity of the earlier acrylate resins. This allows some simple parts to drain in a few minutes, and most parts in a few hours.

However, complex parts resulting in narrow internal passages may take several hours to drain. This is carried out in addition to the normal support removal and finishing processes. During draining, what happens to the part dimensions? QuickCast parts are a concern since they involve thin walls, and *must* be left in the green state throughout this postprocessing period.

While RP&M offers the ability to build complex geometries, curl distortion has traditionally been one of the major disadvantages. When additional layers are cured on top of each other, physical or chemical transformation takes place in the material. For SL, it is a photochemical cross-linking reaction, and for RP&M methods that involve sintering or solidification of a heated material, it is changes in the density of the materials due to changes in phase or temperature. All these changes involve some degree of shrinkage and lead to internal stresses. This internal stress ultimately manifests itself as curl distortion when sections of a part are not supported.

The same internal forces lead to a type of delayed warpage known as green creep distortion (GCD). As discussed earlier, even photopolymers with a tendency to curl can be built virtually without curl distortion when using appropriate support attachments. However, when the part is removed from the supports, as well as the underlying metal platform that restricts the distortion from happening, the residual internal stresses generate GCD.

The GCD test was developed to investigate the dimensional stability of SL parts in the green state. The test procedure[21] is schematically shown in Figure 2-6. The test involves a long thin square cross-section strip 8 inches × 1/4 inch × 1/4 inch (200 mm × 6.4 mm × 6.4 mm). Remember, a thin strip with a high aspect ratio is one of the most difficult parts to accurately build and maintain for virtually all layer-additive RP&M methods.

Initially, the strip is completely supported so that it stays flat on the SLA platform during build. The part is then taken out of the SLA vat and the supports are removed. The SL strip is placed on an electro-optical measuring device, shown schematically in Figure 2-7, and then intentionally allowed to undergo creep distortion in the green state. The maximum deflection at the midpoint of the green strip, or the two ends of the strip, is defined as the GCD. Measured electro-optically, GCD is automatically recorded on a computer over a period of 24 hours.

GCD is plotted in Figure 2-8 as a function of time for the three acrylate resins, SL 5081-1, SL 5143, and SL 5149, all built in STAR-WEAVE, and also for the epoxy resins, SL 5170 and SL 5180, using the QuickCast build style.[21,23] The data

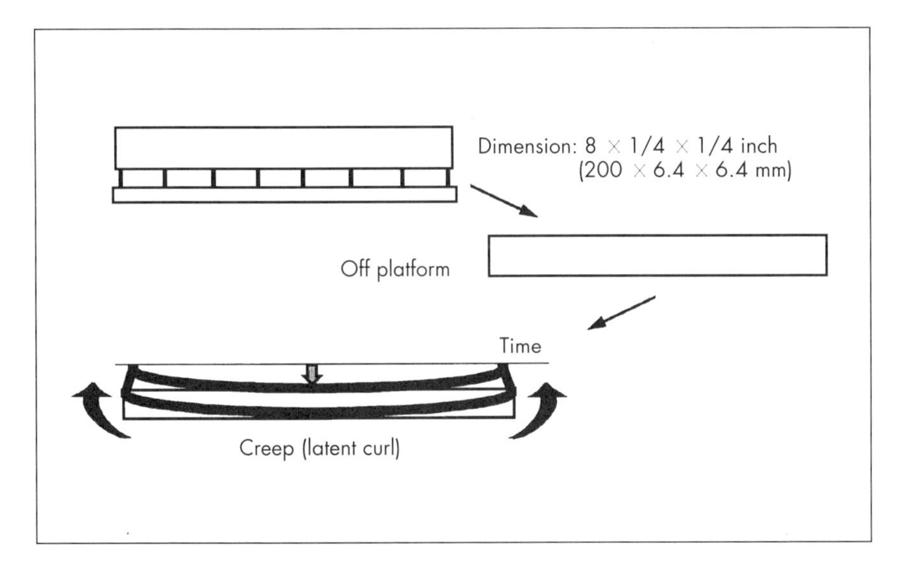

Figure 2-6. Schematic of the green creep distortion test procedure.

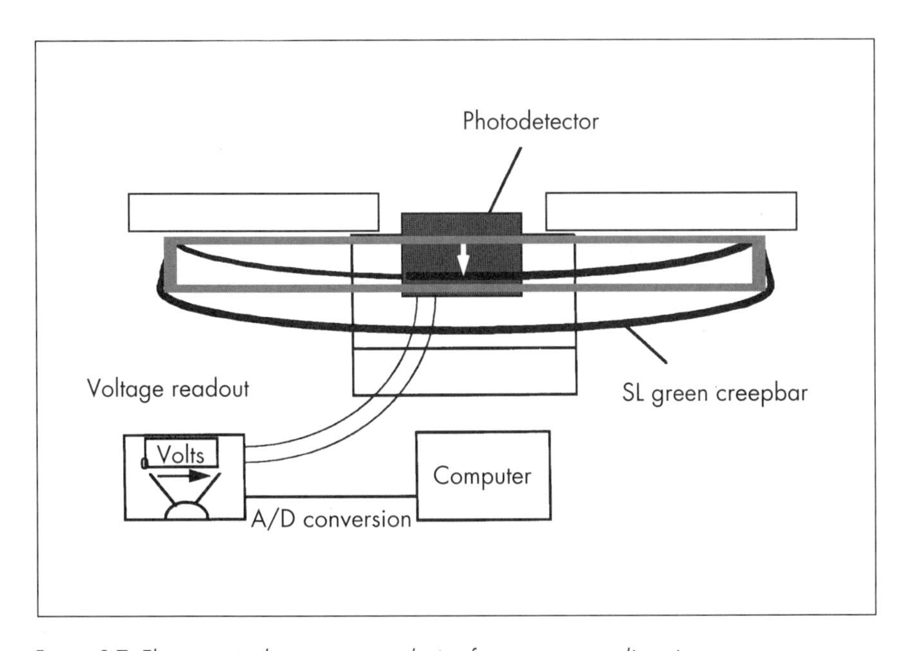

Figure 2-7. Electro-optical measurement device for green creep distortion.

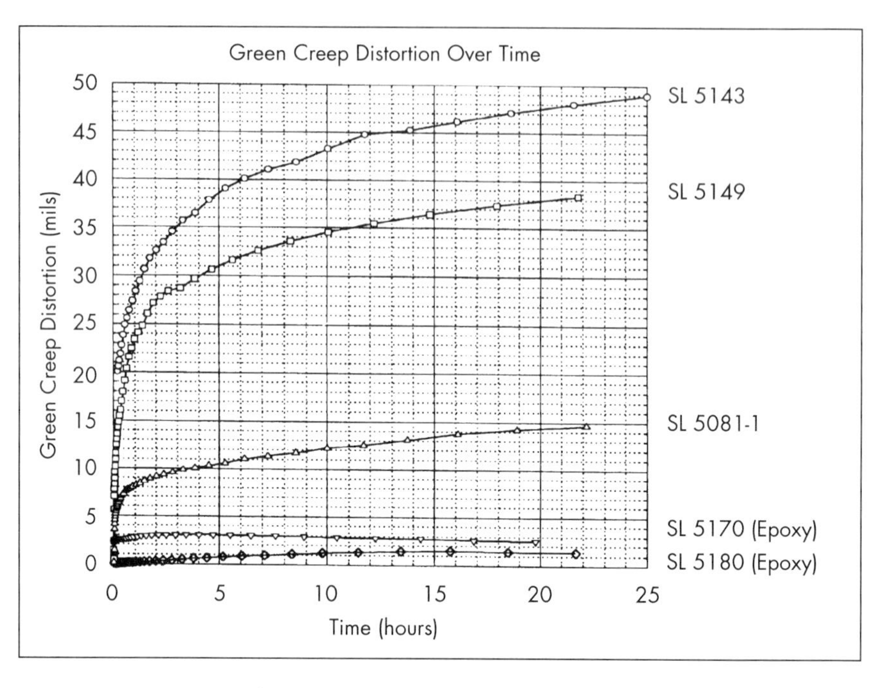

Figure 2-8. Green creep distortion data for a thin strip built in acrylate-based SL 5143, SL 5149, SL 5081-1, and epoxy-based SL 5170 and SL 5180.

for test strips built in the solid ACES style using these epoxy resins was found to be essentially identical. From Figure 2-8, note that the GCD for all resins initially increases quite dramatically. The distortion rate soon slows, approaching an asymptote. For all acrylate resins shown in Figure 2-8, more than 60% of the absolute GCD measured at an elapsed time of 20 hours occurs within the initial two hours. For example, the strip built in SL 5143 distorts to 0.048 inch (1.219 mm) after 20 hours. At two hours, the distortion is already about 0.035 inch (0.889 mm), which is about 73% of the final GCD value after 20 hours. Among acrylate resins, SL 5081-1 has the lowest creep distortion.

This data suggests that cleaning and finishing of acrylate-based SL parts should be performed as quickly as possible, once the restraining support structures are removed. This is especially true if some of the sections have a high aspect ratio, with a long axis in the plane of the liquid resin. It is clear that both epoxy resins SL 5170 and SL 5180 have extremely low GCD. Both epoxy resins have GCD values less than 0.004 inch (0.1 mm) over their full 8-inch (200-mm) length, within the 24-hour test period.

Furthermore, when the GCD data was plotted as a function of the logarithm of time, it was found to be log-linear, as seen in Figure 2-9. Using this characteristic

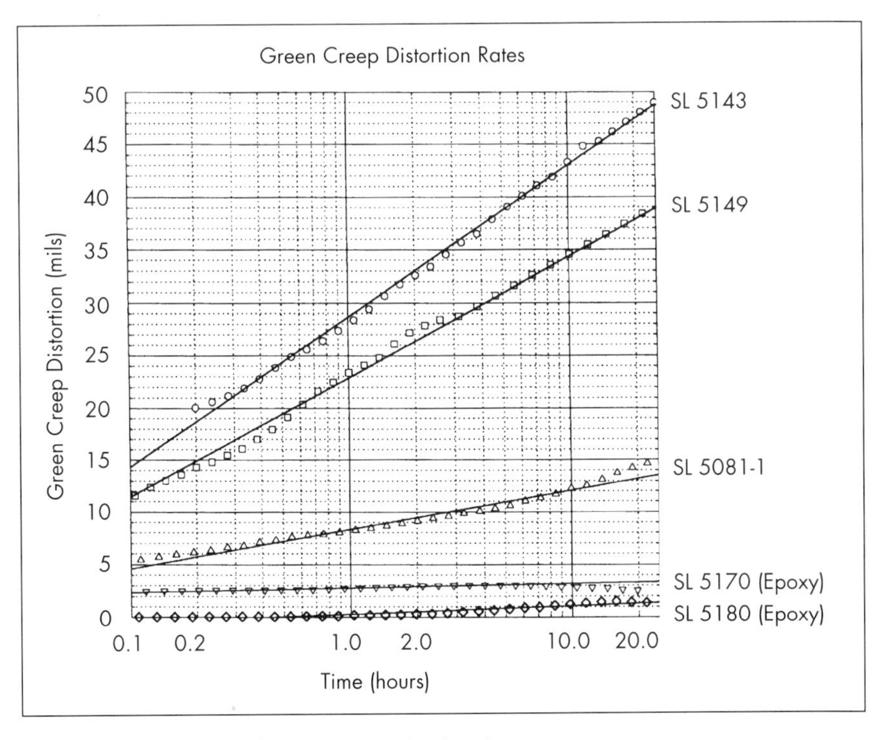

Figure 2-9. Green creep distortion versus log (time).

log-linear creep behavior, a convenient parameter called the green creep distortion rate (GCD rate) was defined, where GCD rate is defined as $dGCD/dlog_{10}t$. This parameter characterizes the rate at which the test part undergoes creep distortion, as indicated by the slope of the curve in Figure 2-9. Thus, a single parameter may be used to describe the creep behavior of SL resins. This allows one to directly compare the dimensional stability of various resins or build styles.

The GCD rates for the SL resins tested are summarized in Figure 2-10. It is apparent that the epoxy resins have substantially lower GCD rates than acrylate resins. The latest results indicate that SL 5180 built in ACES, with a GCD rate of 0.3 mils/log hour, has the highest dimensional stability in the green state, among the SL resins tested. This means that an 8-inch (200-mm) long thin square strip built in SL 5180 will undergo only about eight microns of creep distortion if left in the green state for an additional ten hours. Considering the error-bars for the experiment to be ~±0.0005 inch (±12 microns), test parts in SL 5170 built using either QuickCast or ACES are also comparable.

Compared to SL 5143, the *green creep distortion rate of the epoxy resins has improved more than 40-fold over the past three years*, from 0.0144 inch (366 microns) to about 0.0003 inch (8 microns) for each decade of time. In summary,

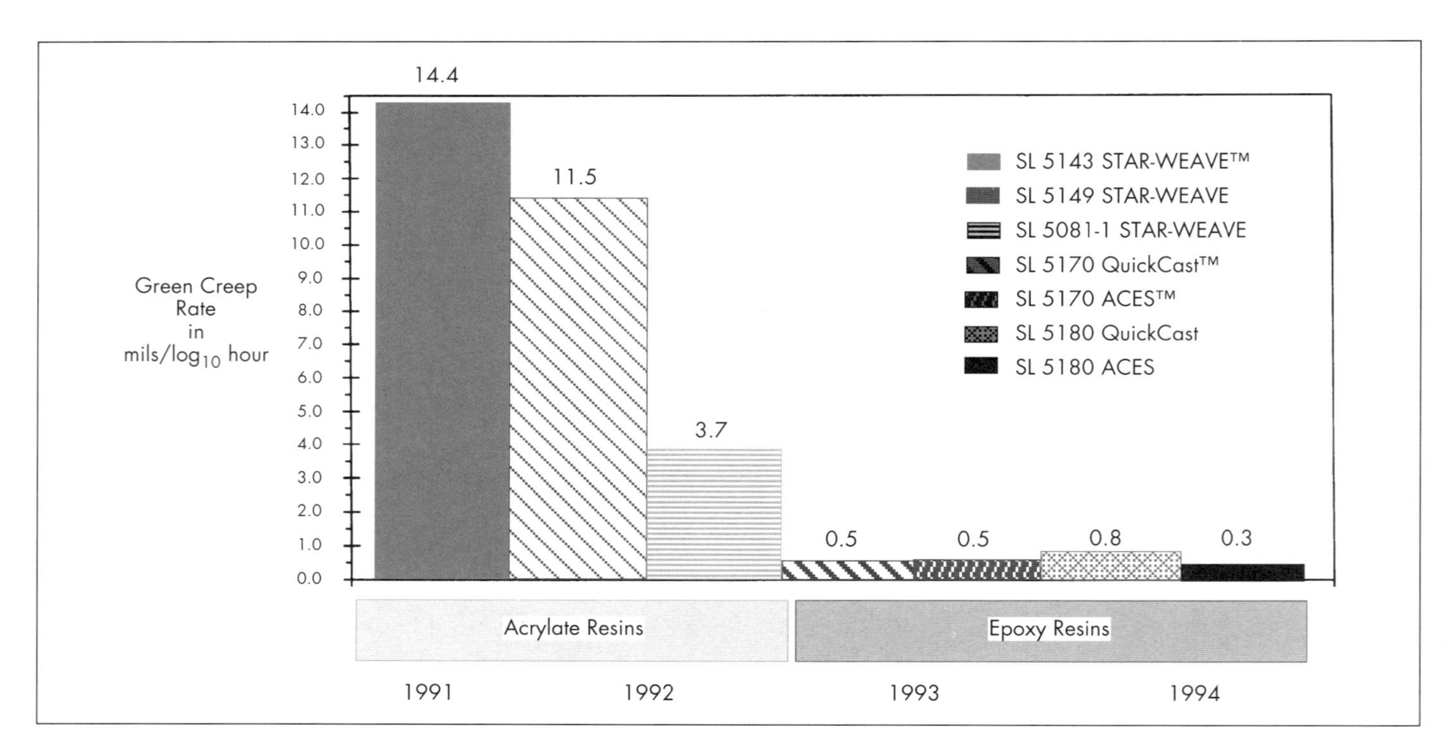

Figure 2-10. Green creep distortion rates for various SL resins and build styles.

the GCD rates of the various Ciba-Geigy SL resins may be ranked in ascending order of GCD rate:

SL 5170 & SL 5180 << SL 5081-1 < SL 5149 < SL 5143.

The dramatically reduced GCD rates of SL 5170 and SL 5180 are key to preserving SL part accuracy.

Part Flatness

Green creep distortion is the result of the residual internal stress generated during the SLA building cycle. Postcuring is not involved. The green creep distortion rate does not tell us what happens when the SL part is postcured. Would an SL part maintain its CAD designed shape even after normal postcuring?

The Slab 6×6 flatness diagnostic test is a more comprehensive test that provides the answer to this question. Slab 6×6 is a simple 6 inch × 6 inch × 1/4 inch (150 mm × 150 mm × 6 mm) thin, nominally flat horizontal slab, built flat on an SLA. The slab is built and then UV postcured for one hour. However, the UV radiation is exposed from one side only, to simulate the worst case scenario. Finally, the postcured slab is aged for one week to observe the postcured creep distortion. The Slab 6×6 distortion test takes into account all the events that a normal SL part would encounter.

While this part may seem simple, it is actually one of the most difficult parts to build accurately in SL, or any other layer-additive RP&M method that involves a phase change, or variation in the dimensions of the solidifying layer during the building process. It should be stressed that parts with vertical walls or other stiffening features on top of a thin horizontal slab tend to increase the moment of inertia of the object, thereby reducing distortion. The end result is that a flat slab section, with added complex components, tends to build flatter than a simple thin horizontal slab. Somewhat ironically, part flatness actually improves with greater geometrical complexity. This is certainly another nonintuitive result, compared with conventional technologies. Conversely, a thin slab is an excellent "worst case" diagnostic to characterize part flatness.

The distortion range (i.e., highest measurement minus lowest measurement) is established using a coordinate measuring machine (CMM) on the upper surface. Generally, the highest values occur at the corners, and the lowest values occur in the middle of the slab. Additional measurements are actually taken between each of the process steps in the Slab 6×6 test procedure.[22] However, for the purpose of this discussion, only the final maximum error value, measured seven days after the build and UV postcure, is reported.

The Slab 6×6 flatness diagnostic test results are summarized in Figure 2-11. In 1989, the Tri-Hatch build style was found to build a highly concave Slab 6×6, having a distortion range of 0.227 inch (5.76 mm). A Slab 6×6, built with the conventional acrylate-based SL resins in this obsolete build style or with resins that yield high curl factors, often came out warped and significantly concave

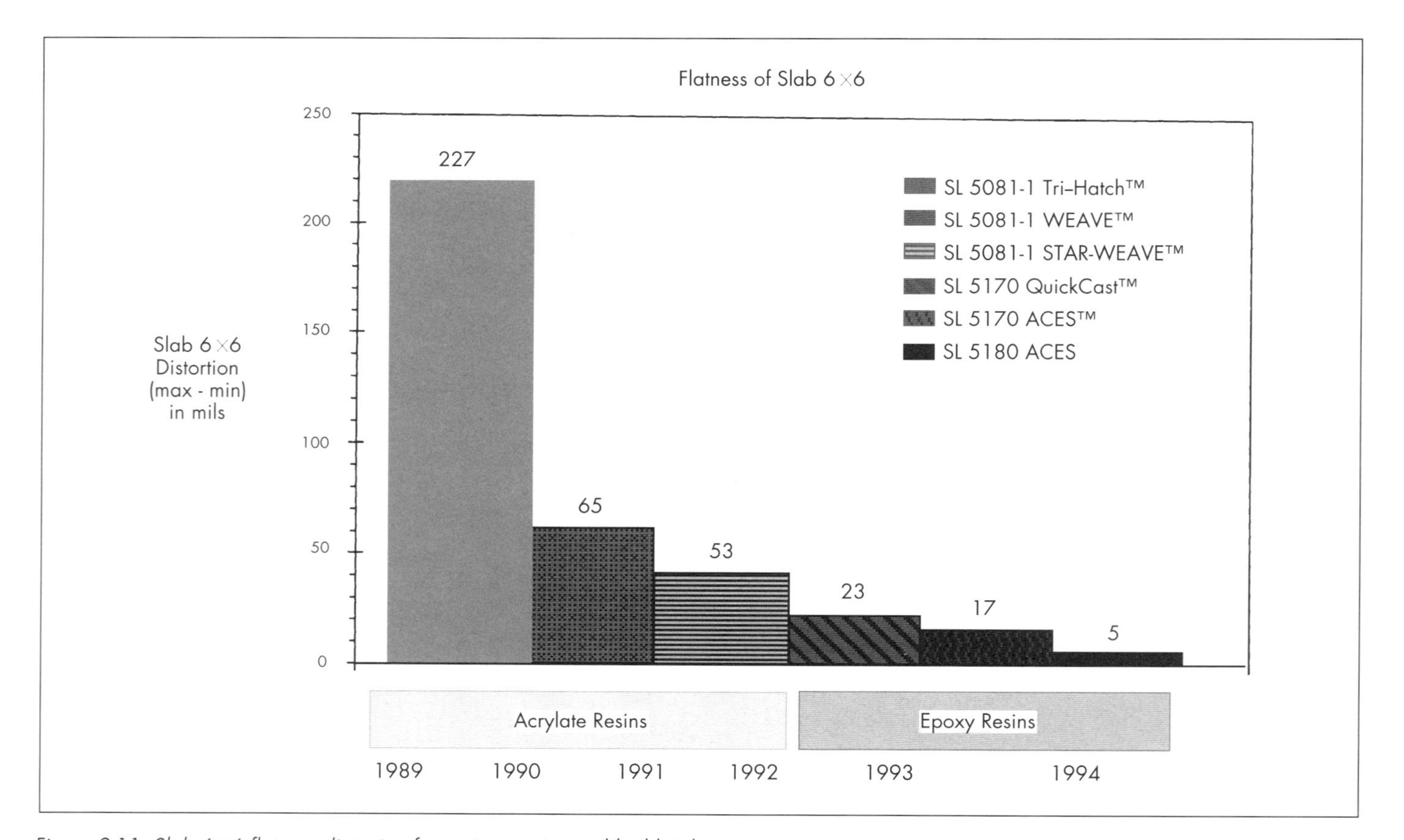

Figure 2-11. Slab 6×6 flatness distortion for various resins and build styles.

upward after the seven-day period, looking like a shallow bowl. Most of the slab distortion occurred in the UV postcuring step.

When the new WEAVE build style was introduced in 1991, the flatness immediately improved by more than three-fold, to 0.065 inch (1.65 mm). After the release of STAR-WEAVE in 1992, the Slab 6×6 distortion range was further reduced by more than four-fold to 0.053 inch (1.35 mm), an accuracy improvement of an additional 18%. The minimum distortion range values in 1992 were all achieved using SL 5081-1 acrylate resin.

Further process modifications on the acrylate resin did not yield better results. However, when the epoxy resin, SL 5170, was introduced in 1993 together with the QuickCast build style, the flatness suddenly improved by more than another factor of two relative to the next best acrylate resin. The measured Slab 6×6 distortion instantly dropped from 0.053 inch to 0.023 inch (1.35 mm to 0.58 mm). With process optimization, and the development of the new ACES build style, the distortion was later reduced to 0.017 inch (0.43 mm).

Now the Slab 6×6s were finally starting to look like flat slabs. Remember, the Slab 6×6 diagnostic test involves the absolute worst case of postcuring from one side only. In real applications, such a structure would normally be postcured from both sides to provide more uniform UV exposure, thereby minimizing postcure distortion.

The resin development program continued, resulting in the 1994 release of SL 5180, an epoxy resin for the SLA-500. Using SL 5180 in the ACES build style resulted in a superior flat slab, with a maximum distortion of only 0.005 inch (0.13 mm). This Slab 6×6 distortion, obtained in 1994, corresponds to an almost 50-fold improvement compared to that of 1989, and a 10-fold improvement from the best Slab 6×6 results in 1992. Now, SLA users can build nearly undistorted flat parts when required, using epoxy resins SL 5170 and SL 5180.

Overall Part Accuracy

All the described diagnostic tests focus on distortions inherent to layer additive RP&M methods, namely, inaccuracies in the Z-direction. All of these distortions may be traced to differential inter-layer shrinkage. While distortions of this type are often the largest sources of part inaccuracy, how shrinkage affects the overall part dimensions in the X-Y dimensions must also be well understood. Accuracy in the Z-dimension depends on the formation of the layers and cure depth control, in the case of SL. This involves the laser scanning velocity, as well as liquid leveling. For SL, the X-Y dimensional accuracy depends strongly on the shrinkage of the resin and the positioning accuracy of the laser scanning system.

User-part Test. The user-part accuracy diagnostic[24,25,26] was designed by the North American Stereolithography Users Group. Each user-part involves the measurement of 170 dimensions of a relatively complex geometry. A CMM is used to measure 78 dimensions each in X and Y, and 14 in Z, for a total of 170 data points per user-part. Geometries include thin, medium, and thick walls; short, medium, and long sections; circular and square holes; and walls intersecting at 45 and 90° angles. The goal of the user-part test is to determine statistically significant accuracy and repeatability estimates through Error Distribution Functions (EDF) generated from these measurements.

Root-Mean-Square (RMS) Error. User-part accuracy is established from the error values associated with the CMM measurements. The root-mean-square (RMS) error is calculated from at least 170 measurements (one user-part). With increased measurements, the statistical significance of the results improves. The RMS error represents the error range (+ or − the nominal dimension) within which 68% of the part dimensions will fall, provided the EDF is Gaussian.

The RMS error values of SL user-parts built during the five-year period from October 1989 to October 1994 are presented in Figure 2-12. This figure clearly shows significant advances in the accuracy of SL user-parts over the last five years. Note that user-part accuracy has improved from an RMS error of ±0.0089 inch (225 microns) to ±0.0018 inch (45 microns). This represents a five-fold advance in user-part accuracy over a period of only five years.

The improvement in overall part accuracy came from a number of factors. Advances in hardware, software, process, and resin all contributed incrementally to better SL part accuracy. However, some factors contributed more than others. For example, all accuracy data between 1989 and 1992, shown in Figure 2-12, corresponds to user-parts built with SL 5081-1 acrylate resin. Most of the advances during this time period, resulting in a two-fold accuracy improvement from an RMS error of ±0.0089 inch (225 microns) to ±0.0045 inch (114 microns), were due to SL process developments. Here the process involves the appropriate selection of parameters that define the SL build style, using the SLA hardware and photopolymer provided. Some of these parameters include laser scanning speed, drawing sequence, border overcure, hatch spacing, layer thickness, part deep-dip distance, etc., as well as leveling and recoating. In particular, the two-fold improvement of the accuracy data for SL 5081-1 resin was mostly due to build style advances from Tri-Hatch*, to WEAVE, to STAR-WEAVE.

However, accuracy improvements from process modifications alone could not be sustained indefinitely. In retrospect, an inherent limitation due to the SL resin systems had been reached. In early 1993, the development program for SL 5170 was underway. A user-part was subsequently built in that resin. The RMS error immediately decreased from a value of ±0.0045 inch (114 microns) using acry-

**Tri-Hatch is one of the earlier SL build styles that resulted in pockets of liquid resin trapped between walls of cured resin. WEAVE and STAR-WEAVE are advanced build styles resulting in minimal internal stress, and maximum laser-cured resin. Reference 26 explains the latter two build styles in great detail.*

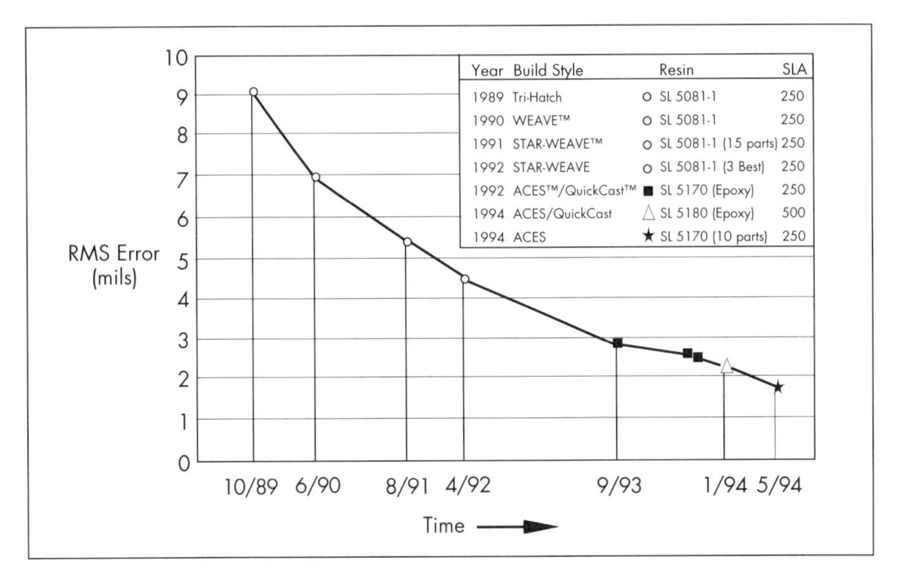

Figure 2-12. User-part improvement from 1989 to 1994. All data beyond April 1992 are those of epoxy-based SL photopolymers.

late resin SL 5081-1 to ±0.0028 inch (71 microns) in the epoxy resin SL 5170, resulting in roughly another 40% improvement in accuracy.

This advance in accuracy was necessary, especially for the QuickCast application, which was also commercially released, together with SL 5170, in July 1993. In this process, an SL pattern generated in the QuickCast build style is converted directly into metal using the shell investment casting technique. With the advent of QuickCast, and the subsequent availability of metal prototypes from SL patterns, part accuracy requirements were pushed substantially. The accuracy specifications for a functional prototype are much more demanding than those for parts that are intended mainly for concept models and design visualization.

In March 1994, the epoxy resin SL 5180 was also introduced for the SLA-500. At this point, hardware and software improvements culminated in the Orion Imaging Technology, as well as the new Diode Laser Leveling system, both developed by 3D Systems. With the help of these advances, user-parts built in SL 5180 on an SLA-500 were now able to achieve an overall RMS error of ±0.0022 inch (56 microns).

Furthermore, the latest results for user-parts built with SL 5170 on an SLA-250 produced the current record of ±0.0018 inch (45 microns) RMS error. It is important to note here that this RMS error value is based on not one, but 10 user-parts (viz. 1700 measurements) built on the same machine over a period of three months. This demonstrates that SLA systems have reached a very high repeatability when parts are built using epoxy resins.

2.6 Other Relevant Properties of SL Photopolymers

Surface Wetting

Surface wetting is also a very important property for epoxy resins. In SL, thin layers of liquid resin are formed on top of cured resin. If the surface tension of the liquid resin is much greater than the surface energy of the cured resin, then the liquid will ball up, much like drops of water on the smooth surface of a teflon-coated frying pan.[27] The equivalent problem that may arise on an SL building surface is called dewetting. If this happens during the SL building process, a number of problems may result. If dewetting occurs on the very last up-facing layer, then the part will have an imperfect top surface. If it occurs on the interior layers, dewetting may lead to layer delamination because the subsequent liquid layer requires twice the original cure depth to achieve layer adhesion. If it occurs at the borders, dewetting often results in line separation, also referred to as "bird-nesting," leading to parts that have fuzzy outside walls.

One of the most important parameters relevant to dewetting prevention is layer thickness. The dewetting process happens mainly due to the difference between the surface tension, or adhesive energies of the liquid, and laser-cured surfaces. The surface of a nicely coated liquid layer can break when the surface energy differences between the liquid and the cured resin are large enough to overcome the gravitational forces that try to flatten the liquid layer.

Obviously, if the layer thickness is large, then the gravitationally induced pressure difference between the liquid surface and the cured surface becomes correspondingly greater. If you keep pouring water into a teflon-coated frying pan, at some point the surface will become smooth and flat, and the balling effect or apparent dewetting, disappears. Of course, water on teflon is an extreme case in which the liquid surface tension is much greater than that of the teflon substrate.

The next important parameter is the surface UV exposure level. The surface tension of the cured SL resin surfaces depends on the degree of cure. If inappropriate exposure levels (i.e., C_d and overcure values) are used, excessive dewetting may occur for some resin formulations.

Swelling

Swelling is another important property. When SL resins are cured in the vat and exposed to liquid resin for an excessive time interval during the building cycle in an SLA, part dimensions may increase due to the infusion of the liquid monomers into the semicured green part. A diagnostic test for swelling was established some time ago.[22]

For a given resin, swelling is primarily a function of the wetted liquid surface to cured volume ratio, degree of cure, and immersion time inside the liquid resin. Greater wetted liquid surface to cured volume ratio, reduced degree of cure, and longer immersion time all result in increased swelling. This can become a measurable source of dimensional inaccuracy if the SL resin swells excessively, especially when parts are built with a quasi-hollow internal structure.

The results for some of the SL resins are shown in Table 2-3, when built in the most susceptible quasi-hollow build styles such as Tri-Hatch or QuickCast. Note that QuickCast build styles are not available for acrylate resins. Thus, the data comparison is not altogether straightforward. However, the QuickCast build style results in a greater wetted liquid surface to cured volume ratio, because cured lines are jogged between several layers, while the Tri-Hatch build style does not offset the hatch vectors. In the Tri-Hatch build style, all hatch lines are drawn on top of each other throughout the entire part. For this reason, the QuickCast build style is expected to result in greater swelling than Tri-Hatch for a given resin. Nevertheless, even with this handicap, the epoxy-based SL 5170 and SL 5180 resins show minimal swelling. This ensures that inaccuracies due to swelling in the vat are negligible, and is yet another reason why these resins build accurate QuickCast patterns.

Table 2-3. Swell Tower* Diagnostic Test

Resin	Type	Build Style	Swelling (mil/inch)
SL 5081-1	acrylate	Tri-Hatch	2.2
SL 5143	urethane-acrylate	Tri-Hatch	10.8
SL 5149	urethane-acrylate	Tri-Hatch	9.6
SL 5170	epoxy	QuickCast	0.0
SL 5180	epoxy	QuickCast	0.6

*2×2-inch 100-mil-thick-walled, 5-inch tall rectangular tower built over 24 hours.

Humidity and Temperature Effects

Environmental conditions are also important when using epoxy-based SL photopolymers. Epoxy SL photopolymers SL 5170 and SL 5180, just like conventional epoxy resins, tend to be mildly hygroscopic. If liquid resin is left in contact with water or a high-humidity atmosphere, it will absorb a finite amount of water. To prevent excessive water absorption, liquid epoxy resin exposed to air should be kept in a room at moderate to low relative humidity levels. Liquid epoxy SL resin exposed to excessive humidity for prolonged periods may contain sufficient water that its photospeed ultimately becomes impaired. If cured SL parts built in epoxy resin are left in a high-humidity environment, they will soften with time.

Furthermore, dimensional change also occurs at moderate to high relative humidities. To prevent these inconveniences, store and ship SL parts made from epoxy resins inside plastic containers with desiccant sacks included therein. Also, maintain the room humidity as low as possible to prevent part softening in the vat. The humidity levels must be kept under control to assure accurate, strong epoxy parts.

QuickCast patterns are significantly more susceptible to softening and dimensional expansion from exposure to high humidity than solid ACES parts, due to their relatively greater surface to volume ratio. Constantly maintaining the relative humidity of the room under 50% at 72° F (22° C) has been confirmed to bring these adverse humidity effects under control.

The temperature effects on part dimensions are often overlooked. The coefficient of thermal expansion (CTE) of organic polymers such as cured SL resins are generally between 30 to 110 ppm° F (50 to 200 ppm° C) for temperatures below the resin glass transition temperature. If the temperature is not controlled, especially in the room in which the part is used for the particular application, the dimensional disparity from the nominal dimensions can become unacceptable. The part size and the absolute dimensional error are also related.

For example, for a 1-inch (25.4-mm) part made of a material having a CTE of 100 ppm° C, an increase of 1.8° F (1° C) would result in an expansion of only 0.0001 inch (2.5 microns). For a 10-inch (254-mm) part, the change becomes 0.001 inch (25 microns). Furthermore, a 9° F (5° C) increase, which is realistic in non-temperature controlled areas, would correspond to 0.005 inch (125 microns) of expansion. The RMS error of a user-part built in epoxy resin SL 5170 can currently be held to less than 0.002 inch (50 microns). However, if the temperature is not carefully controlled, part accuracy could be compromised.

2.7 Photopolymer Photospeed and Laser Scanning Velocity

SL resin photospeed is directly associated with the amount of laser exposure necessary to achieve a prescribed cure depth, C_d. SL photopolymer photosensitivity, which is often used interchangeably with the term "photospeed," also implies wavelength sensitivity. Nonetheless, the impact of the resin photospeed on the time required to build a part is the most relevant property for SLA users. The primary resin photosensitivity parameters (i.e., the penetration depth, D_p, and the critical exposure, E_c), are certainly not intuitively obvious quantities. However, these resin parameters can be directly associated with the much more physically significant and intuitively obvious "laser beam scan velocity," V_s, at the free liquid resin surface, necessary to generate a specific cure depth C_d.

The photosensitivity of SL resins is established using the WINDOWPANE™ procedure,[22] which involves exposing the resin to a laser beam scanned at a prescribed series of different velocities. The resulting C_d values are then plotted as a function of the logarithm of the laser exposure, E_{max}. An example is shown in Figure 2-13 for SL 5149. The dependence of cure depth as a function of exposure is called the "resin working curve," and is generally log-linear. This log-linear response to actinic radiation is one of the most fundamental SL photopolymer characteristics.[6,22,28]

The slope of the working curve is D_p, and the X-intercept is E_c. The latter corresponds to the gel point of the photopolymer. These are the fundamental pa-

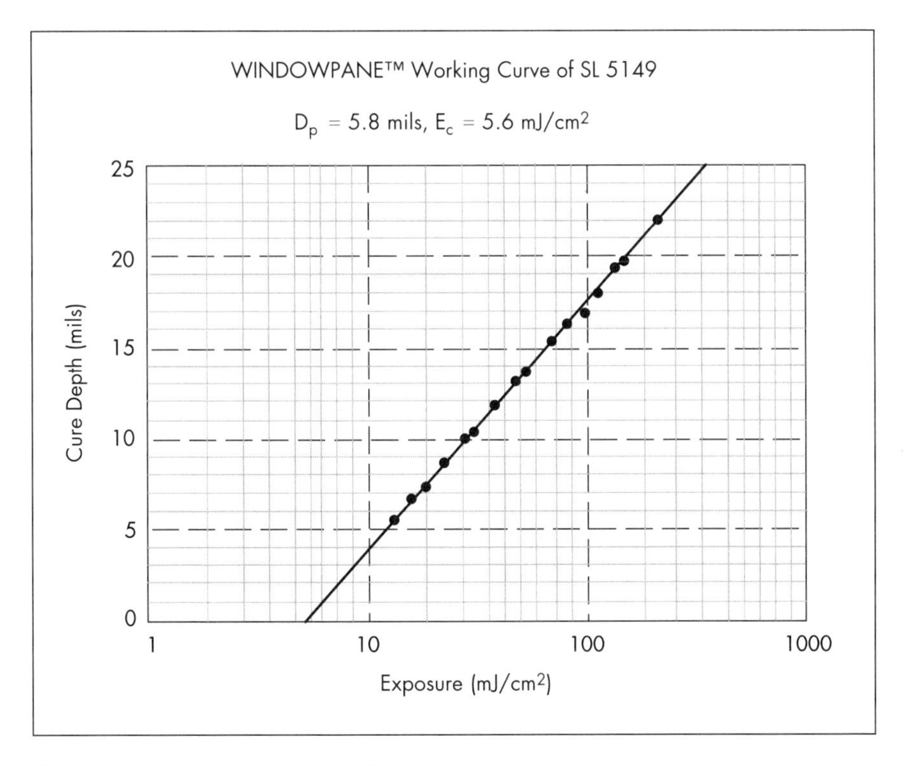

Figure 2-13. Working curve for SL 5149 resin.

rameters that define the photosensitivity of an SL resin when exposed to an actinic laser radiation source. Neither D_p nor E_c alone define the photospeed, but both parameters collectively define the generated cure depth, C_d, according to the fundamental working curve equation,[28]

$$C_d = D_p \, \mathrm{Ln} \, (E_{max}/E_c). \tag{2-1}$$

Estimating the actual laser scan velocity, V_s, required to achieve a given C_d from the values of D_p and E_c for a specific resin, is not intuitive at all, owing to the logarithmic exposure dependence. The photospeed is often erroneously quoted to be a simple function of E_c only. In reality, it is a function that must be defined by both D_p and E_c.

The most intuitive way to describe photopolymer photospeed is to convert these parameters to V_s. The greater the scan velocity, the faster the photospeed of the resin. The laser scan velocity can easily be calculated if we assume a Gaussian laser irradiance distribution. Even though the laser beams used in SLAs are not perfectly Gaussian, the following equations provide an excellent first-order approximation.

Using the equations derived previously for the line spread function of a scanned Gaussian laser beam,[29] the maximum exposure incident on the resin surface E_{max} is given by

$$E_{max} = (2/\pi)^{1/2} [P_L / W_0 V_s] \quad (2\text{-}2)$$

where P_L is the laser power incident on the photopolymer, W_0 is the $1/e^2$ Gaussian half-width, commonly referred to as the beam radius, and V_s is the laser scanning velocity at the liquid photopolymer surface.

Now, solving the working curve equation (2-1) for E_{max} gives,

$$E_{max} = E_c \exp (C_d/D_p). \quad (2\text{-}3)$$

Equations 2-2 and 2-3 can be combined and rearranged to provide the fundamental stereolithography scan velocity equation as a function of the resin parameters D_p and E_c, as well as the laser power P_L, and the desired cure depth C_d.

$$V_s = (2/\pi)^{1/2} [P_L / W_0 E_c] \exp (- C_d / D_p) \quad (2\text{-}4)$$

For a typical beam diameter of $2W_0 = 0.009$ inch = 230 microns, this equation can be numerically reduced to

$$V_s = 69.8 [P_L / E_c] \exp (- C_d / D_p) \quad (2\text{-}5)$$

where P_L is expressed in mW, E_c is in mJ/cm^2, and V_s is in cm/sec. As an exercise, let us calculate V_s to cure $C_d = 3$ mils with SL 5170 resin, when $P_L = 30$ mW. This corresponds to drawing the hatch vectors for a 6-mil (0.006-inch≈0.15-mm) layer in the ACES build style. The D_p and E_c of SL 5170 are 4.8 mils and 13.5 mJ/cm^2, respectively. Substituting into equation (2-5), we obtain $V_s = (69.8) [30/13.5] \exp (-3/4.8) = 83$ cm/sec, or about 33 inches/sec.

It is important to note that V_s is directly proportional to the laser power P_L, and is inversely proportional to both W_0 and E_c. Thus, doubling E_c will cut the scan velocity in half. However, an increase in the cure depth C_d decreases the laser scan velocity V_s exponentially. As the ratio C_d/D_p increases from 1 to 2, the exponential factor in equation (2-5) decreases by a factor of $e = 2.718 \ldots$. For example, doubling C_d slows down the laser scan velocity by a factor of 2.718 . . . resulting in a drawing speed only about 37% of the original value!

However, the actual build time depends not only on the laser power, layer thickness, or overcure values applied for the particular part, but is also strongly influenced by part size and geometry, as well as the recoating parameters and the build orientation. Nonetheless, SL users should find equation 2-4 very useful to compare the relative drawing speed of one photopolymer system with another, provided that the C_d values and the associated build style are specified.

2.8 Mechanical Properties

Green Strengths

As stated previously, cured SL resin must possess sufficient mechanical strength to support itself in the laser-cured, green state. The relevant mechanical properties are termed green strength. This includes the elastic modulus, also known as Young's modulus, which is a measure of resistance to deformation, as well as the tensile or flexural strengths of the green SL part. Without sufficient green strength, cured layers may be physically displaced by the viscous forces generated during SL part building.

Furthermore, parts without sufficient green strength may be dimensionally unstable after removal from the SLA platform. Such green parts may sag under the influence of gravity or collapse under their own weight. Support structures may become impossible to remove without damaging or distorting the part geometry.

To determine the green strengths of an SL part, a diagnostic test strip 2 inches long, having a cross section 0.4 inch × 0.1 inch (50 mm × 10 mm × 2.5 mm), is built on an SLA. These strips are then tested by measuring the resistance to bending under applied load. The green strength, or more strictly, the green flexural modulus (GFM), is measured on a Lloyds tensile test system using a compressive bending rate of 10 mm/min, at a fiber strain of 1%.

In general, the GFM values of SL parts were found to be strongly dependent on build style, including the variation in the laser beam diameter, B, hatch spacing, h_s, layer thickness, a, and cure depth, C_d. A phenomenological equation relating B, h_s, and C_d with the GFM has been proposed for various acrylate-based SL photopolymers.[31]

$$GFM = k_{WV} (C_d - C_{d0}) (B/h_s)^2 \tag{2-6}$$

where k_{WV} is the resin green strength constant for the WEAVE build style, and C_{d0} is the extrapolated X-intercept of the curve.

According to this equation, the GFM of acrylate-based SL photopolymers are directly proportional to C_d, and h_s^2, when built in the WEAVE build style. This is apparent from actual GFM data for the SL 5131 and SL 5154 resins plotted and fitted to equation 2-6 in Figures 2-14A, B, and C, and as a function of C_d and $(B/h_s)^2$, as seen in Figures 2-15A, B, and C. The GFM dependence on h_s is presented numerically in Table 2-4, together with some of the GFM data[17] for epoxy-based SL photopolymer systems.

These results indicate that the green strengths of test strips built in the epoxy-based photopolymers SL 5170 and 5180 are much greater than those of the acrylate-based resins, even when the epoxy resins were generated from thinner a = 0.006 inch (0.15 mm) layers, while those of acrylate resins were built using a = 0.010 inch (0.25 mm) layers. *The GFM of the epoxy-based SL 5180 is 5 to 12*

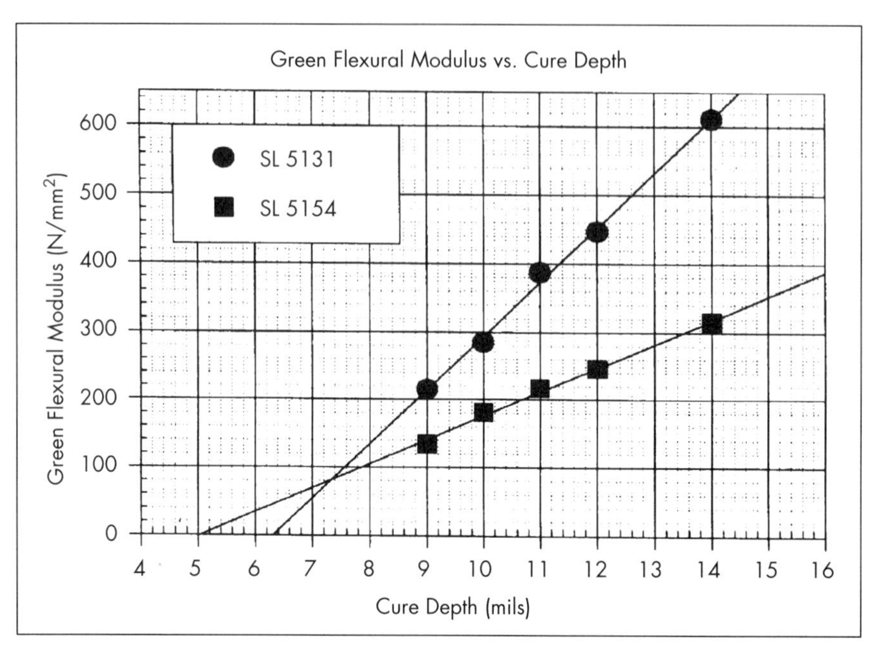

Figure 2-14A. Green flexural modulus dependence on various cure depths for specimen built in 10-mil layer WEAVE, using a beam diameter of 10 mils.

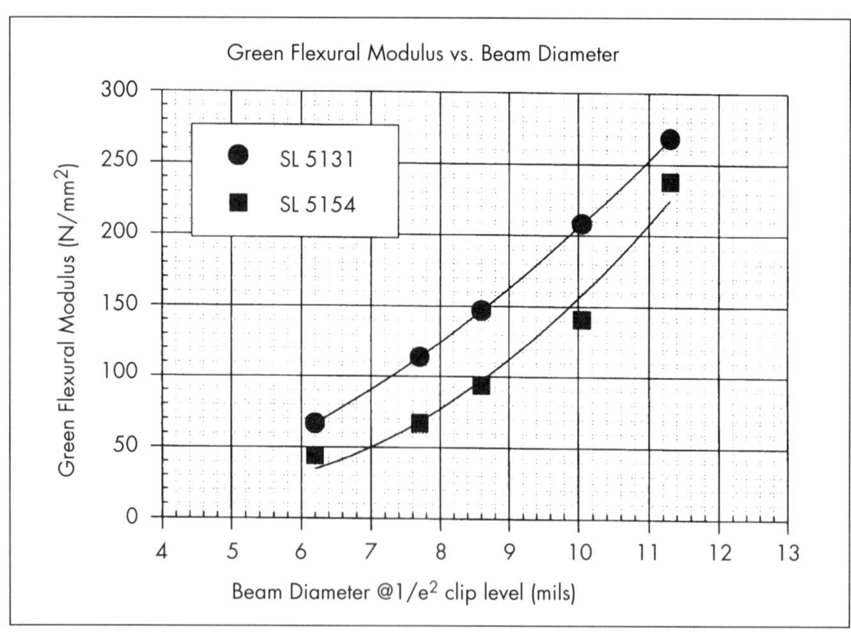

Figure 2-14B. Green flexural modulus dependence on beam diameter, using a hatch spacing of 9 mils.

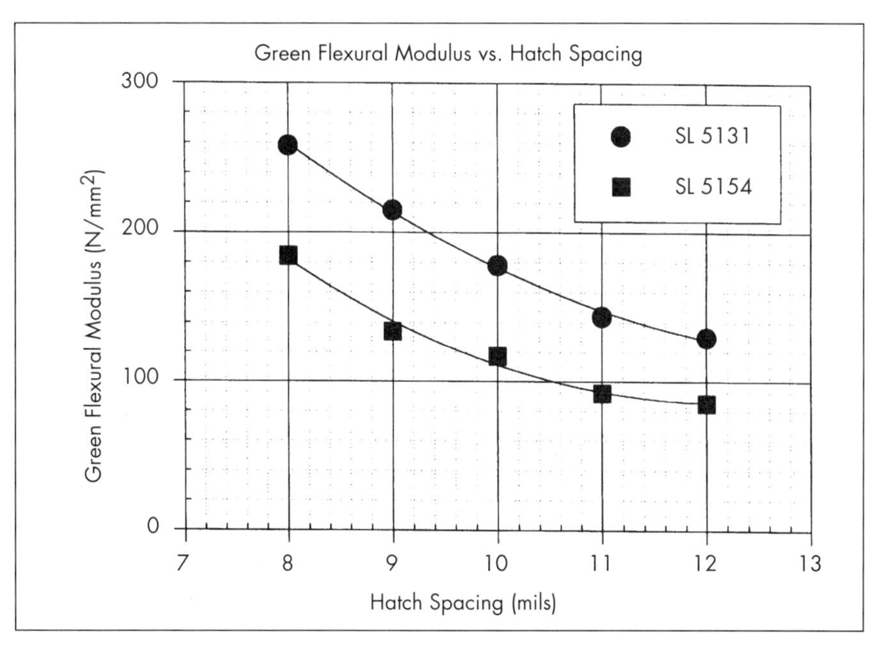

Figure 2-14C. Green flexural modulus dependence on hatch spacing, built with a beam diameter of 10 mils.

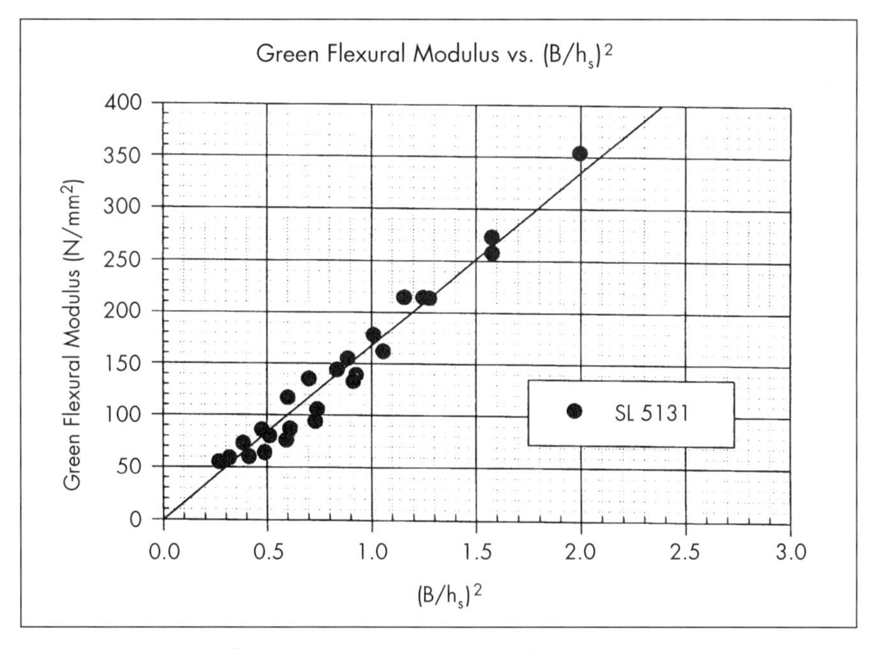

Figure 2-15A. Green flexural modulus versus $(B/h_s)^2$ for acrylate-based AL 5131 photopolymer.

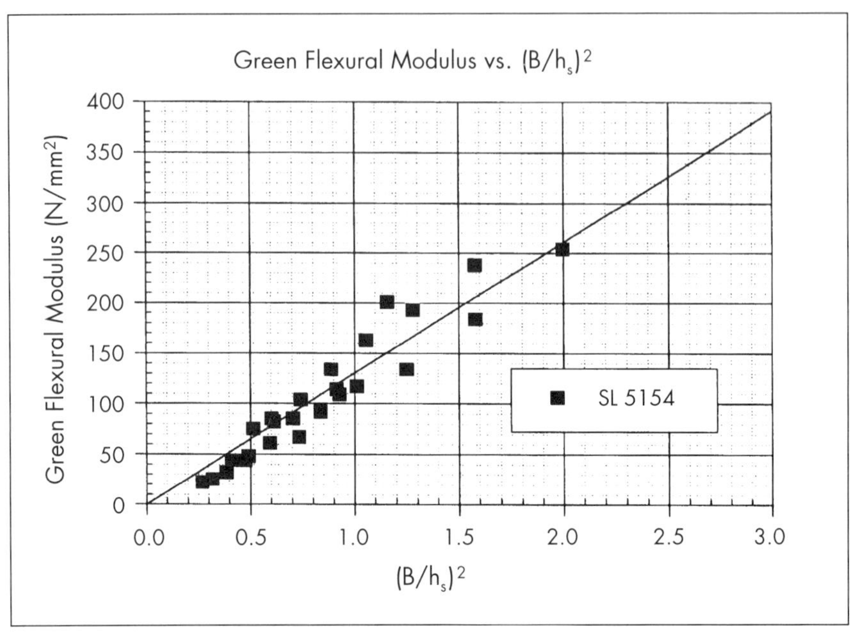

Figure 2-15B. Green flexural modulus versus $(B/h_s)^2$ for acrylate-based SL 5154 photopolymer.

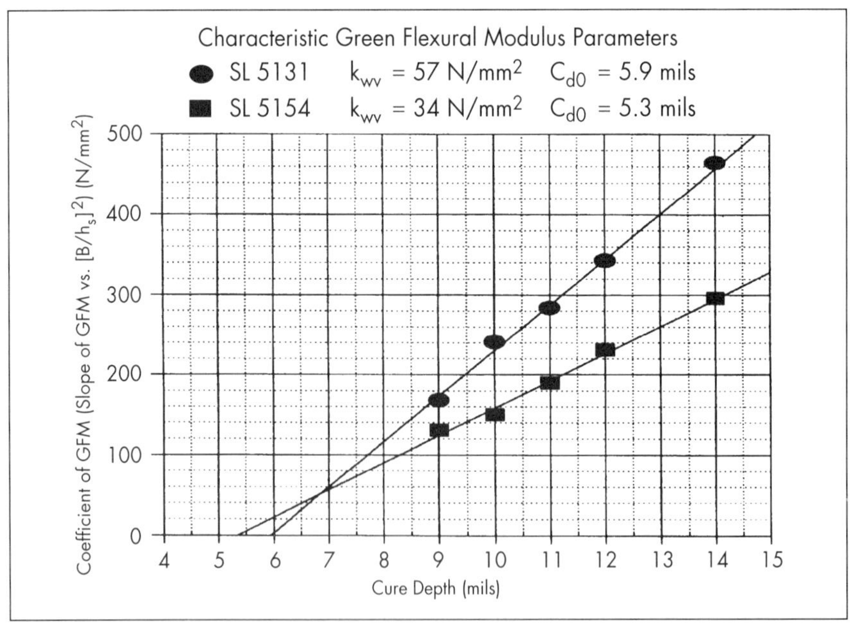

Figure 2-15C. Coefficient of green flexural modulus versus C_d for acrylate-based SL 5131 and SL 5154 photopolymers, used to calculate green strength parameters k_{wv} and C_{d0}.

Table 2-4. Green Flexural Modulus* of Stereolithography Strip Test Part (MPa)

Hatch Space (mil)	Epoxy Based (Built in 6-mil Layers)		Acrylate Based (Built in 10-mil Layers)	
	SL 5180	SL 5170**	SL 5131	SL 5154
8	700	1846	139	109
9	601	- -	94	67
10	570	1845	76	61
11	- -	- -	64	48
12	530	1827	60	45
ACES™	1141	1784	N/A	N/A

* Laser Beam diameter approximately 8.5 mils at the $1/e^2$ clip level of 13.5%; for epoxy resins SL 5170 and 5180, GFM was measured one hour after SL building. See text for details.

** For SLA-250. Other resins, SL 5180, SL 5131, and SL 5154 are all SLA-500, argon ion formulations.

times greater than the corresponding values for the acrylates. In addition, the GFM of SL 5170 is about three times greater than SL 5180. Both epoxy-based resins are significantly less dependent on h_s within the tested range of 0.008 to 0.012 inch (0.2 to 0.3 mm). The new ACES build style improves the GFM of SL 5180 to about 64% of the value for SL 5170, compared with only 38% for WEAVE using 8-mil (0.2-mm) hatch spacing.

Because of continued thermal polymerization, the GFM of these epoxy-based systems further increases significantly over time. The GFM increase is shown in Figure 2-16. Therefore, to validate the comparison, the GFM values reported in Table 2-4 were all measured one hour after build.

All these characteristics of the epoxy-based photopolymers are advantageous for SL. The small dependence on hatch spacing provides a wider process latitude for users, thereby minimizing concern about building soft parts. Greater green strength can now be achieved where it is most needed, for larger parts that take longer to build. Also, high green strength is one of the key properties necessary for success with QuickCast. When built in the quasi-hollow QuickCast style, resins with low green strength are simply too weak to withstand cleaning and handling, nor would these resins be able to maintain dimensional accuracy or stability in the green state.

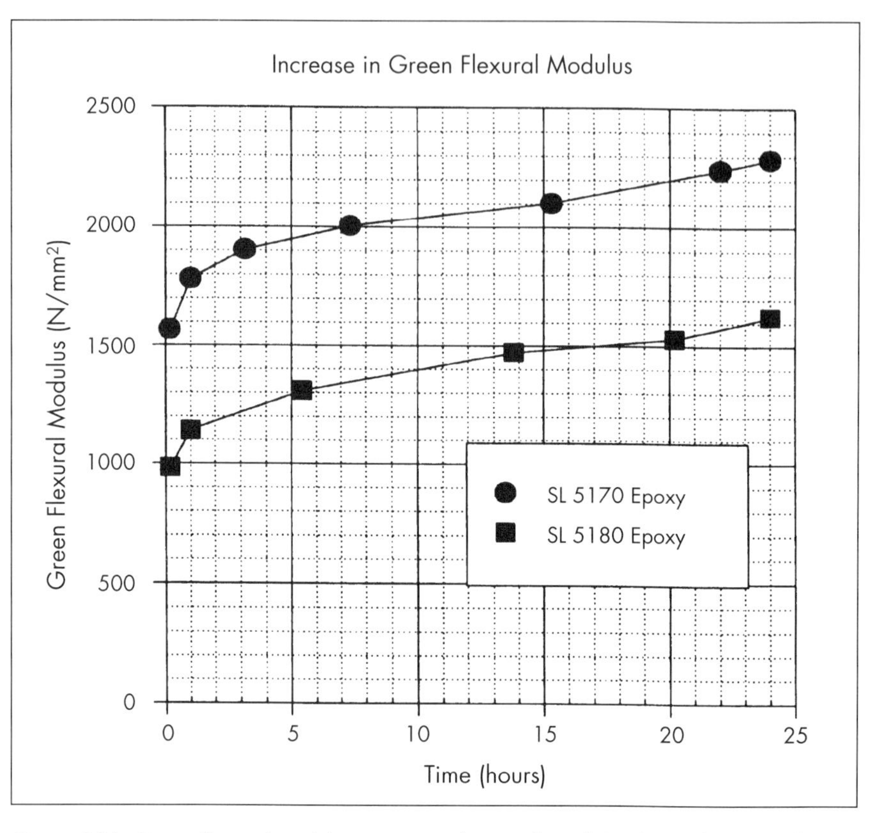

Figure 2-16. Green flexural modulus increase of epoxy-based SL photopolymers with time after the strip part is completed on an SLA.

Postcured Mechanical Properties

How do the mechanical properties of the epoxy resins compare to those of acrylate-based SL resins, and also to other common thermoplastics? For mechanical testing, the SL resins were all built on the appropriate SLA using the recommended solid build styles, and were postcured normally. For the epoxy resins SL 5170 and SL 5180, the ACES build style was used. The other SL resins were built in the recommended STAR-WEAVE style.

The American Society for Testing and Materials (ASTM) mechanical tests parts are schematically described in Figure 2-17. The experimental test results for these SL resins are plotted in Figures 2-18 through 2-24, together with the literature values for acrylic plastic and medium impact polystyrene. In these figures, the darkened tip areas of the horizontal bars provide the range of the measured test values. For the epoxy resins SL 5170 and SL 5180, the range is based on the error bar associated with testing multiple samples. In all cases, at least five samples were evaluated.

Mechanical properties of the SLA-250 and SLA-500 versions of the acrylate-based SL resins have been paired. Therefore, in addition to the normal measurement errors, the range for the acrylate resins includes a small but nonnegligible variation between the SLA-250 and SLA-500 versions of similar resins.

For the acrylic and medium impact polystyrene data listings, the literature values presented[30] include the range of values corresponding to multiple grades of commercially available materials described as acrylic plastic and medium impact polystyrene. Obviously, there are a number of different manufacturers as well as various grades of these common thermoplastics. The listed range of the mechanical property values is often very large.

The applications of these materials are many and varied. Acrylic plastics are typically used for transparent aircraft enclosures, radio and TV parts, lighting equipment, and goggle lenses. Medium impact polystyrenes are most commonly used for radio and TV cabinets, toys, containers, and packaging.

Tensile and Flexural Strengths

Tensile testing of SL resins is performed according to the ASTM D638 method. The tensile strength data, given in Figure 2-18, demonstrates that the stiff, but relatively brittle, acrylate resins, SL 5081-1 and SL 5131, have the highest tensile strengths in this group. Their tensile strengths are 7,200 to 11,500 psi (50 to 80 MPa). The urethane acrylate-based SL 5143, SL 5149, and SL 5154 have the lowest tensile strengths, with values around 5,000 psi (35 MPa). The SL 5170 and SL 5180 resins have tensile strengths in the range of 6,000~9,000 psi (40~62 MPa). These values are in the same range as those of acrylic plastic. The tensile strength of SL 5180 is comparable to medium impact polystyrene. Furthermore, SL 5170, with a tensile strength of about 8800 psi (60 MPa), is almost 50% stronger than medium impact polystyrene.

Flexural strength is precisely defined as the stress measured at a fiber strain of 5%, in accord with the criteria for flexural testing designated by ASTM D790. For the great majority of RP&M applications, flexural strength is actually more relevant because parts are more often subjected to bending loads rather than being pulled along their long axis. The flexural strength data is presented in Figure 2-19. The urethane acrylates have the lowest strengths, for both the tensile and flexural tests. Generally, flexural strength scales with tensile strength. However, an interesting trend is observed between the acrylate and epoxy-based resins. In the case of acrylate resins, the flexural strengths are equal to, or slightly greater than, their tensile strengths. For example, the maximum tensile strength for SL 5081-1 is 11,600 psi (80 MPa), whereas the flexural value is 14,000 psi (98 MPa), an increase of about 20%.

However, epoxy-based resins have significantly higher flexural strengths than tensile strengths. For example, the flexural strength of the epoxy SL 5180 is 12,700 psi (88 MPa), which is more than 100% greater than its tensile strength of 6,100 psi (42 MPa). Thus, while the tensile strength of SL 5180 is comparable to that of medium impact polystyrene, its flexural strength is superior by about 20%.

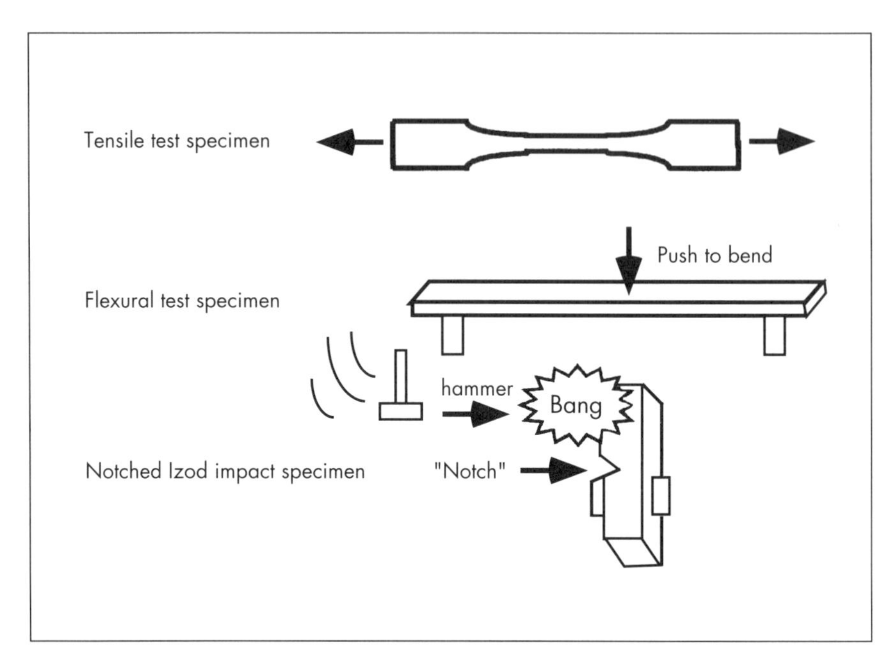

Figure 2-17. ASTM test methods for plastics.

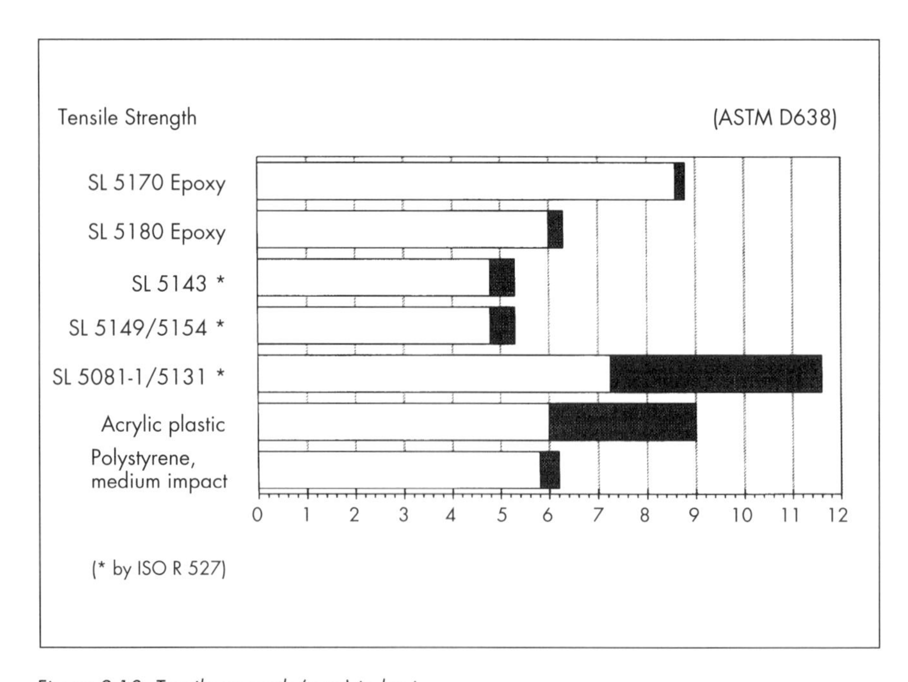

Figure 2-18. Tensile strength (max) in kpsi.

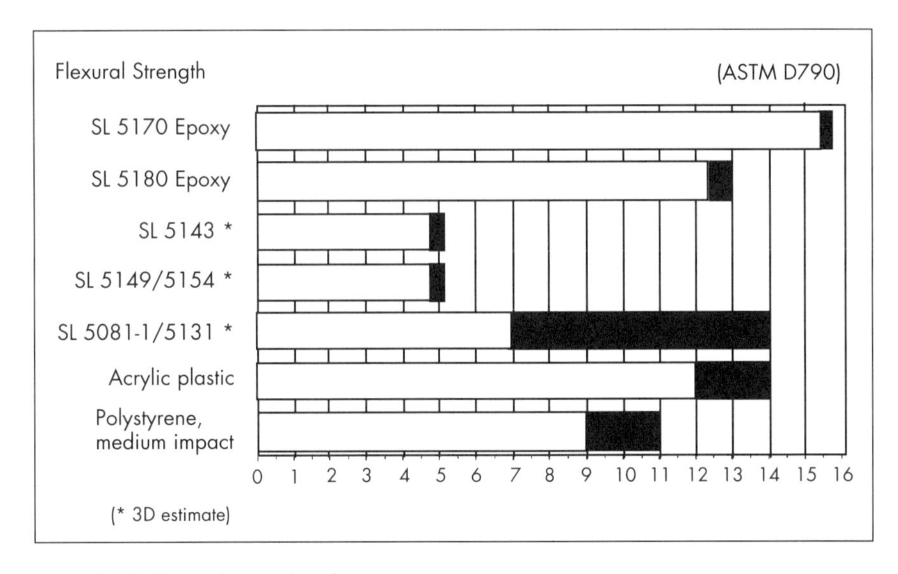

Figure 2-19. Flexural strength in kpsi.

Similarly, for SL 5170, the increase from tensile to flexural strengths is from 8,800 to 15,600 psi (60 to 108 MPa), an increase of almost 80%. Furthermore, SL 5170 has a higher flexural strength than the strongest acrylate resin, SL 5081-1, and has an even greater flexural strength than commercially available versions of either acrylic plastic or medium impact polystyrene.

Tensile and Flexural Modulus

Modulus is a measure of how much stress is generated in a material as it undergoes elastic deformation, and is determined by the initial slope of the stress vs. strain curve for that material. This is one of the most important mechanical parameters, and is often referred to as the rigidity of the material. Objects with low modulus feel rubbery. The tensile modulus is measured in compliance with the test protocol defined by ASTM D638, and for flexural modulus data, in accord with ASTM D790.

The tensile modulus of SL 5081-1 resin, given in Figure 2-20, is relatively high among the SL resins, with a maximum value approaching 600,000 psi. The urethane acrylates, SL 5143 and SL 5149, have relatively low tensile modulus values, ranging from only 100,000 to 160,000 psi, compared to the tensile modulus for acrylic and medium impact polystyrene, which is about 390,000 to 470,000 psi. However, the epoxy-based resins range from 400,000~600,000 psi, with SL 5180 being comparable to the thermoplastics. *On the other hand, SL 5170 has a tensile modulus almost 30% greater than either acrylic or medium impact polystyrene.*

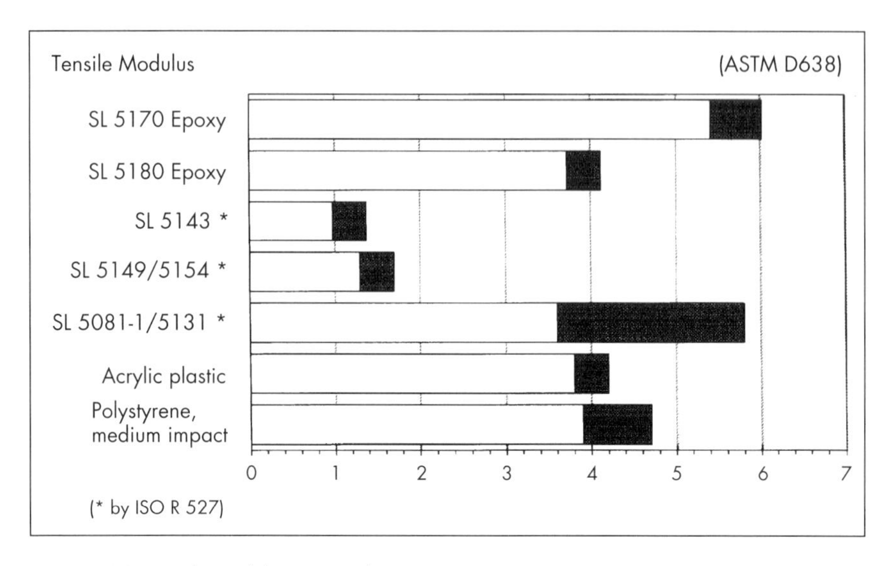

Figure 2-20. Tensile modulus in 100 kpsi.

The flexural modulus, shown in Figure 2-21, is a measure of bending, or flexural stress, when the material is subjected to a strain perpendicular to the long axis of the test sample. The flexural modulus test was carried out in accord with the ASTM D790 procedure.

Tensile and Flexural Resistance. This type of bending deformation results in shear forces within the part, and is a common deformation mechanism in complex geometries. Tensile and flexural resistance is the force needed to elongate or bend a sample with a given modulus. There is a marked difference between tensile and flexural resistance in terms of their dependence on the sample geometry. Tensile resistance increases linearly as the cross-sectional area increases. However, flexural resistance increases as the cube of the part thickness in the direction of the bending action. When deflecting a sample of a different material with a finger, as many RP&M users often do, remember that the sample thickness becomes a predominant factor. A part that is twice as thick requires eight times the force to achieve the same bending deflection as the thinner part.

The trend for flexural modulus of SL resins in Figure 2-21 is similar to that for tensile modulus. Notice, though, that the epoxy-based resins are substantially more rigid in bending than the acrylate-based resins. Even the high strength acrylate resins SL 5081-1 and SL 5131 have only about 75% of the flexural modulus of SL 5170 and SL 5180. The difference between the modulus of the epoxy resins and those of the urethane-acrylate resins SL 5143 and SL 5149 is much greater. SL 5170 and SL 5180 epoxy resins are almost three times as flexurally rigid as the urethane-acrylate SL resins. As mentioned before in the case of

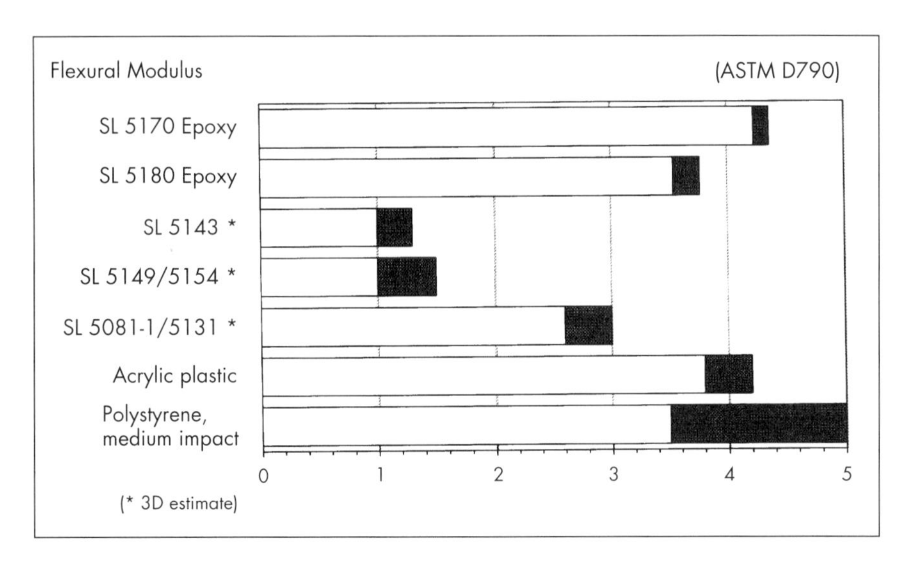

Figure 2-21. Flexural modulus at 100 kpsi.

green parts, high modulus in the postcured state is also a key to improved dimensional stability.

Elongation to Break

The elongation-to-break data, obtained per ASTM D638, is shown in Figure 2-22. The value for the flexible urethane-acrylate resin, SL 5143, is the greatest among the Ciba-Geigy resins. The range of elongation-to-break values for medium impact polystyrene is extraordinarily large, ranging from as small as 3% to as large as 40%, depending on the grade of material. SL 5081-1 has a maximum elongation to break of only 3%. The epoxy resins SL 5180 and SL 5170 have values as great as 16 and 19%, respectively. These values are significantly greater than 7% for acrylic plastic, or about 11% for the urethane acrylates, SL 5149 and SL 5154.

The Stress-Strain Curves of SL 5170 and SL 5180

The experimental tensile stress data for SL 5170 and SL 5180 is plotted as a function of elongation in Figure 2-23. The tensile modulus values, defined as the slope of the initial linear portion of the stress vs. strain curve, are slightly different between the two epoxy-based resins. SL 5170 has a greater slope than SL 5180, corresponding to the difference in the tensile modulus value of 570,000 psi (3.95 GPa) versus about 400,000 psi (2.77 GPa), respectively.

The measured stress increases as the tensile sample is strained further, until an elongation of about 3% has been reached. Since the stress becomes maximum at

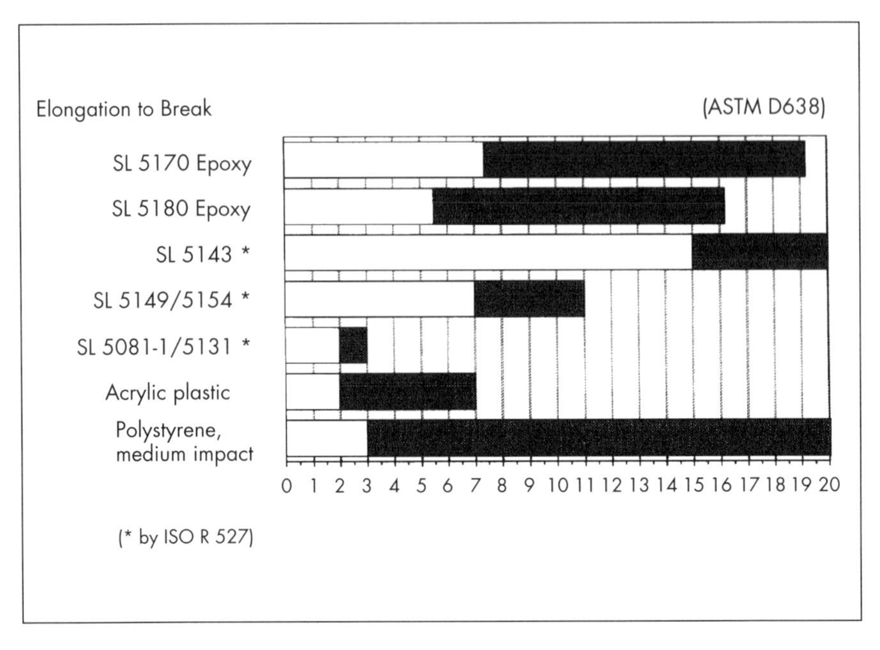

Figure 2-22. Elongation to break in percent (%).

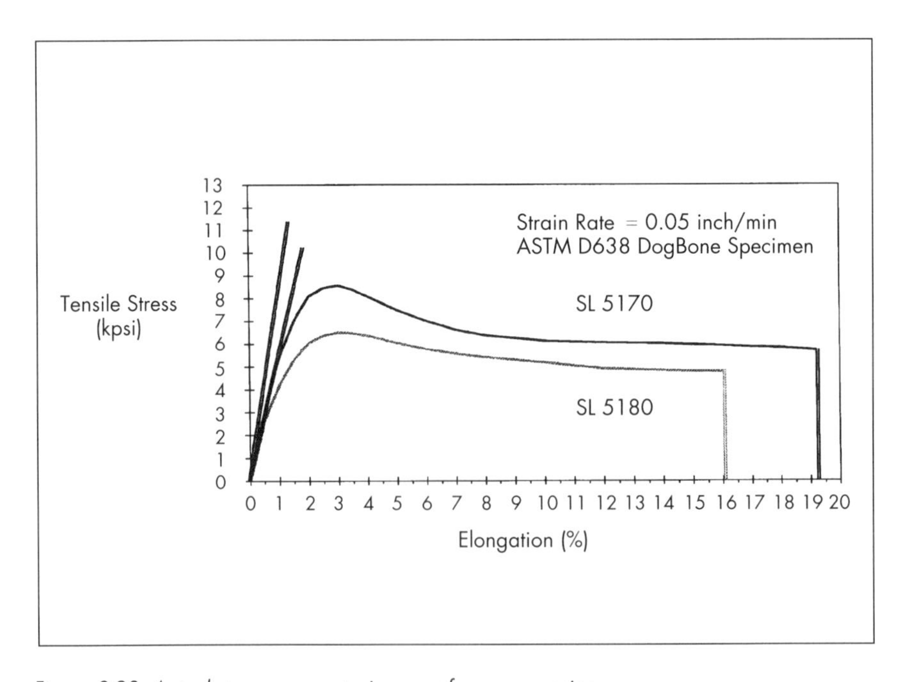

Figure 2-23. Actual stress versus strain curve for epoxy resins.

this point, then by definition, these stress values are assigned as the tensile strengths for the two resins. The materials then pass beyond the yield point, and go into plastic deformation before finally breaking at an elongation of about 16 to 19%. *Such a high degree of plastic deformation is usually NOT a characteristic of highly cross-linked polymer systems.* Perhaps the effective cross-link density, which has not been measured, may be relatively low. The photo cross-linked epoxy-based resins SL 5170 and SL 5180 do indeed undergo substantial elongation, well beyond the yield point.

Material toughness is defined as the area under the stress vs. strain curve. According to this view, SL 5170 and SL 5180 would be considered tough materials. Admittedly, there are many high-molecular weight engineering thermoplastics that are considerably tougher than either SL 5170 or SL 5180. However, these epoxy-based SL resins have proven to be quite rugged for numerous applications, based on SLA user surveys carried out in late 1993. (SL 5180 had not been released at the time of this survey, but it had been in Beta testing.)

Furthermore, SL 5170 and SL 5180 parts have survived many functional tests that include spinning propellers at high speeds, exposing models to high velocity flow in wind-tunnels, snap-fits for such parts as telephone and computer housings, and fluid flow testing both with and in liquids, to name a few. Many of these functional tests could not have been performed with the earlier acrylate resins.

Impact Strength

Finally, the impact strengths of SL resins, shown in Figure 2-24, were measured according to the methods of ASTM D256. The impact samples were notched in the CAD data, such that no machining was involved. Thus the SL parts were immediately ready to be tested. The width of the impact test sample was 0.25 inch (6.4 mm): a thicker sample, instead of the thinnest allowed by ASTM at 0.125 inch (3.2 mm). This was done because thin samples are known to result in higher impact strength values due to the flexing and multi-nodal bending of the sample during the impact. Thus, thin samples dissipate the energy proportionally more than thick samples.[32] Consequently, such values from thin samples may not be representative of the actual impact strength of the resin. A thicker sample can concentrate the energy of the impact in one direction. While the numerical result may be smaller, it is more representative of the material.

It is important to note that the impact strengths of the epoxy resins are comparable to medium impact polystyrene, and are slightly greater than acrylic plastic. Also, SL 5170 and SL 5180, having impact strengths of 0.5 to 0.8 ft-lb/inch (27 to 43 mJ/mm), are 3 to 4 times more impact resistant than the acrylate resins SL 5081-1 and SL 5131, which have impact strengths of 0.1 to 0.2 ft-lb/inch (5 to 10 mJ/mm). SL 5180, with a maximum value of 0.8 ft-lb/inch (43 mJ/mm), has the greatest impact strength of the resins tested. Within experimental error, the impact strengths of SL 5170 and SL 5180 are comparable to medium impact polystyrene and acrylic plastics.

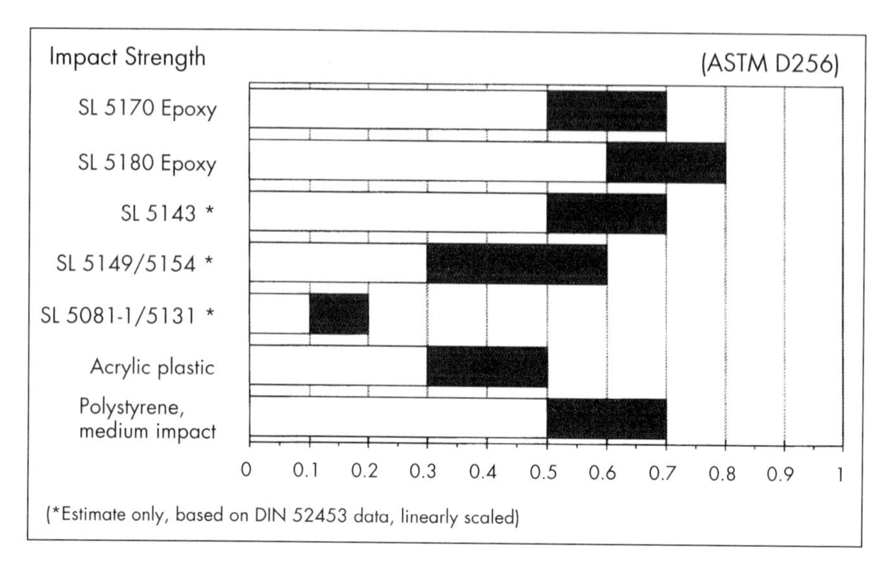

Figure 2-24. Impact strength in foot-pound/inch.

Furthermore, it has been discovered that the epoxy resins become even more impact resistant with aging. The impact strength of SL 5170 and SL 5180 steadily increases over a period of one month, at which point the impact strength values have increased three-fold to as high as 2 ft-lb/inch (108 mJ/mm).

Thermal Properties

Two thermal properties have been reported for SL resins. One is the glass transition temperature, T_g, and the other is the heat deflection temperature. T_g is the temperature at which large scale molecular motion ceases in an amorphous material when it is cooled from a high to a low temperature.[33] In general, T_g is related to the hardening or softening of an amorphous material.

T_g is usually also a point at which the heat capacity changes upon cooling, as measured with a differential scanning calorimeter (DSC). T_g can also be measured mechanically using a dynamic mechanical analyzer (DMA) by oscillating the polymer as the temperature is varied. T_g is basically a rate dependent parameter, involving both the rate of cooling and the mechanical oscillation rate in the case of DMA. A faster rate of cooling results in a higher T_g.[33]

Using the ASTM D648 procedure for heat deflection temperature, a sample 5 × 0.5 × 0.125 inch (127 × 12.7 × 3.2 mm) is placed under three point flex loading at 66 psi (0.46 MPa) in a silicone oil bath. The bath temperature is then increased at 3.6° F/minute (2° C/minute). The temperature at which the sample deflects 0.010 inch (0.25 mm) is called the heat deflection temperature.

The thermal properties of SL resins are generally quite low. For example, all Ciba-Geigy SL resins have T_g values of about 175° F (80° C), except for SL 5081-1 and SL 5131 acrylate resins, which have glass transition temperatures of roughly 300° F (150° C). However, both of these resins exhibit moderate curl and are extremely brittle.

The epoxy resins SL 5170 and SL 5180 have T_g values of about 175° F (≈80° C), and heat deflection temperatures of about 120° F (≈50° C). Such materials are generally inadequate for thermally demanding conditions. Nevertheless, direct injection molding of polyurethane and polystyrene have been successfully carried out briefly at over 390° F (200° C) in a simple core and cavity mold made from SL 5170 resin, built in ACES.[34]

SL resins with higher glass transition and heat deflection temperatures are desired for direct injection molding applications. Unfortunately, both of these values are relatively low for existing SL resins. Thus, it is common for users to cast SL patterns into polyurethanes via secondary tooling operations using silicone RTV rubber. If the parts are accurate, they can be converted into other materials using secondary processes.

2.9 Applications of Epoxy-based Photopolymers

Direct functional testing of SL parts built in the solid ACES build style is attractive due to the remarkable optical clarity and improved surface finish, in addition to the enhanced mechanical properties.

The material properties of prototypes need not be limited by those of the SL parts themselves. Using secondary processes, a wide variety of materials can be used in place of SL photopolymers. Master pattern transfer techniques such as silicone RTV molding are established processes from which various grades of castable polyurethane and epoxy thermoplastic parts can be obtained. Surface finish and dimensional accuracy are the most important factors when secondary processes are considered. SL parts made from the epoxy resins SL 5170 and SL 5180 are widely used for such applications.

For applications that require greater mechanical strengths than SL 5170 and SL 5180, precision metal parts can be obtained using the QuickCast process. The release of epoxy resins SL 5170 and SL 5180 was essential to enable the QuickCast process.[35] A number of properties of the epoxy resins were responsible for achieving this capability. QuickCast has already succeeded in generating about 5,000 precision shell investment cast metal parts directly from SL 5170 or SL 5180 epoxy resin patterns.[36,37]

An advanced application of QuickCast is emerging, involving the fabrication of metal tooling cast from QuickCast patterns. The core and cavity negatives of a part are produced in metal using QuickCast. This was demonstrated by the successful generation of QuickCast tooling in A-2 tool steel.[38] Using the metal core and cavity pair prototype, eventually production functional parts may be injection molded in the desired engineering thermoplastic material. This allows the user to test a component design quickly and in the material of his or her choice, not limited by the properties of currently available SL photopolymers. See Table 2-5 for data on currently available photopolymers for SLA.

2.10 Industrial Hygiene and Safe Handling of SL Resins

The author would like to remind SL resin users to read the MSDS, and to practice appropriate industrial hygiene at all times. As in handling other chemicals, it is best to control the chemical hazard, which is proportional to both the toxicity and chemical exposure. The toxicity of a substance is established in terms of the amount of material that can be physiologically tolerated by humans. The smaller the amount tolerable, the higher the toxicity. The exposure refers to the amount of material that actually comes into contact, or is ingested or inhaled over a given amount of time. The most effective method for a safe chemical environment is to decrease the hazard both by using materials that are less toxic, and also by avoiding exposure. By preventing chemical exposure, the hazard can be practically zero, even if the toxicity is high. While the toxicity of SL photopolymers can vary, proper industrial hygiene, such as wearing impervious gloves, can decrease the chances of chemical exposure tremendously.

2.11 Future SL Photopolymer Systems

There are shortcomings and disadvantages for each SL resin system. However, many problems will be solved as new resin systems are developed, tested, and introduced into the market. SL photopolymers with improved properties will open new applications. Resins capable of forming thin layers will be valuable for high accuracy rapid tooling applications where stair-stepping is intolerable. A temperature and flame resistant resin would allow direct application of SL parts in various under-the-hood evaluations of automobile components. Superfast resins could increase the build speed, further increasing productivity.

Electrically conductive resins might make direct fabrication of electrical circuit boards possible. Photoelastic resins already allow the detection of critical stress points in static tests. Resins that have modified surface properties may enable facile metal coating techniques that, in turn, could lead to increased thermal and abrasion durability during applications involving tooling for injection molding. Tooling resins capable of withstanding high temperatures may eventually be used for direct injection molding of engineering thermoplastics. Indeed, ACES multi-part tooling for wax injection molding, from which hundreds of metal components are then investment cast, has already been accomplished.[39]

SL curable ceramics will allow the generation of intricate 3-dimensional ceramic parts not accessible with conventional methods. Tough resins similar to ABS would increase the use of SL parts for direct functional testing. Optically transparent resins might be applied in generating headlight lenses for vehicles. Perhaps biocompatible SL resins may ultimately be implanted directly into humans. Solvent resistant resins could be used for harsh chemical environments. Finally, composite resins with dramatically improved mechanical properties could also enable direct functional testing of SL parts intended for challenging conditions.

2.12 Summary

RP&M technology must provide adequate and repeatable dimensional accuracy suitable for various end use applications. For RP&M technology to advance, two simple, but fundamental, requirements must be met. On one hand, users must understand the accuracy and tolerances required for their specific applications. On the other hand, the RP&M machine supplier has the responsibility to present statistically significant data adequate to define the dimensional accuracy and repeatability that can be achieved by that RP&M machine. All these properties must be determined for each RP&M build material, and should be reported to the users for comparison. At 3D Systems, new SL resin systems are taken through extensive parametric and diagnostic testing to map their overall performance. Test results for SL 5170 and SL 5180 epoxy resins have been presented in comparison to the earlier acrylate resins.

In July 1993, the first epoxy-based SL photopolymer, Cibatool SL 5170, became available for the SLA-250. In March 1994, SL 5180 resin became available for the SLA-500. These epoxy resin formulations have demonstrated numerous significant advantages over the earlier acrylate SL resin systems. Parts built in these resins show a high level of overall accuracy, negligible curl, improved flatness, and minimal green creep distortion. Also, parts built in these resins exhibit substantial improvements in their mechanical properties, relative to the earlier SL acrylate photopolymer systems.

Overall dimensional accuracy, measured by building not one, but ten sets of user-parts in SL 5170 on a single SLA-250, has been documented. The measured RMS error, based on 1700 data points, is ±0.0018 inch (±45 microns). User-parts in SL 5180, built on the SLA-500, are nearly identical, with an RMS error of ±0.002 inch (±50 microns). The results from a total of 20 user-parts built in the two epoxy resins demonstrated not only excellent overall accuracy, but outstanding repeatability as well.

Flatness

The Slab 6×6 flatness diagnostic test showed that these epoxy-based resins are now capable of producing extremely flat parts that were formerly a challenge with the acrylate-based resins. The maximum Slab 6×6 distortion in epoxy resin SL 5180 is only +0.005 inch (+0.127 mm). This occurred even when the slabs were intentionally postcured from one side only, to simulate the worst-case scenario. It is now possible for SLA users to build very flat parts with confidence.

Table 2-5. Photopolymer Systems for SLA and Their Typical Properties

Manufacturer	Ciba-Geigy Cibatool								Du Pont SOMOS Photopolymer		Allied Signal EXactomer	
Name	5081-1	5131	5143	5149	5154	5170	5177	5180	2100 or 2110	3100 or 3110	2201	2202 SF
Chemistry Type**	A	A	UA	UA	UA	Epoxy	UA	Epoxy	A	A	VE	VE
Laser	HeCd	Ar^+	HeCd	HeCd	Ar^+	HeCd	HeCd	Ar^+	Ar^+ or HeCd	Ar^+ or HeCd	HeCd	HeCd
SLA Type	250	400 500	250	250	400 500	190 250	250	500	250 400 500	250 400 500	190 250	190 250
Photosensitivity Parameters												
D_p (mil)	7.5	5.7	5.7	5.8	5.1	4.8	10.6	5.2	8.5 or 4.7	7.4 or 5.0	7.0	6.6
E_c (mJ/cm^2)	6.6	3.9	4.3	5.5	4.2	13.5	11.2	16.2	2.9 or 3.5	4.0 or 2.5	27.0	8.5
Liquid Properties												
Viscosity (centipoise) @ 30° C	2400	2000	2000	2000	2000	180	2000	187	3800	1000	205	230
Density (g/cm^3) @25° C	1.14	1.14	1.11	1.12	1.12	1.14	1.12	1.15	1.16	1.13	1.13	1.13
Cured Solid Properties												
Density (g/cm^3)	1.21	1.21	1.20	1.20	1.20	1.22	1.20	1.22	1.20	1.21	1.16	
Tensile Strength (k lb/in^2)	8.6	10.2	5.1	5.1	5.1	8.7	4.3	6.2	1.0	3.1	8.0	9.0

Tensile Modulus (k lb/in²)	435	508	138	160	160	573	154	391	5.4	120	211	201
Elongation to Break (%)	2.5	2.5	20	10	10	19	12	16	46	9.2	6-10	7
Flexural Strength (k lb/in²)						15.6	8.6	12.7	0.082	3	6.2	
Flexural Modulus (k lb/in²)						430	203	366	2	94	325	
Impact Strength (ft-lb/in)	0.06	0.07	0.75	0.42	0.42	0.60	0.31	0.70	2.9	0.28	0.40	
Hardness (Shore D)	89	89	80	78	78	85	78	84	41	80	80	
T_g (°C)	150	150	80	83	83	83	72	80			65	
Heat defl. Temp. (°C)						49	42	42			46.5	
Key Properties	Hard, brittle, temp. resist	Hard, brittle, temp. resist	High elongation	General purpose	General purpose	High accuracy, QuickCast, ACES, optically clear	Express mode	High accuracy, QuickCast, ACES, optically clear				

**A=Acrylates, UA=Urethane Acrylates, VE=Vinylethers.

Less Green Creep

Parts built in the epoxy resins exhibit considerably less green creep, as demonstrated by the green creep distortion diagnostic test. The results presented in this chapter show that the green creep distortion rates for SL 5170 and SL 5180 are very small, with values from 0.0003 inch (8 microns) to about 0.0008 inch (20 microns) for every decade in time. This is more than a 40-fold improvement over the earlier acrylate resins. Reduced creep distortion in the laser-cured state is very important for most SL applications. However, it is critical for parts generated using the QuickCast build style.

Furthermore, the overall mechanical properties of these epoxy resins, measured according to ASTM standards, were found to be comparable to, or exceed, those of thermoplastic materials such as acrylic plastic (PMMA) and medium impact polystyrene.

Substantial Improvements

With respect to the earlier acrylate resins, the SL 5170 and SL 5180 epoxy resins excel in almost every category of measured mechanical properties. It is worth noting that the improvements are not merely incremental, but rather are quite substantial. For example, the flexural strength of the epoxy resins is 2 to 3 times greater, and the tensile and flexural modulus values are more than four times greater than the earlier urethane-acrylate SL resins. The elongation to break for the epoxy resins is more than six times greater than that for SL 5081-1. Finally, the impact strength is comparable to the flexible urethane acrylates. At the same time, however, the impact strengths for SL 5170 and SL 5180 are 4 to 7 times greater than that of either SL 5081-1 or SL 5131. The epoxy resins are considered by many users to possess the best combination of mechanical properties for various end-use applications.

In a nutshell, the significance of epoxy resins for SLA users may be best summarized by dramatically reduced effective linear shrinkage, increased overall accuracy, improved flatness, negligible curl distortion, low swelling during build, greatly reduced green creep distortion, optical clarity when built in ACES, excellent mechanical properties, enhanced process latitude, and ease of use coupled with a high probability of successful part building the first time.

Major characteristics of the epoxy resins that enabled QuickCast are: (1) high green strength and low swelling that make it possible to generate wide hatch spacings and hence a larger, more easily drained cell structure; (2) low viscosity for faster and more complete pattern drainage; and (3) improved dimensional precision so that QuickCast patterns can be directly shell investment cast into accurate functional metal prototypes.

The epoxy resins also enabled the ACES build style that led to the following properties of SL parts:

1. high optical clarity due to superior laser-cured homogeneity;
2. minimal mechanical anisotropy;
3. high green strength and dimensional stability;

4. greatly improved flatness;
5. strong, accurate, dimensionally stable thin walls;
6. very few build failures;
7. greatly reduced curl; and
8. electroless metal coatings can be applied.[40]

Significance of QuickCast

Probably the greatest significance of the epoxy resins is associated with the QuickCast application. Here, the high inherent strength of the resin is essential to prevent premature deformation. QuickCast patterns necessarily require a quasi-hollow internal structure to prevent breakage of the ceramic investment casting shells during the burnout cycle. The toughness, rigidity, and impact strength of QuickCast patterns made from SL 5170 and SL 5180 are much greater than those of waxes commonly used for investment casting. This allows foundries to shell investment cast thin walls and delicate features that were once thought to be impossible for wax patterns.

Future developments in SL photopolymers will continue to advance new applications by achieving improved electrical, optical, thermal, or mechanical properties. However, accuracy and dimensional stability, coupled with good overall mechanical properties, continue to be keys to expanding SL into new applications, such as generating prototype and, eventually, production rapid tooling. Plastic parts will then be quickly and even more economically manufactured by injection molding in the desired engineering thermoplastic materials.

References

1. Wilson, J., *Radiation Chemistry of Monomers, Polymers, and Plastics*, Marcel Dekker, New York, 1974.
2. Lawson, K., *UV/EB Curing in North America*, Proceedings of the International UV/EB Processing Conference, RadTech International North America, Orlando, Florida, May 1-5, 1994, Vol. 1, p. 298.
3. Hoyle, C. and Kinstle, J., *Radiation Curing of Polymeric Materials*, Chapter 1, American Chemical Society, 1990.
4. Hug, W., "Lasers for Rapid Prototyping & Manufacturing," Chapter 3, Jacobs, P., *Rapid Prototyping and Manufacturing: Fundamentals of Stereolithography*, Society of Manufacturing Engineers, Dearborn, Michigan, 1992.
5. Phillips, R., *Sources and Applications of Ultraviolet Radiation*, Academic Press, New York, 1983.
6. Hunziker, M. and Leyden, R., "Basic Polymer Chemistry," Chapter 2, Jacobs, P., *Rapid Prototyping and Manufacturing: Fundamentals of Stereolithography*, Society of Manufacturing Engineers, Dearborn, Michigan, 1992.
7. Plews, G. and Phillips, R., *Journal of Coating Technology,* 1979, Vol. 51, No. 648, p. 69.

8. Pappas, S.P., "Photoinitiators for Radical, Cationic, and Concurrent Radical-Cationic Polymerization," Chapter 1, *UV Curing: Science and Technology*, Technology Marketing Corporation, Norwalk, Connecticut, 1985, Vol. II, p. 13.
9. Odian, G., "Ring-Opening Polymerization," Chapter 7, *Principles of Polymerization*, Wiley-Interscience Publication, New York, 1981.
10. Licari, J. and Crepeau, P., North American Aviation, U.S. Patent 3,205,157, 1965.
11. Crivello, J. and Lam, J., *Journal of Polymer Science*, Polymer Symposium, (56) 383, 1976.
12. Crivello, J., Lam, J., and Volante, C., *Journal of Radiation Curing*, 1977, Vol. 4, No. 3, p. 2.
13. Crivello, J. and Lam, J., *Journal of Polymer Science*, Polymer Chemical Edition, 1979, Vol. 17, No. 4, p. 2.
14. Watt, W., *Epoxy Resin Chemistry*, Bauer, R. ed., American Chemical Society, Symposium Series, 1979, Vol. 17, p. 114.
15. Lapin, S., *Radiation Curing: Science and Technology*, Pappas, P., ed., Plenum Press, New York, 1992, p. 241.
16. Lapin, S., Noren, G., and Schouten, J., Proceedings of the International UV/EB Processing Conference, RadTech International North America, Boston, April 26-30, 1992, Vol. 1, p. 167.
17. Pang, T. and Kwo, K., "Increase of the Green Flexural Modulus of Epoxy Photopolymers SL 5170 and SL 5180 Over Time," *3D Systems Report*, December 22, 1994.
18. Odian, G., "Chain Copolymerization," Chapter 6, *Principles of Polymerization*, Wiley-Interscience Publication, New York, 1981.
19. Sitzmann, E., Barnes, D., Anderson, R., Dinkel, E., Srivastava, R., Haynes, R., Green, G., and Krajewski, J., "EXactomer™ Resins for Stereolithography," Proceedings of the Rapid Prototyping & Manufacturing Conference, Society of Manufacturing Engineers, Dearborn, Michigan, 1992.
20. Allied Signal Preliminary Test Results for EXactomer 2202 SF, October 1993.
21. Pang, T., "Stereolithography Epoxy Resin Development: Accuracy and Dimensional Stability," Proceedings of the Solid Freeform Fabrication Symposium, University of Texas, Austin, Texas, August 8-11, 1993, p. 11.
22. Nguyen, H., Richter, J. and Jacobs, P., "Diagnostic Testing," Chapter 10, Jacobs, P., *Rapid Prototyping and Manufacturing: Fundamentals of Stereolithography*, Society of Manufacturing Engineers, Dearborn, Michigan, 1992.
23. Pang, T., "Stereolithography Epoxy Resin Development: Accuracy, Dimensional Stability, and Mechanical Properties," Proceedings of the Solid Freeform Fabrication Symposium, University of Texas, Austin, Texas, August 8-10, 1994, p. 204.

24. Gargiulo, E. and Belfiore, D., "Stereolithography Process Accuracy: User Experience," Proceedings of the Second International Conference on Rapid Prototyping, University of Dayton, Dayton, Ohio, June 1991, p. 311.
25. Gargiulo, E. and Belfiore, D., "Photopolymer Solid Imaging Process Accuracy," Intelligent Design and Manufacturing for Prototyping, Winter Meeting, ASME, December 1-6, 1991, Vol. 50.
26. Richter, J. and Jacobs, P., "Accuracy," Chapter 11, Jacobs, P., *Rapid Prototyping and Manufacturing: Fundamentals of Stereolithography*, Society of Manufacturing Engineers, Dearborn, Michigan, 1992.
27. Adamson, A., *Physical Chemistry of Surfaces*, 1982.
28. Jacobs, P., "Fundamental Processes," Chapter 4, Jacobs, P., *Rapid Prototyping and Manufacturing: Fundamentals of Stereolithography*, Society of Manufacturing Engineers, Dearborn, Michigan, 1992.
29. ibid., p. 88.
30. "1994 Materials Selector Issue," *Machine Design*, December 1993, Vol. 65, No. 26.
31. Pang, T., "Green Strengths of Stereolithography Resins: Phenomenological Green Flexural Modulus Equations I & II," *3D Systems Report*, February 25, 1992.
32. ASTM D 256- 92, "Standard Test Methods for Impact Resistance of Plastics and Electrical Insulating Materials," Note 11, p. 63.
33. Aklonis, J. and Shen, M., *Introduction to Viscoelasticity*, Inter-Science Publishing, 1982.
34. Schulthess, A., *Direct Injection Molding of Engineering Thermoplastics in a Stereolithography Cavity made in SL 5170*, 1994.
35. Pang, T. and Jacobs, P., "Stereolithography 1993: QuickCast™," Proceedings of the Solid Freeform Fabrication Symposium, University of Texas, Austin, Texas, August 8-11, 1993, p. 158.
36. Jacobs, P., 3D Systems, Kennerknecht, S., Cercast Group, Smith, J., and Hanslits, M., Precision Castparts Corporation, and Andre, L., Solidiform, Inc., *QuickCast™: Foundry Reports*, 3D Systems, Valencia, CA, April 1993.
37. Discussion with Jacobs, P. and Fedchenko, R.
38. Denton, K., Ford Motor Company, Jacobs, P., 3D Systems, "QuickCast™ Tooling: A Case History at Ford Motor Company," Proceedings of the Rapid Prototyping and Manufacturing Conference, Society of Manufacturing Engineers, Dearborn, Michigan, April 26-28, 1994.
39. Dzugan, R., "ACES Rapid Tooling for Wax Injection Molding," North American Stereolithography Users Group Meeting, Tampa, Florida, 13 March, 1995.
40. Schulthess, A., "Electroless Metal Plating of Stereolithography Parts," Proceedings of the Rapid Prototyping Conference, Society of Manufacturing Engineers, Dearborn, Michigan, April 24-26, 1994.

Acknowledgment

I would like to thank Doctors Adrian Schulthess, Bettina Steinmann, Manfred Hofmann, and Max Hunziker, as well as their technical staff members, Christian Bovet, Jean Francois Broillet, Daniel Chambertaz, Veronique Charriere, Daniel Fankhauser, and Beatrice Maire—all of them from Ciba-Geigy—for advancing SL resin formulation development. My appreciation also goes to Charles Hull and Doctors Paul Jacobs and Richard Leyden of 3D Systems for providing a fertile environment for enhancing SL process and chemistry. Last, but not least, I thank (also from 3D Systems) Thomas Almquist, Bryan Bedal, Michelle Guertin, Kelly Kwo, Hop Nguyen, and Dr. Jouni Partanen for their technical contributions and support in SL resin development.

Thomas Pang

CHAPTER 3

Stereolithography Hardware Technology

The inquiring mind is never satisfied with things as they are. It is always seeking ways to make things better and do things better. It assumes that everything and anything can be improved.

Harlow H. Curtis

All stereolithography (SL) systems, whether manufactured by 3D Systems or others, share the same core elements or subsystems: the input data type, part preparation phase, layer preparation functions, and the actual laser scanning, or imaging, of each part cross-section.

The input data is a CAD solid, fully surfaced CAD description, or series of contours representing a 3-dimensional object, translated most commonly into a format called the STL file. Other file formats are used, such as SLC, though less often. Detailed information on STL and SLC file formats is contained in Chapter 4. Additional file formats are under development.

The part preparation phase uses software algorithms to translate the input data into a sequence of instructions that the stereolithography apparatus (SLA) uses to generate the physical model. These instructions include operator-controlled optimization parameters, which are useful in tailoring the output of the SLA to the specific end-use application of the model. See Chapter 4 and/or Reference 1 for more information on part preparation.

The layer preparation phase establishes a thin layer of liquid photopolymer resin, typically in the range of 0.004 to 0.010 inch (0.10 to 0.25 mm) thick. This is accomplished using various mechanical and electrical subsystems, as well as a number of software control algorithms.

The imaging phase selectively solidifies, or cures, the layer of resin at each part cross-section. The system's laser, along with various mechanical, optical,

By Diana Kalisz, Director of Engineering, and Jeff Thayer, Chief Engineer, 3D Systems Corporation, Valencia, California.

electrical, and electronic elements, directed by control algorithms, is used to image each layer in sequence.

When designing the hardware and software for each SL basic element, the effectiveness of any subsystem must be maximized, while at the same time minimizing deficiencies in implementation. RP&M is a new industry, undergoing continuous improvement. At any instant, the hardware and software in the field represent the best solutions for each element at the time they were developed. Meanwhile, our understanding of the processes continuously improves, and there is recognition that basic elements can be further enhanced. Work then goes on to develop designs, which once again reflect an improved solution for the new slice in time.

3.1 Layer Preparation

Preparing the surface of the liquid resin in an SLA is analogous to preparing the foundation of a house. Errors in that foundation will be reflected throughout the entire structure, adversely affecting the structure's resistance to outside influences that might distort it. The liquid resin surface is the foundation of each layer of the SL model. The quality of the resin surface directly impacts the model's quality. For example, a rippled resin surface can result in layer-to-layer delamination or, if it occurs on the top surface of the part, poor surface quality.

The importance of surface preparation is evident in all SL machines. Recoating blades, fluid pumps, surface-finding lasers, and robotics have all been directed at achieving a uniform resin surface. Requirements for this process are simple in concept, but represent a daunting task in implementation. The system must simultaneously achieve the following goals.

- The resin surface must be maintained at the focal plane of the imaging system.
- The resin surface must be uniformly flat, level, and free of extraneous features.
- The resin surface must be a controlled distance above the previously drawn cross-section of the part.

Making these tasks even more difficult is the necessity to accomplish them as quickly as possible, in order to maximize system productivity.

Resin Surface Location

The resin surface defines the drawing plane of the imaging system. It has a fundamental influence over the quality of the solidified resin, the basic building block of an SL model. It also impacts calibration of the imaging system, which is the "road map" transforming the CAD input data into the model cross-section drawn at the resin surface. Additionally, location of the resin surface affects the quality of the "line" of cured resin that is drawn.

Changing Spots. The initial diameter of the laser beam is reduced in size by focusing optics in order to control the width of a cured line of resin, and to enable imaging of reasonably small features. When using a fixed focal length optical system, the diameter of the focused laser beam, or "spot," is a function of the distance from the focusing elements to the resin surface. Thus, the spot's shape and size, incident on the resin surface, changes as the surface is displaced up or down.

As the spot size varies, the irradiance distribution changes. It is this irradiance distribution that determines the photopolymer curing characteristics, such as how deeply the resin is cured, the width of the cured line, and how strongly a given line attaches to the one below. To obtain a model that most closely matches the original mathematical representation in CAD, this curing must be closely controlled. Keeping the laser spot consistent in both size and shape is vital.

Depth of Focus. The free liquid resin surface defines a horizontal plane, while the motion of the focused laser spot describes a portion of the surface of a sphere, as shown in Figure 3-1. Thus, the theoretical best point of focus does not intersect the resin surface over the entire image area. Fortunately, keeping the laser spot a consistent size and shape throughout the working envelope can be achieved by taking advantage of depth of focus, an optical phenomenon well known in photography.

Significant depth of focus is possible due to the nonlinear Rayleigh propagation of a focused Gaussian laser beam. This perhaps nonintuitive behavior results in a beam diameter in the vicinity of the focal point that does not behave geometrically, as seen in Figure 3-2.[2] Thus, the focus, both upstream and downstream, of the optimum focal point does not degrade in the manner that would be predicted by the theories of physical optics. Simply stated, there is enough depth of focus available that acceptable beam size and shape can be maintained across the flat resin surface. The gradual convergence and divergence of the laser beam in the region near the optimum focus is referred to as the beam "waist," and is similar in shape to a gradually tapering hourglass.

The resin is positioned such that the spherical surface of theoretical optimum laser focus is bisected by the free resin plane. If the resin surface is placed significantly above or below this expected level, the system begins to operate outside the range of acceptable focus, and beam shape is different across the resin surface, depending on the point at which the laser impinges. Thus, significant displacement of the resin surface relative to its ideal position results in varying cured linewidth and depth across the working envelope. This would make it difficult, if not impossible, to find a global set of parameters that would provide proper resin curing across the entire imaging plane. Figures 3-1 and 3-2 also illustrate the impact on beam shape as "defocusing" is attempted. When defocusing, the beam waist actually moves either toward or away from the laser. Thus, the apparent diameter may change at any plane bisecting the beam path, but the beam waist has simply translated along its path. This produces a situation where a nonoptimal energy distribution is being used for imaging, and astigmatism (variance between perpendicular axes) may develop.

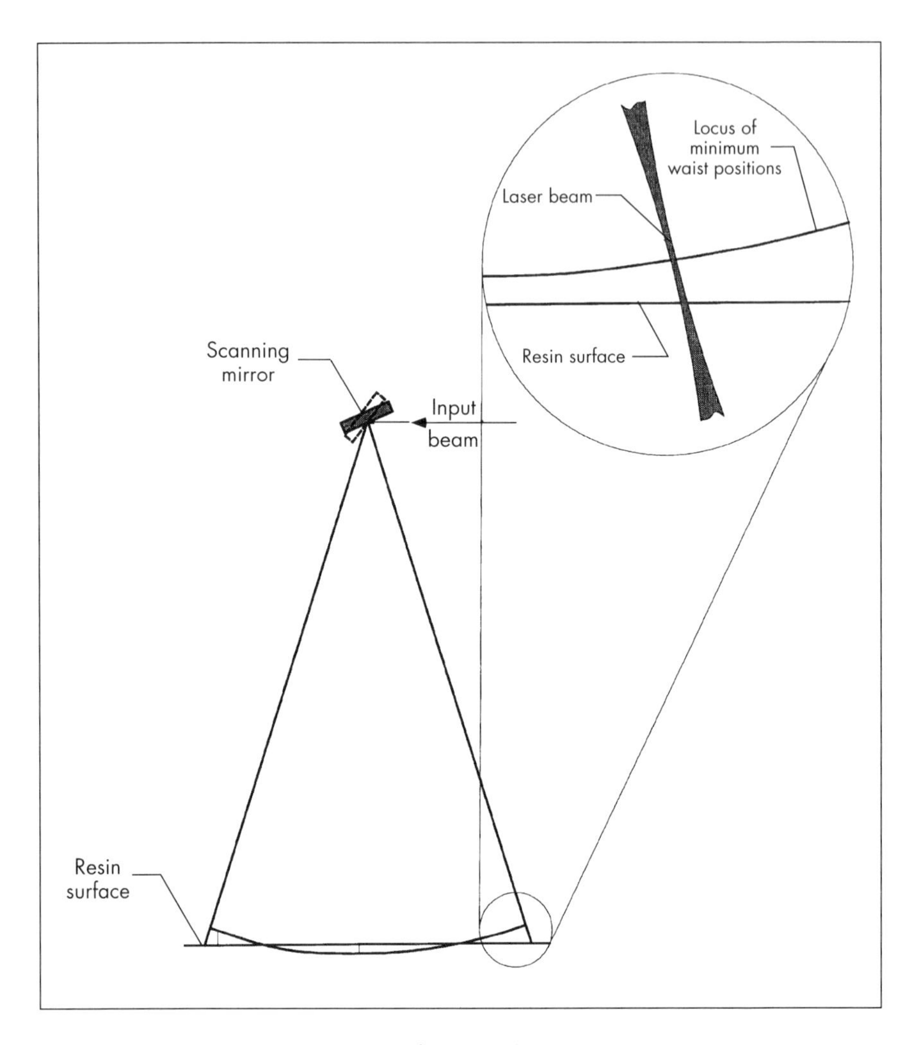

Figure 3-1. Spherical beam sweep over flat resin plane.

Imaging System. The resin surface's position also affects calibration of the imaging system. This calibration provides a map that translates the CAD model data into the data space of the SLA. Since the imaging system of the SLA describes a portion of a spherical surface, and the free liquid resin surface is planar, the scanned laser's spherical coordinate system must be translated into a planar coordinate system. Another way to express this concept is that the CAD coordinates must be translated into angular coordinates appropriate for the scanning system.

The spherical-to-planar mapping is only accurate for the specific distance between the imaging system and the resin surface for which it was generated, within

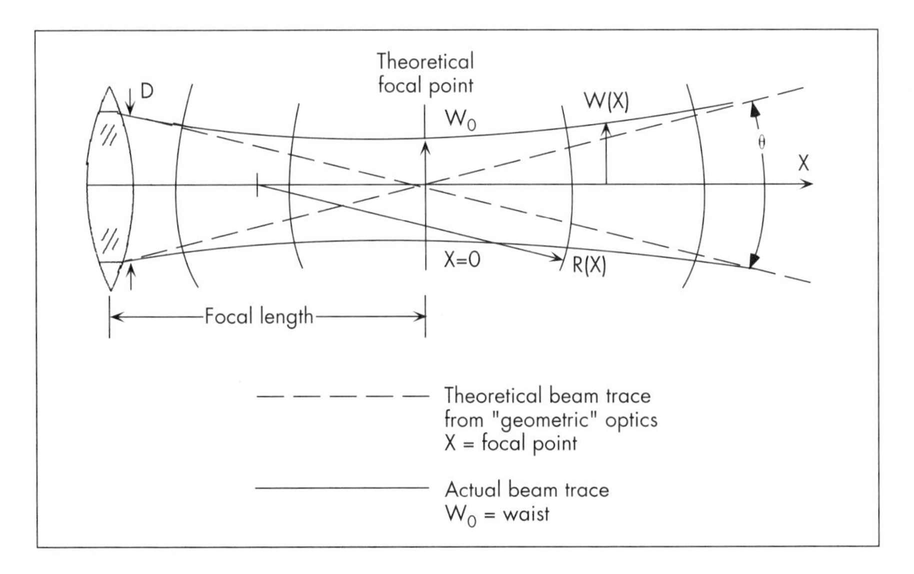

Figure 3-2. Beam waist, compared with theoretical focus.

a narrow tolerance band. Small errors in positioning the resin surface result in errors in model scaling. Larger errors in resin surface positioning can result in obvious distortions in both shape and scale of the model.

Flat and Level Resin Surface

The flatness of the resin surface influences two critical qualities of the finished model: layer-to-layer adhesion and dimensional stability. Both flatness and distance from the surface to the previously drawn layer are controlled with a recoater system.

If there are areas of the resin surface not uniformly flat, or if a layer is thicker than intended, the laser exposure provided by the scanning system may be insufficient in that area to produce adequate layer-to-layer adhesion. This can result in partial or complete layer separation, the outcome being poor surface quality of the model or even failure of the system to complete the building process.

Conversely, if the layer has areas that are thinner than intended, resin exposure, and thus resin cure, may be excessive. This can result in unacceptable residual stresses, which are responsible for both short- and long-term distortion of the model.

Recoating and Leveling Systems. The recoating and leveling systems work together attempting to achieve the goal of providing a flat layer of liquid resin at the proper thickness. These systems execute a number of operations: lowering the part in the vat, wiping away excess material, and smoothing the remaining material to provide the desired thickness of liquid resin above the previously solidified layer. As the desired layer thickness decreases, the difficulty in achieving a flat, level coating of liquid resin increases. While a recoating error of 0.001 inch (0.025 mm) is relatively minor when building with 0.020 inch (0.51 mm) layer thickness, the same error becomes very significant when building with 0.004 inch (0.10 mm) layer thickness. The recoater mechanisms must be controlled precisely to achieve the required layer thickness and uniformity, since any errors in this step are directly reflected in the next layer image.

If left to settle and smooth under the influence of gravity, the resin surface will eventually come to equilibrium as a flat plane, with its normal vector passing directly through the center of the earth. In practical terms, the time required to reach this state of equilibrium could require many minutes or even hours, as shown in Figure 3-3. Depending on resin viscosity and surface tension, along with desired resin smoothness and uniformity, these time periods can easily remove the R from RP&M. Thus, the practical objective of the recoater system is to achieve a reasonable approximation of this flat and level state within a few seconds per layer.

Most recoater systems used in stereolithography employ mechanisms that traverse the surface of the resin, and through direct or combined mechanical, hydraulic, or other motive forces, produce a smooth and level surface. Any nonuniformity in the resin surface results either from inadequacy in the recoater system or from the fundamental properties of the resin.

Controlling Resin Volume

Even when a smooth, uniform, and accurate coating of liquid resin has been achieved over the previous layer of the model, layer preparation is still unfinished. The system must establish the correct volume of resin within the vat necessary to begin part building. "Correct" in this case is defined as sufficient resin so that, as the model undergoes shrinkage during curing, there is adequate resin reserve to compensate for the resulting volume reduction. At the same time, the resin level must be maintained at the focal plane of the imaging system.

Opposing decreased resin volume due to shrinkage is volume increase from resin displacement. As the part becomes taller and taller, the descending elevator mechanism adds its volume to the vat. The resin displacement influences the required volume, but only to a small extent. The resin volume requirements are driven more by the necessity that the free liquid surface not become superelevated either beyond the limits of the imaging system focal plane or, worse, beyond the top of the resin vat. Satisfying these requirements bounds the leveling system between having enough, but not too much, resin on hand for any build.

3.2 Imaging System Objectives and Requirements

The imaging system in an SLA is responsible for translating the abstract data of the CAD world into specific patterns of UV exposure on the surface of the photopolymer. In large measure, it is the quality of these patterns and the speed at which they are drawn that characterize the productivity of the SLA. The image

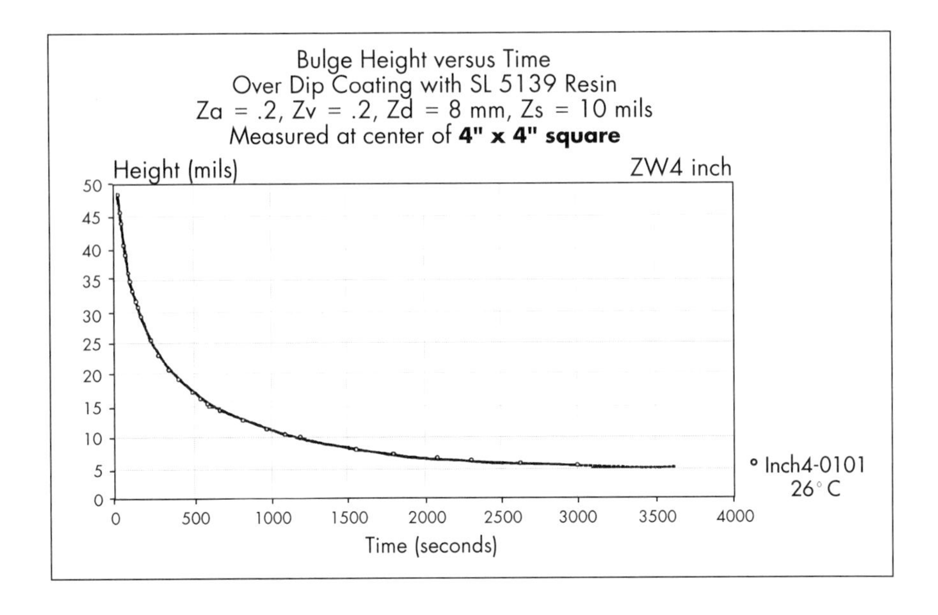

Figure 3-3. (Above) Leading edge bulge decay with time, for a 4" × 4" square. (Below) Leading edge bulge decay with time for a 1" × 1" square.

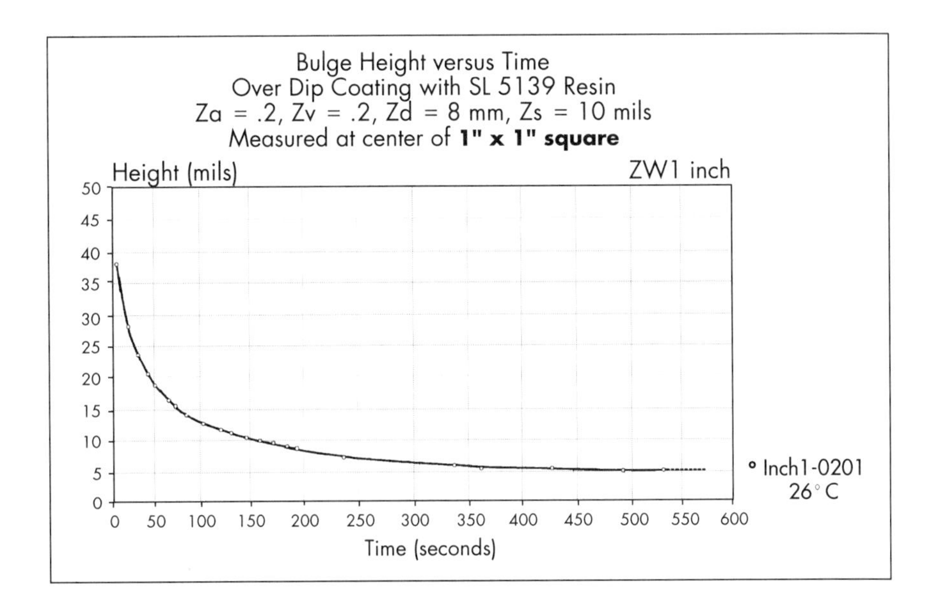

generated must accurately, precisely, and repeatedly reflect the geometry of the CAD data, while additionally responding to the constraints of the overall technology of which it is a part.

As an analogy, the SLA imaging system acts much as your television monitor. A TV receives abstract input signals in the form of a time-varying voltage waveform (the nondigital broadcast signal). The TV interprets and acts on these input signals to produce a pattern of glowing phosphor on the screen, which we see as our favorite show. Similarly, the SLA receives vectors representing the part cross-section. It then translates this information, and produces an image on the resin surface that will become, through polymerization, a single cured cross-section of the model. The performance achieved by the imaging system translates directly into various aspects of model quality such as accuracy, surface finish, and mechanical properties such as green strength.

The general requirements for an SL imaging system are not unlike those for other imaging systems. The system must faithfully represent the desired shape by mapping the input data (vectors, usually in CAD coordinates) onto the resin surface (usually called vat space) in such a manner that the spatial relationships in the data are preserved. The system must also control the laser exposure received by each segment of the model being built.

In SL, "faithfully representing the desired shape" has a specific meaning. The imaging system must take the data that describes the geometry of interest, and translate it onto the surface of the polymer. To accomplish this, the system must perform data transformation, geocorrection, drift correction, linewidth compensation, and numerous calculations to provide the appropriate exposure. These processes will be described in detail in the following sections.

Data Transformation

The CAD data contains a description of the geometric relationship between one point of a part and every other point. Typically, each point is represented as three 32-bit floating point numbers (X, Y, and Z). The CAD data's resolution in this case is one part in 1.7×10^7. Together, these three numbers describe the location of that point in CAD space. These points and their geometric relationships must be mapped into the image space of the SLA. This mapping requires a number of manipulations.

- The origin of the coordinate system must be established. The image space of the SLA is likely to have a different "origin" than the CAD world. The problem is compounded when two or more objects are imaged simultaneously, each with potentially different CAD coordinate systems and/or origins.
- Coordinate system resolution conflicts must also be resolved. The CAD data may have a different resolution/representation than the SLA. For example, one SLA may use a 16-bit integer (one part in 2^{16}), another a full 32-bit float to represent space. Further complicating this issue is the fact that the original CAD data may be in either metric or English units.

- Additional coordinate system transformations may be required by the user. The part may require scaling, translation, or rotation in image space in order to accommodate user needs.

Geocorrection

Geocorrection is a term given to the process of compensating for the "pin-cushion" image distortion that occurs when the planar coordinate space of the data is imaged onto the flat resin surface using spherical coordinate system-based scanning mirrors.

The imaging system on most SL equipment directs the laser beam, using orthogonally arranged, servo-controlled, galvanometer-driven scanning mirrors. These mirrors use a rotation sensor for position feedback, enabling their control subsystem to point to a particular Cartesian coordinate in the SLA image space. This requires mapping from the flat CAD space into the rotational space of the scanner position sensor. Geocorrection is the term adopted by SL to describe this mapping.

Suppose the system is instructed to draw a vector whose end points in CAD coordinates are denoted (x_1, y_1) and (x_2, y_2). In order to image this vector with the scanning system, the coordinates must be translated into angular positions (θ_1, ϕ_1) and (θ_2, ϕ_2) on the mirror encoders. There must be a means of relating the values of (θ_1, ϕ_1) to (x_1, y_1) with some accuracy. The problem is compounded by a substantially nonlinear relationship between angle and projected distance along the image axis, as shown in Figure 3-4. This coordinate system transformation is called "geocorrection," since it corrects the distortion introduced by the spherical- (geo-) based coordinate system of the scanners, interacting with the flat resin surface.

Since geocorrection is strictly a geometric correction, it is theoretically calculable. Some systems on the market do employ a theoretical coordinate transformation. The process is a function of the image field size, its distance from the scanning mirrors, the centering of the scanners over the image field, and finally the mirror encoder details. A reasonably accurate estimate of the correction can be made by measuring a few points in the image field and relating those points to the mirror position. A curve is fit to the points and a resulting correction is obtained.

Several factors make strict reliance on the purely analytical approach unadvisable. Encoders are not perfect, and their response may have a complex and nonlinear relation to a number of parameters including absolute position. The intervening optics in the system between the scanners and the image plane can also introduce additional perturbations, which must also be corrected if a quality model is required. Finally, any errors in establishing the data points on which the theoretical geo curves are based will result in degradation of image accuracy and, thus, model quality.

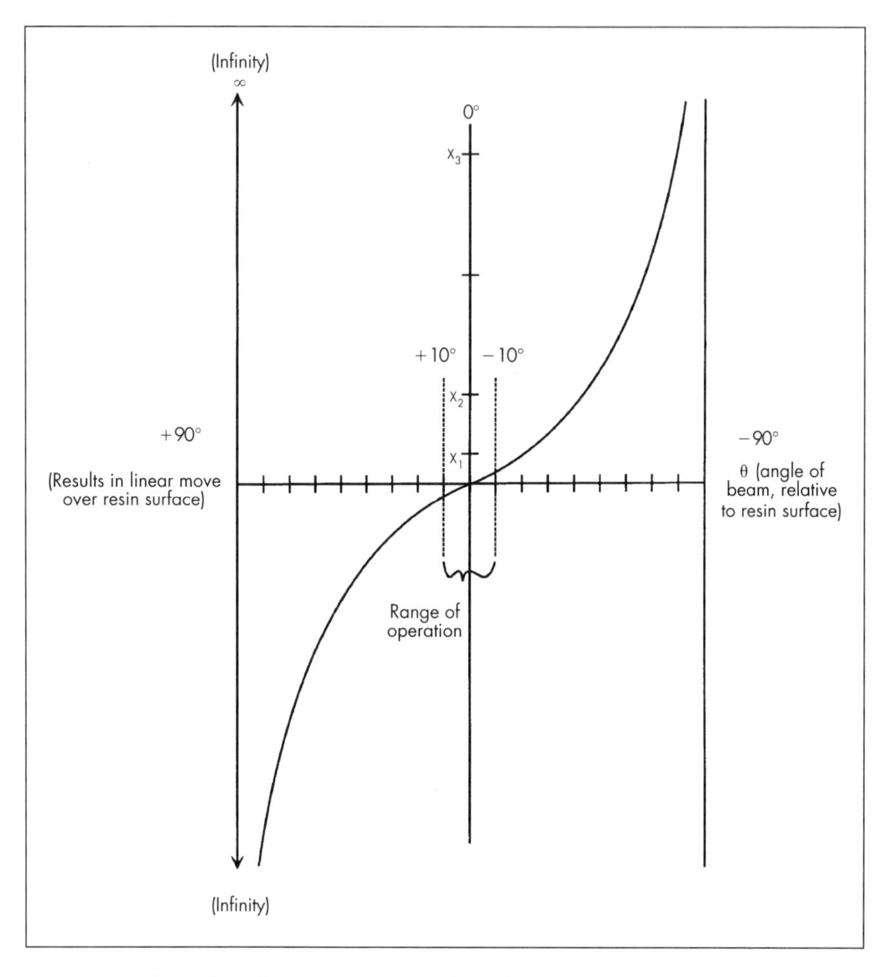

Figure 3-4. Relationship of scanning mirror angle and projected distance along resin surface.

Drift Correction

After data translation and geocorrection, the CAD data can be faithfully imaged onto the surface of the polymer . . . once. In SL, however, the accurate generation of hundreds or thousands of images onto the resin surface is required. The image drawn at the start of a part build must be precisely registered with the last image drawn, even though many hours may have elapsed from start to finish. A change in any of the metrics that contribute to mapping the CAD space into the vat space will result in image misregistration or image drift. A seemingly simple vertical wall, 10 inches (254 mm) high, may require as many as 2,500 repetitions of the same vector—drawn one on top of another. Even tiny deviations, fractions of a vector width from vector to vector, will result in poor model surface finish.

Image Drift Causes. The causes of image drift are many, and some are quite difficult to control. The temperature of the machine and therefore its physical size and shape may fluctuate with changes in room temperature. This will alter the location of geocorrection points, or vat space, relative to the scanning mirror locations. The response of the scanning mirror position sensors may drift as a function of operational conditions. As a laser ages, or with changes in ambient environment, the laser beam may wander slightly, which in turn will displace the position of the scanned laser spot in the image space. Any or all of these effects occur and must be accommodated. The approach taken in SLA systems is to limit the variations as much as possible, and additionally to measure and correct for the small remaining sources of image drift.

SLA systems use a pair of reference points located in the image space. The scanning system records and keeps track of the location of these reference points before each layer is built, and interprets changes in their location as an indication of drift. The drift is quantified and input to an algorithm, which then corrects the next image appropriately for the encountered drift.

Linewidth Compensation

Another element in mapping CAD space to image space relates to the finite width of the laser beam. In CAD space, a vector has no width. The boundary of the part is mathematical, not physical. Yet to image that boundary or vector, be it on a cathode ray tube (CRT) or on the surface of the photopolymer of an SLA, it will be imaged with finite thickness. On the CRT, this is the pixel size. On an SLA, it is the diameter of the laser spot, typically about 0.010 inch (0.25 mm). If the SLA were to image the vector precisely over its zero-width location, the resulting cured line of photopolymer would be mislocated as shown in Figure 3-5. To compensate for this, the SLA offsets the vector's zero-width location by the half-width of the cured line. This process is called *linewidth compensation*. This step is performed in the part preparation element, but some discussion is included here in the imaging section to provide a complete review of CAD to image space translations.

Exposure Control

The system must also control the laser exposure received by each segment or vector of the model being built. Accurate laser exposure is necessary in order to produce a high-quality model. Underexposure may result in delamination, while overexposure can lead to excessive curl, or other photopolymer dependent effects, such as dewetting. Exposure also affects both the cured linewidth and depth of polymerization.

The laser exposure influences the width of the polymerized line, as a result of the profile or irradiance distribution of the laser beam focused onto the resin surface. A typical laser beam profile is shown in Figure 3-6. Note that the width of the profile is a function of the point along its height where it is measured. For

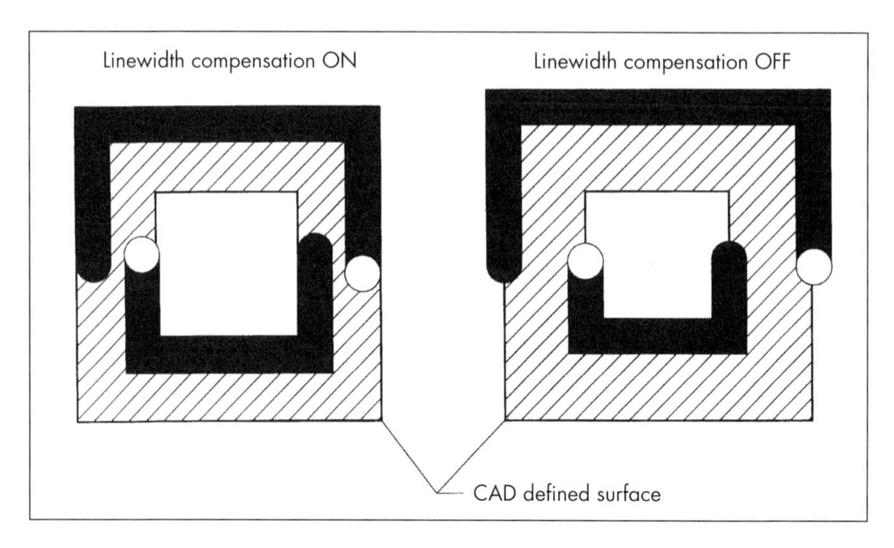

Figure 3-5. Linewidth compensation.

a Gaussian laser beam, the so-called half-width (W_0) is typically measured at its $1/e^2$ point (e.g., 100 μm @ 13.5% of peak amplitude). In SL, the cured polymer linewidth is calculated and used in the imaging process. From Reference 3, equation (4-38), we find:

$$L_w = W_0 \sqrt{\frac{2C_d}{D_p}} \tag{3-1}$$

Based on this equation, as the desired cure depth, C_d, increases, so will the linewidth. Note that the cured linewidth is a function of W_0, which is measured at a point rather low on the beam profile curve.

As a vector is drawn with increasing exposure, the depth to which the liquid polymerizes also increases, up to the point where it reaches the previous solidified layer. At that point, additional exposure increases the extent of polymerization within the new layer of resin and also increases the bond to the previous layer. Further polymerization can be considered beneficial in general, since material green strength and interlayer adhesion are improved. But increases in polymerization are not without some detrimental side effects.

Residual Stress. First and foremost, as the rate and amount of polymerization increase with exposure, so does the shrinkage associated with polymerization, as noted in Chapter 2. Briefly, the dynamics of resin modulus development as a function of increased exposure, known as "photomodulus,"[3] can lead not only to increased strength but also to internal stresses. The common term for this is re-

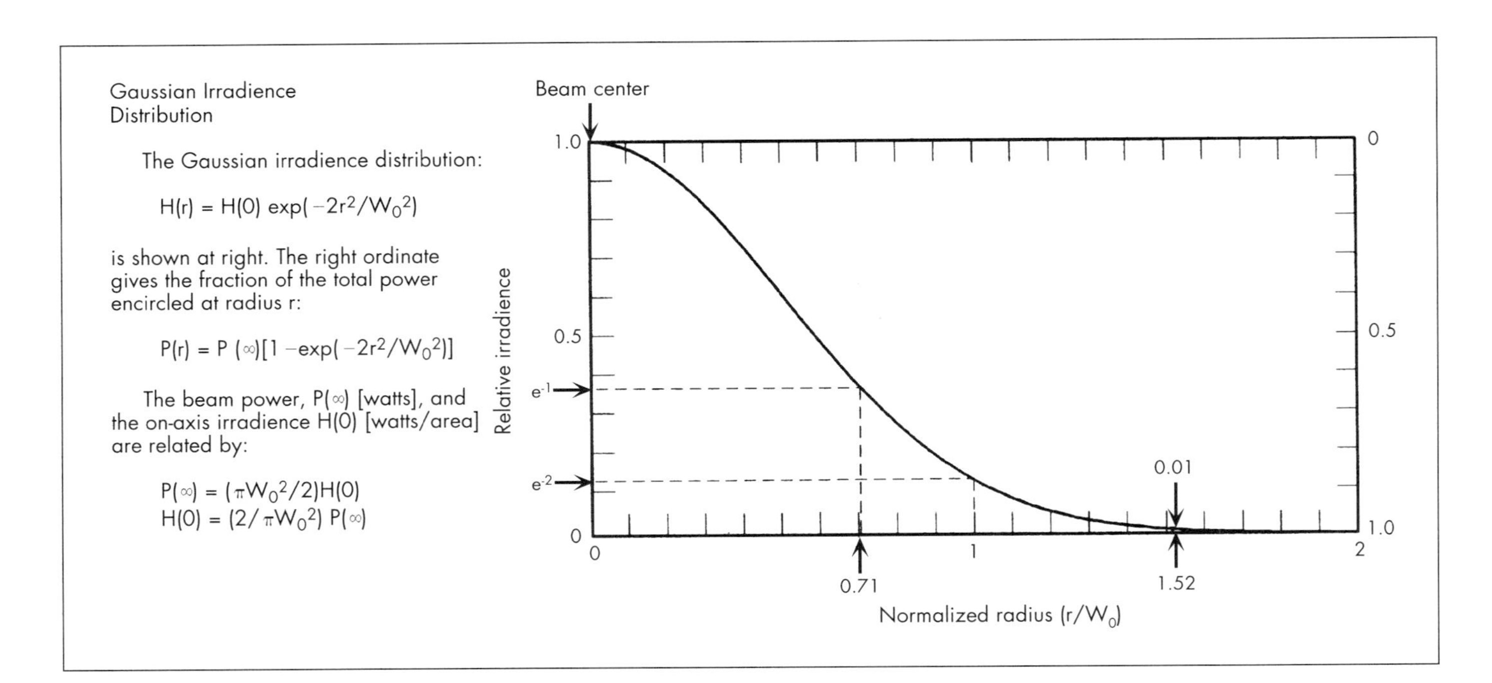

Figure 3-6. Typical laser beam profile.

sidual stress, and it will be relieved over time. This may be a matter of minutes, days, or even weeks. It is relieved by moving the molecules that make up the polymer, through a process called strain. The strain appears on a macroscopic scale as a gradual change in the shape and size of the model, and is commonly referred to as "creep," which can lead to warpage.

Reducing Stress. Much of the effort in developing SL in recent years has been directed at processes that reduce stress buildup in the model. Most of these process developments rely primarily on very precise control of exposure, involving:

- how the interior geometry of the model is converted into individual vectors;
- how the vectors are arranged in time and space; and
- how the vectors interconnect both within a layer and to previous layers.

A large class of stress control techniques rely on limiting the overlap and interconnection of vectors in the model interior. One technique, known as WEAVE, controls the gap between parallel vectors and the depth of polymerization. The vectors should be as close together as possible, without touching. This is achieved through precise servo control of the position of the scanning system.

When employing this technique, the vectors should be cured as deeply as possible, without effecting a bond to the previous layer. This requires precise control over the integrated exposure received by the photopolymer, as the laser spot scans the surface and draws the vector. This is achieved through precision servo control of the velocity of the scanning system.

These requirements necessitate both a velocity priority and a position priority servo control of the scanning system. Both are generally viewed as mutually exclusive control regimes. In SL, they must work in concert. Fortunately, a reasonable compromise can be achieved between the two control regimes, and a scanning system can be, and has been, designed to simultaneously achieve both good position and good velocity control.

The imaging system is also responsible for its own metrology related to exposure. It must provide measurements of the energy within the focused laser spot, as well as its distribution. This information is necessary to implement requests from the part preparation element regarding cured linewidth and depth, as well as other desired model properties. The metrics must be taken at the commencement of the build process, compared to nominal values and, if necessary, corrective action taken or warnings provided, should the metrics fall outside acceptable limits. The metrics must also be monitored continually during the model building process. In this way, the apparatus can adjust for minor fluctuations or arrest the build and alert the operator if uncommon conditions occur that indicate a state outside its accommodation range.

3.3 General System Implementation

The various machine subsystems have evolved over the past seven years. They resulted from an increased understanding of the processes involved, leading to improved part qualities for an expanding number of applications. Differences in overall model size also result in distinct implementations of the major elements. In concept, a large part is simply generated by a bigger version of an SLA that builds smaller parts; in implementation, the differences are significant.

The following sections describe the historical and current implementation of each subsystem, and are organized from the point where electricity flows into the SLA, through the laser and optical system, into the electronics, the mechanisms, and finally the control of the part building process.

Power, Environmental Control and Electronics

Electrical power is connected from the customer's facility to the SLA. Power specifications range from standard U.S. 110 volt "wall" power, to 208-volt, 3-phase high amperage requirements for the larger SL machines. This variation is based primarily on the laser in use, as the laser is by far the greatest consumer of power in an SLA.

Maintaining a specific temperature within the build area makes a major contribution to ensuring good model output. The photochemistry of a given resin performs best at a certain temperature within a small tolerance band.[4] Heaters, fans, and sensors are provided to maintain the desired temperature.

The laser employed in the imaging system requires cooling to maintain proper operation. Depending on the type of laser, cooling may be simple air-to-air heat exchange using fans, water-to-air using a radiative heat exchanger, or water-to-water in either a closed or open-loop system. The method selected is based on the amount of waste heat to be removed and the recommendations of the laser manufacturer.

Each SLA contains an X86-based control computer that accepts the build file from the computer used for the part preparation element. The controller regulates the part building processes, monitors machine interlocks, tracks the positions of various components, supervises laser power, and regulates the temperature in the part building area. Separate electronic boards control functions such as network communication, movement in the X, Y, and Z axes, coordination of other mechanisms, and scanning mirror operation.

Lasers and Optics

The laser chosen for the system must be appropriate for the resin used. Wavelength, output beam shape, and available power are important characteristics. If the resin and laser are wavelength incompatible, the resin will not be properly solidified. If the laser output power is insufficient, throughput of the SLA will be compromised by slow drawing speeds.[3]

Optical elements are used to fold the laser beam path, helping to keep overall machine size as compact as possible, and providing an appropriately sized laser spot on the surface of the resin, as discussed in Section 3.2. The optics are coated to either reflect or transmit the wavelength in use, depending on the function of each optical element in the system.

The scanning mirrors then position the laser spot at the desired location on the resin surface. Special consideration is given to minimizing the error that would be introduced by the elliptical sweep of the laser spot over the flat resin surface. In addition, dynamic mirror drift and laser power variations over time are recognized and corrected.

Resin Vat, Part Movement and Leveling Systems

Vats of various sizes are used in different SL machines. The vat is simply a container to hold liquid resin. To allow for resin shrinkage from the increased density of cured regions in a part, and also for resin displacement by the platform and its supporting structure, a subsystem is included that senses the resin height and requests adjustment, when required. Implementation of this level-finding system ranges from a float mechanism to a light-bounce system, both of which detect the current resin height, electronically report the information, and provide software interpretation of the resin level.

To change the resin level once it has been determined, several methods may be employed. Two have been used thus far in SLA machines, a displacement device positioned in the resin, and the direct method of moving the entire vat of resin up or down. To ensure uniform layering, the resin surface must be as featureless as possible. Waiting for the surface to become level is effective, but, depending on resin viscosity, surface tension, and the desired surface uniformity, this method can adversely affect part building speed. To hasten this process, most SLAs use a recoater (or "doctor") blade that sweeps the surface of the resin a selected number of times at a specified height above the last cured layer. As each layer is solidified, the part is moved down into the resin vat a distance equal to the thickness of the next layer by a stepper-motor driven stage.

Imaging System Implementation

The earliest SLA systems employed an X/Y pen plotter directed, fiber-optic coupled UV energy source to image data onto the surface of the resin. The systems were crude, but operative. They were also slow, inaccurate and, given the thousands of images required to build a complete SL model, prone to breakdown.

Second generation SLA machines employ imaging systems built around scanning mirrors supplied by General Scanning, Inc. (GSI) of Watertown, Massachusetts. This imaging system, used in SLA-190/250 machines of today, employs two orthogonally-mounted, servo-controlled, galvanometer-driven mirrors that direct the laser energy onto the surface of the vat. The laser spot position is monitored using a rotary encoder attached to each mirror. The servo loop is closed in a purely analog circuit, and a digital-to-analog interface is provided between the control electronics and the GSI equipment. This approach proved to be not only faster than the X/Y pen plotter system, but more accurate, and far more reliable.

Address Locations. Commands sent to the GSI system are simply an X and Y position in the form of two 16-bit words, one for X and one for Y. The system directs the mirrors to the point in the image field represented by the address numbers. The image space of the SLA is divided into $2^{16^2} = 2^{256} \approx 10^{77}$ address locations, which is the number of possible combinations of a 16-bit word. By controlling the time interval between sending subsequent address locations to the scanners, the rate at which the laser spot moves across the resin surface can be controlled. In this way, the SLA images vectors under both direction and speed control.

The 16-bit space of the GSI-based system is spread over an angular extent of 20 degrees. In the SLA-250, the distance from the laser source to the resin surface is approximately 30 inches (760 mm), which results in an effective resolution of about 0.00037 inch (9.4 microns) per address step, equivalent to a system resolution of 2,700 steps per inch. There exists no manipulative velocity control over how rapidly the system reaches the commanded address. To assure adequate exposure, the imaging system is commanded to a location no greater than one-half beam width away from the last commanded point. Using this method, the average laser exposure incident on the resin is controlled over the distance of one beam diameter, though the instantaneous speed and, therefore, exposure are not.

Servo Lag. One issue concerning most, if not all, imaging systems is the finite time required to reach the next address point. As the rate of commanded addresses is increased beyond the time required to reach the commanded point, the system begins to fall behind. Consequently, the actual location of the laser spot lags behind the theoretical, commanded address. This is known as servo lag.

In certain respects, servo lag is not a bad characteristic. In SL, this lag helps smooth the speed ripples so that very uniform instantaneous speed/exposure can be achieved. Unfortunately, when the commanded address changes direction, the actual position cuts the corner as shown in Figure 3-7. This effect produces rounded corners, resulting at the very least in small errors in reproducing the exact part geometry, and in extreme cases, geometry not being drawn at all, as seen in Figure 3-8.

In SLA machines using tens of milliwatts of laser power, the 10-20 inch-per-second (ips) (0.25-0.51 meter-per-second [mps]) drawing speeds are sufficiently low that servo lag is not observed. The system's simple interface, robust design, and stable performance are a great advantage in present day SLA equipment.

Faster Imaging Systems

When the SLA-500 system specification was being developed, it became clear that the required drawing speeds of a several hundred milliwatt SLA system were beyond what the existing analog imaging system could deliver. Angular scan rates on the order of 1.25 radians per second (100 ips for SLA- 500, or 35 ips for SLA-250) were required. The GSI systems available at the time exhibited servo lag at an angular scan rate of about 0.75 radians per second (60 ips for SLA-500, or 20

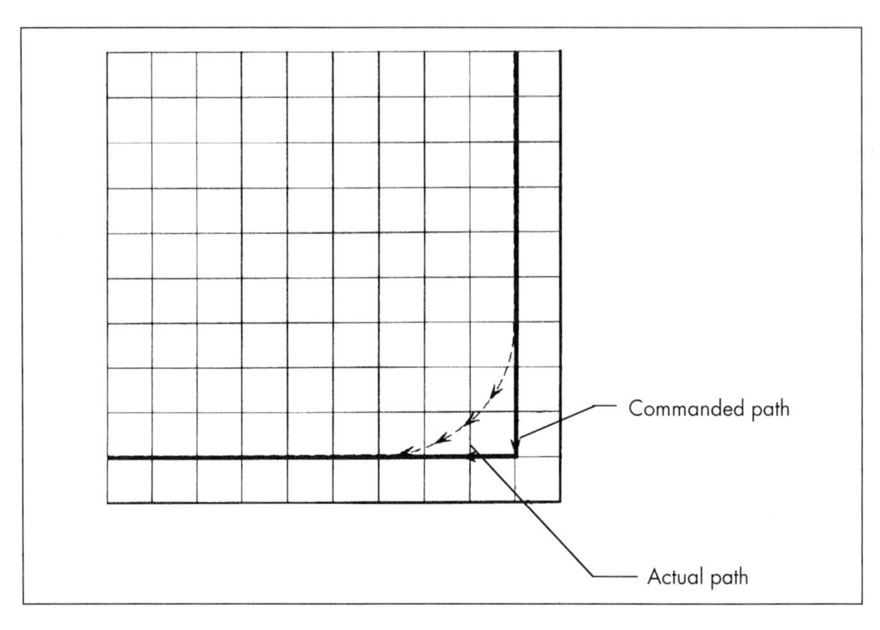

Figure 3-7. Position error due to servo lag combined with directional change.

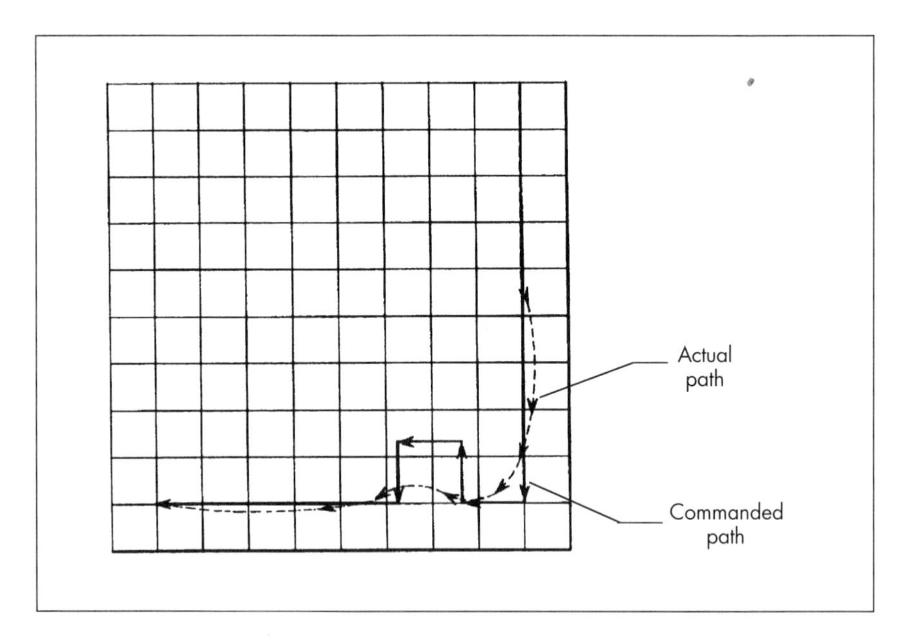

Figure 3-8. Extreme position error due to servo lag combined with directional change.

ips for SLA-250). The performance characteristics of all commercially available scanning mirror systems had either too much inertia and/or insufficient motor torque. At the same time, systems that could attain high drawing speed did not possess sufficient optical aperture to accommodate the focused laser beam. Simply stated, the mirrors were too small.

Digital Servo Control. A completely different approach was needed. The apparently fundamental contradiction of a large optical aperture, which creates high inertia of the motor/mirror combination, and fast and accurate system performance, had to be overcome. Digital servo control offered a possible solution.

In order to appreciate the novel approach taken in developing the scanning system for the latest SLA-500, a basic understanding of servo control theory is necessary. At the heart of the scanning system is the control loop. The loop attempts to maintain the actual position of the laser spot at the commanded position, or command address in the case of the SLA-250. As the commanded position is changed (the system is disturbed from equilibrium), the loop compares the actual position to the commanded one, and calculates a difference or error. The servo then moves the actual position to minimize the error. The time required to minimize the error is related to a quantity known as "bandwidth."

Measuring the Bandwidth. A standardized measure of the performance of a servo control system is the speed (or frequency) at which the actual location follows the command to within $\sqrt{2}/2$, or about 70%, when the command varies as a sinusoid. This frequency is known as the bandwidth of the servo system. Typical measures of bandwidth are found on "Bode" plots. A Bode plot graphically shows the amplitude and phase of a system responding to a sinusoidal time varying command.

Imagine a vector scanning system is given a sine wave to draw across the image field. At low frequencies, one spot is seen traversing the field. The commanded position and actual position are the same. But as the frequency (speed) increases, a second spot (the actual) emerges, lagging behind the first (the commanded). As the frequency increases still further, the actual position gets so far behind that the commanded position has already reversed direction before the actual position has reached its peak. The amplitude of the actual path begins to decrease as it lags further and further behind, until the commanded path has reached the negative peak when the actual path is just reaching a positive peak, as shown in Figure 3-9.

Increasing the Bandwidth. The key to an excellent servo system for SL is to increase the bandwidth as much as possible . . . without going too far. Increasing the bandwidth of the system results in the actual path following the commanded position much more closely. Too far, and the system may actually have difficulty moving slowly—for example, slowly enough to cure very deeply.

In classic, "reactive" servo control theory, a system can be tailored to be either fast or smooth, but not both. Digital servo control offers the potential of achieving both by implementing two control systems in one, with a means of switching

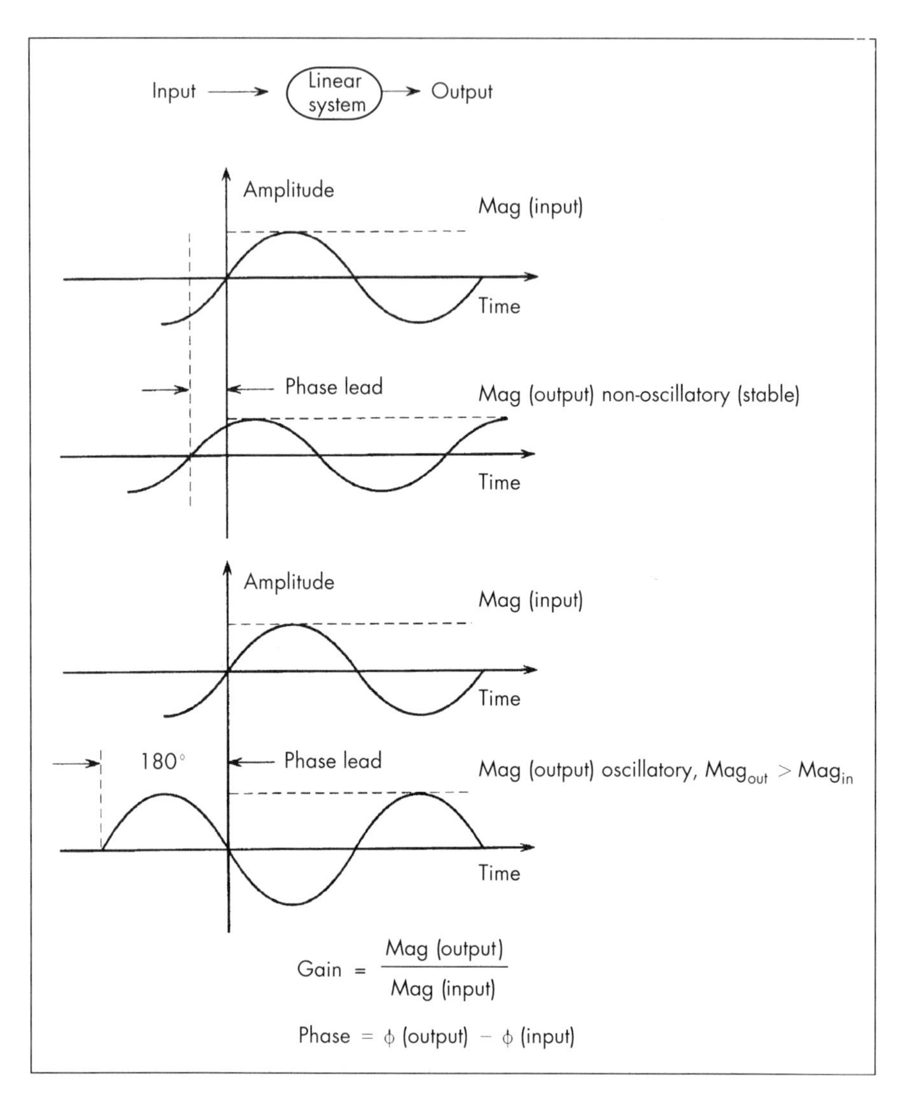

Figure 3-9. Amplitude of the actual path begins to decrease as it lags further and further behind, until the commanded path has reached the negative peak, when the actual path is just reaching a positive peak.

control between them, depending on the system demand at any given time. This is the approach chosen for the SLA-500.

The original SLA-500, and later the SLA-500/20, used digital control based upon an early generation digital signal processor (DSP) from AT&T. These early DSPs afforded performance not previously achieved in servo control, through a conceptually simple planning technique. The system would "look into the future" by examining the command for the next vector. Based on the knowledge of what was coming, the system could plan actions that would enable it to achieve the next desired position.

As the driver of a car on the highway, you are, in effect, the servo control system. The road is analogous to the commanded position. If you are driving a car on a twisting mountain road in the fog, you slow down to protect against errors in following the commands of the road. You slow down because you realize that you have limited knowledge of what is coming. In this way, almost automatically, you allow for your fixed available response time. This response time is analogous to the bandwidth of a servo system.

As the fog lifts, you can increase speed, because your vision of the road ahead improves and your knowledge of the next road command has been enhanced. Though your bandwidth hasn't increased, you are now able to apply correction just a bit before it is needed. This system of looking ahead in the command string is called "feed forward," and is used in the SLA-500 to achieve performance that would normally be beyond the limitations indicated by the bandwidth of the servo system.

Fast Shutter. After the incorporation of this feed-forward technique, the greatest limiting factor in the imaging system was not servo control, but a component in the optics train, specifically, the fast shutter. This component was used to turn the UV energy reaching the vat on and off, at the beginning and end of vectors. The technology available at the time that best fits all requirements for the system was a mechanical device requiring about one millisecond (ms) to change state.

A one ms state change seems fast, until you consider the distance a laser spot, moving at 100 ips (2.5 mps), has traveled while the shutter is opening or closing. The spot moves 0.1 inch (2.5 mm) in one ms. Without correction, this results in vectors being too short at the start and too long at the end. The resulting part accuracy and surface finish would have been unacceptable. Compounding the difficulty, performance of the mechanical shutter changed over time due to temperature, aging, and other factors.

Given all these issues, a scheme of correction was implemented. Much of the computing power of the DSP was therefore dedicated to predictive work, in order to accommodate the relatively slow "fast" shutter. Algorithms were implemented that "looked ahead" for required shutter action, and then calculated backward in time appropriately. These algorithms placed the shutter command at a location in the command string so that the actual opening or closing would occur reasonably near the correct point on the drawn vector.

3.4 Advances in Leveling and Resin Interchange

As our understanding of all SL processes improves, implementation of any subsystem or group of subsystems can also be improved. The next two sections discuss the major elements of the systems where enhancements have resulted not only in better reliability or manufacturability, but in improvements in the quality of the model or the interaction of the user with the system.

Level Sensing Advances

The resin level sensing system for the first commercial SL machines was based on detecting the beam of a helium-neon (HeNe) laser reflected off the resin surface onto a silicon bi-cell detector, as shown in Figure 3-10. If the signal from the bi-cell indicated that the resin level was too high or too low, the system would call for the resin surface to be displaced accordingly, until the level was within an acceptable tolerance band. This "light-bounce" system lends itself nicely to stereolithography, as it is reasonably accurate, fairly fast, noncontact, and insensitive to the density differences that may occur between various resins.

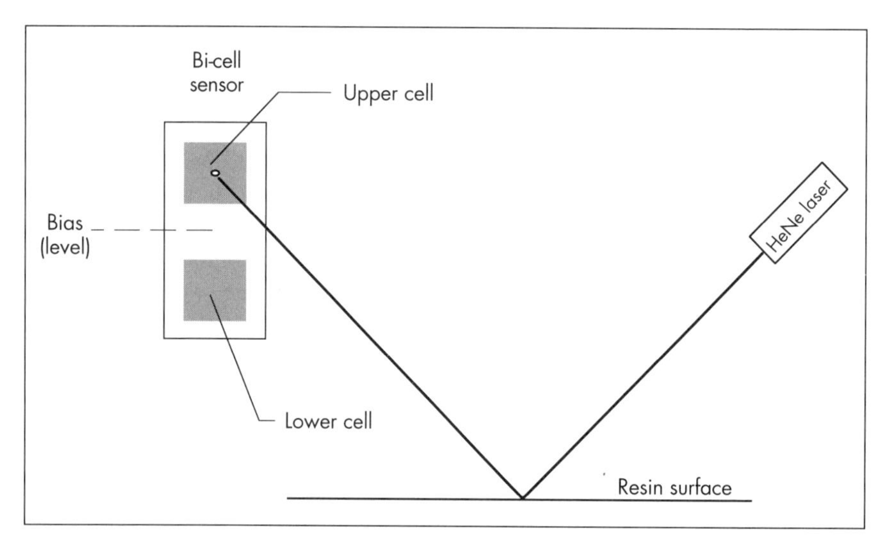

Figure 3-10. The original SL light-bounce leveling system.

Evaluating Light-bounce. One important consideration for the light-bounce leveling system is that any significant discontinuity where the laser spot is incident on the resin surface can result in the resin level being sensed improperly. Small waves or bubbles that might be created when the part is moving in the resin vat must be kept out of the area where the laser reflects off the resin surface. A baffle has proven satisfactory in recent implementations, but other methods have been used, such as an ante chamber whose surface is isolated from the part building area.

Spurious signals caused by electronic noise from the HeNe laser could generate a request for resin displacement when none was needed or, even worse, induce resin movement in the wrong direction. This would then have to be corrected during the next level check and resin displacement. This type of "hunting" behavior results in significantly slower performance of the leveling subsystem than would be possible if the electronic noise were not present.

Also, data from SLA machines in the field has shown that HeNe leveling lasers have an effective lifetime of approximately 3,000 hours. While this may seem reasonable, it falls short of what is desirable for SL machines, which are routinely operated in excess of 6,000 hours per year. This limited lifetime results in relatively frequent HeNe laser replacements, along with the accompanying system downtime.

Overall, the HeNe bounce system works quite well. However, as technologies outside the SL arena have advanced, components have become available that improve SLA accuracy, repeatability, reliability, and even extend the functionality of the original system.

Enter InGaAIP. The commercialization of low-cost solid state lasers was a welcome event for the SLA leveling system. An indium gallium aluminum phosphide (InGaAIP) visible diode laser was chosen to replace the HeNe leveling laser for the SLA-250. One reason for this choice is that the InGaAIP diode laser has an indicated lifetime of 50,000 hours, compared to about 3,000 hours demonstrated for the HeNe. The resultant decrease in SLA downtime is substantial, even if the demonstrated lifetime of the diode laser proves to be somewhat less than indicated by the manufacturers.

In addition, the electronic noise generated by the diode laser is approximately an order of magnitude lower than that for the HeNe. Leveling speed, accuracy and repeatability have all been improved due to the absence of significant noise. Figure 3-11 shows 500 layers of a test involving the HeNe leveling system. In this segment, the HeNe system exceeded the allowable ±0.25 mil (~6 micron) limits 29 times. Figure 3-12 shows the same test of the diode system, where the limits were never exceeded. The new diode laser leveling (DLL) systems have demonstrated leveling accuracy of ±0.0003 inch (±7.5 microns), compared to ±0.0006 inch (±15 microns) for the HeNe system. In contrast, a human hair is generally about 50 microns in diameter, and SL layer thicknesses in common use are now in the range of 100 to 200 microns, so leveling accuracy is very respectable at 3 to 7% of typical layer thicknesses. In addition, the DLL system repeatability equals its accuracy, at 7.5 microns.

The original leveling system for the SLA-500 involved a mechanical float that moved up and down with resin level changes, providing signals that triggered appropriate resin surface movement. While this method was simple, it proved to be susceptible to changes in resin density, hysteresis resulting from sticky bearings, and other mechanical effects.

Following the development of the DLL for smaller SL machines, an enhanced light-bounce system was developed for larger SLAs. While based largely upon the proven system from the smaller SLAs, the silicon bi-cell detector was replaced by a silicon linear cell on larger machines.

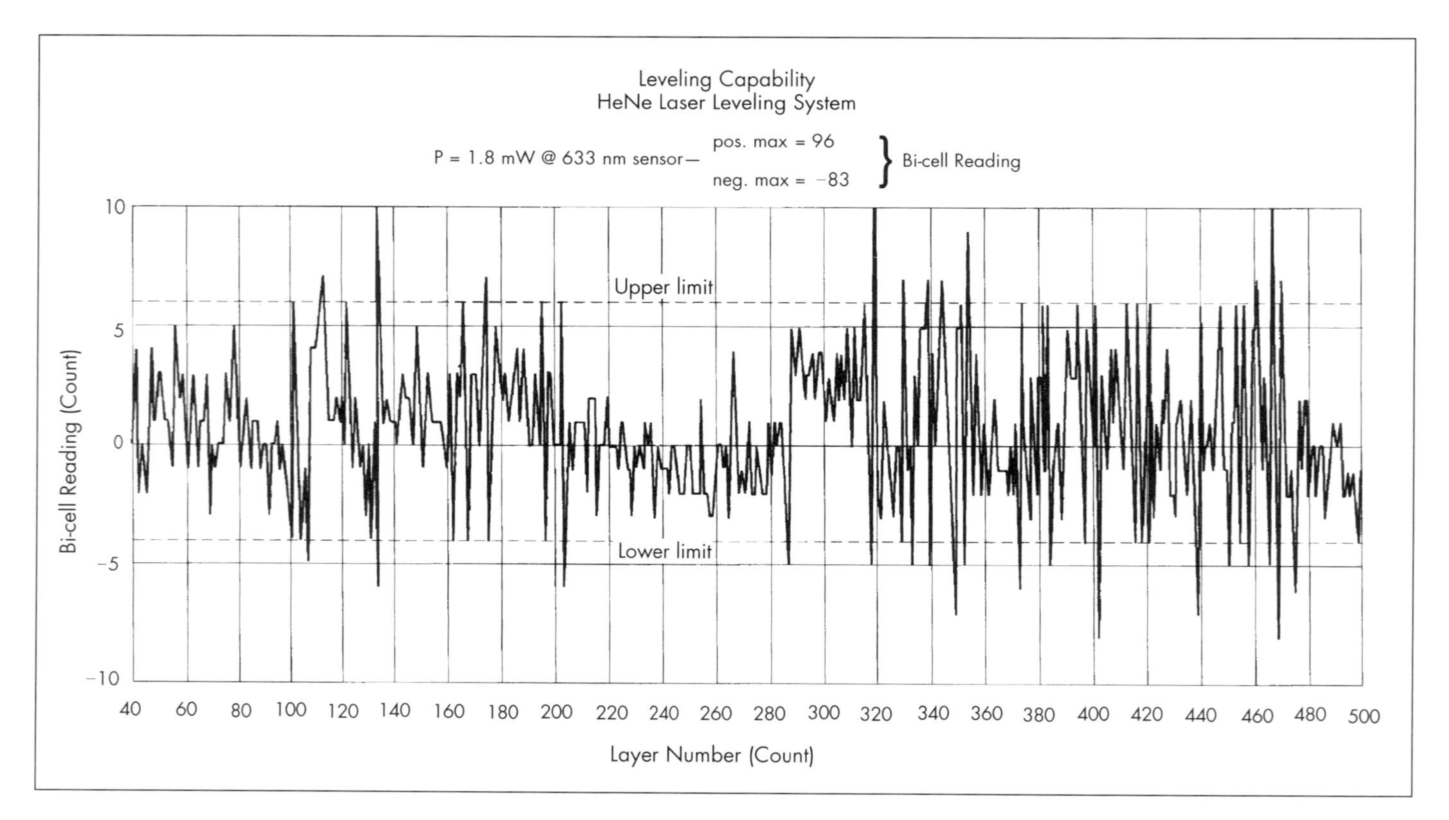

Figure 3-11. Five-hundred layers of a test involving the HeNe leveling system. Here, the HeNe exceeded the allowable limits 29 times.

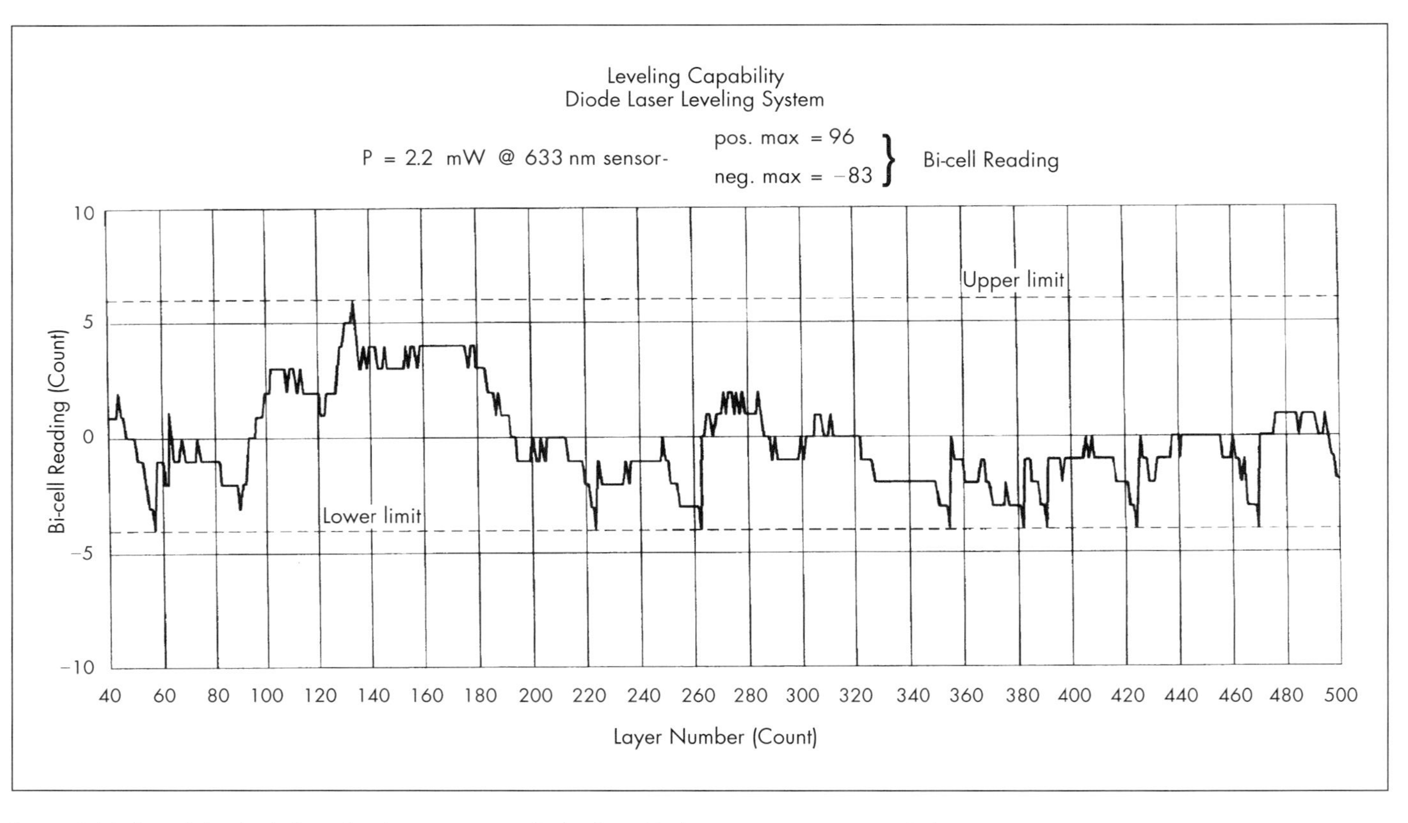

Figure 3-12. Test of the diode laser leveling system in which allowable limits were never exceeded.

One feature of the linear cell in the new SLA-500/30 system is that it not only reports if the resin is too high or too low, it also provides a definitive measure of how far away the free resin surface is from the desired level. Depending on where the light from the diode hits the linear cell, the difference between voltage outputs from each end of the cell can be calculated as shown in Figure 3-13. With this information, the system is directed to move the resin surface within the resin level tolerance band. An advantageous feature of the linear cell is that the slope of the voltage comparison is linear within the operating range of the system, as shown in Figure 3-14. This allows the software, which "zeros in" on the ideal level, to be both simpler and swifter. The software algorithms used in this leveling system are described in Chapter 4.

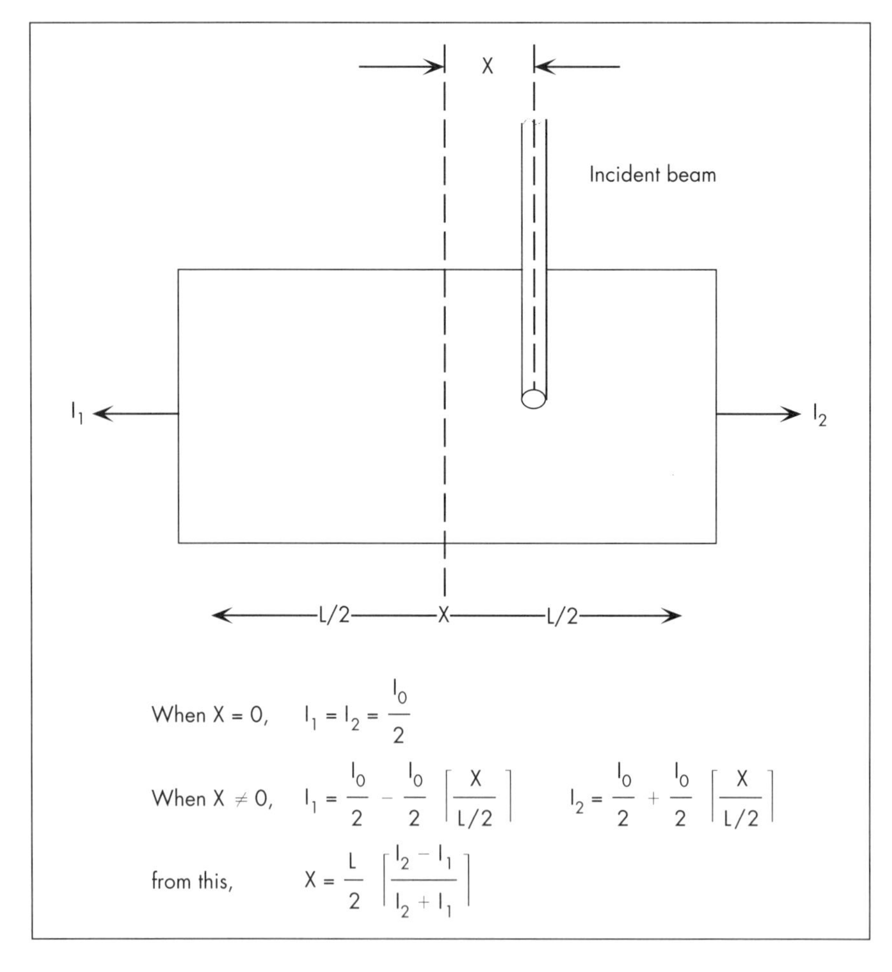

Figure 3-13. Linear cell voltage output difference calculation—find the difference between voltage outputs from each end of the cell.

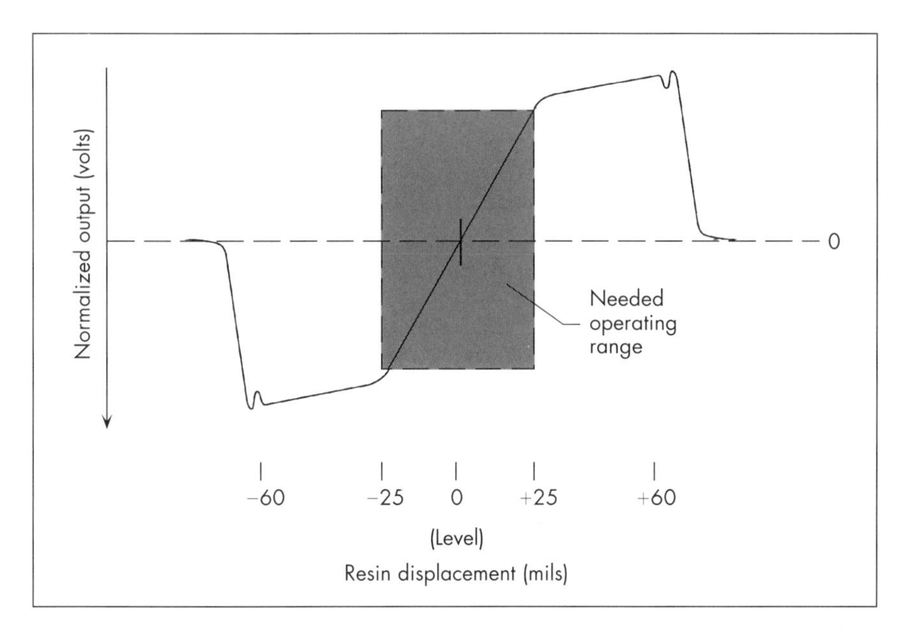

Figure 3-14. Linear voltage comparison slope, within sensing system operating range, for the new silicon linear cell.

To account for as much volume change as possible (e.g., for maximum resin shrinkage during solidification), wide dynamic range in the system is also desirable. The linear cell is approximately 0.197 inch (5 mm) in length by 0.080 inch (2 mm) in width, and the laser spot about 0.008 inch (0.2 mm). This provides not only good dynamic range, but targeting the relatively small laser spot on the large linear cell makes producing, installing, and aligning the system very simple.

Offset Procedure. Improved speed, accuracy, repeatability and reliability have been achieved with the new DLL system using the linear cell. In addition, expanded functionality has also become feasible. Different resins and part drawing styles often require different spacing between the resin surface and the recoater blade. Creating a robust and simple mechanical system to enable the SLA user to change this "blade gap" presents significant difficulties. When employing the linear cell, a simple software offset can perform this function with almost identical results. This technique cannot be used to simulate a blade gap beyond certain limits, due to the dynamic range limits of the sensor and the allowable offset of the resin surface. Nonetheless, the software-generated offset procedure has proven extremely valuable for the range of blade gaps currently used in part building on SLA machines.

This offset strategy works as follows. Whenever a part is built, the SLA software accesses a file containing various resin-specific parameters. The SLA recoater blade is set at a "baseline" blade gap, and the system then offsets the resin level up or down the appropriate distance from that baseline value to comply with the blade gap parameter contained in the selected resin file. Thus, the blade gap can be changed in software to suit the specific requirements of a given resin, or even on a build-by-build basis. The software implementation of this effective blade gap is described in Chapter 4.

The offset in the resin level, relative to the baseline, does change the magnification of the system.[5] The maximum change is approximately 0.02%, or 4 mils (0.1 mm) over a 20-inch (508 mm) dimension. This is truly a scaling effect, equal in all areas, as shown in Figure 3-15 using similar triangles.

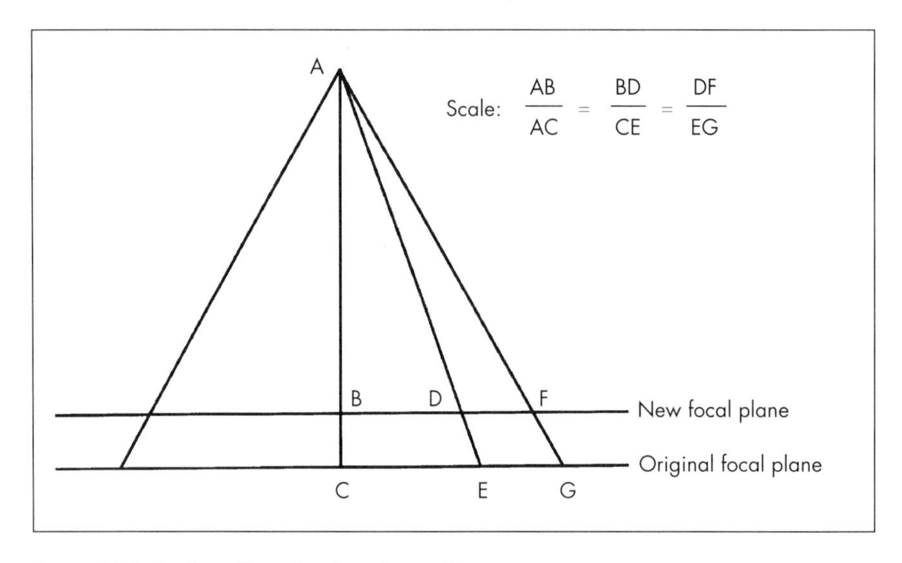

Figure 3-15. Scaling effect of resin surface offset.

An automatic software correction for this magnification change has significant potential to confuse the operator, and is therefore not incorporated. Instead, this and other scaling errors can be compensated for in one step, by building and measuring a suite of test parts. This is often performed by users in order to "fine tune" their SLA equipment. Alternatively, the correction can be calculated, and added to the resin shrinkage compensation factor. An example follows.

Actual blade gap (factory set)	0.025 inch (0.635 mm)
Preferred blade gap (from resin file)	0.006 inch (0.1524 mm)
Resulting change in focal distance	0.025 - 0.006 = 0.019 inch (0.4826 mm)
Correction (0.02%)	0.019/98.0 = 0.0002 inch/inch or 0.0051 mm/mm
Resin shrinkage factor (from data sheet)	0.003
Final shrinkage compensation factor	0.003 + 0.0002 = 0.0032

Consequently, the amount of compensation required for the range of blade gaps currently in use results in such a very small change that it is not generally noticeable in the completed model.

Advances in Resin Interchange

Conceptually, resin vats have always been interchangeable, the primary issues being the amount of time, expense, and difficulty involved. As different resins with unique properties have been introduced, the need has intensified for quick and easy resin exchange.

There are several issues that come into play when interchanging vats. Is the leveling system compatible with the new resin, especially in regard to reflectance and density? Is there ancillary equipment attached to the vat that is sensitive to being removed and replaced? Will the machine calibration be disturbed by exchanging the vat?

Previously, most SLA machines had one or more obstacles to "true" vat interchangeability. Several parallel advances, including DLL systems, have contributed to streamlined vat interchange.

Some early SLA-250 designs positioned the leveling system and recoater blade directly on the resin vat. This arrangement functioned well in normal operation, but presented difficulties when switching vats. Several fundamental mechanical alignments required readjustment when a new vat was installed, and the system also required a field service call for thorough recalibration.

A new hardware design, based on concepts used in the large-frame SLAs, positions the tolerance-critical components on a stationary rim assembly as shown in Figure 3-16. This rim serves as the reference plane for all machine axes and many other mechanical components. The new design allows the vat to be a simple resin container that is easily removed and sealed.

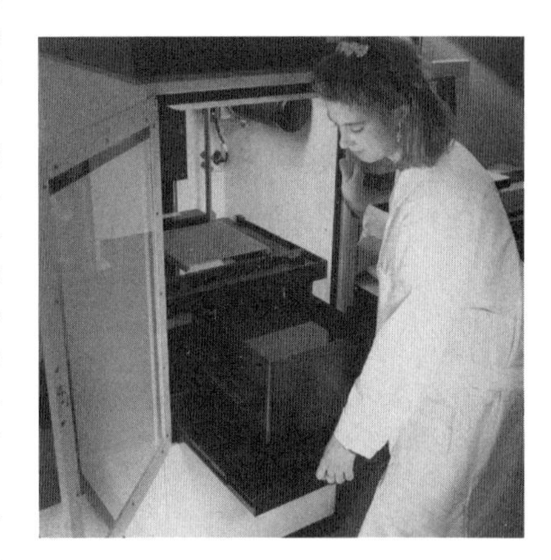

Figure 3-16. A new SLA-250 hardware design positions tolerance-critical components on a stationary rim assembly.

Rather than the fluid displacement device used previously, resin surface positioning in this system is now accomplished by raising and lowering the vat itself, using a single-ballscrew hoist system. This hoist system is also used to lower the vat to an "unload" position from which the user pulls the vat out of the machine on its drawer slides.

Designing the system so that all critical components remain within the SLA during vat exchange, and permitting the vat to be a simple resin "bucket," facilitates quick exchange of resin. The user is now free to build with the resin most appropriate for the part application at hand, without concern for excess machine downtime or the expense of a service visit associated with complete system recalibration.

3.5 Advances in the Imaging System

Major advances in SLA-500 imaging system performance and reliability have been achieved through the introduction of a modern DSP, a truly fast shutter, and a new laser focusing system. While sharing the same scanning mirrors, laser, and optical components with the SLA-500/20, the combination of new hardware and control software results in significant documented improvements in speed and accuracy. The SLA-500/30 system provides enhanced replication of the desired vector image, both in terms of placement and exposure. Furthermore, it does this image after image, part after part.

The Acousto-Optic Modulator

The original electro-mechanical "fast" shutter implemented on the SLA-500/20 has been replaced on the SLA-500/30 by a solid-state Acousto-Optic Modulator (AOM), operating on the principle of acousto-optic birefringence. These devices alter the trajectory of coherent radiation through a solid medium, by introducing sound into that medium. A radio frequency (RF) sinusoidal signal is applied to a piezo-restrictive device bonded to the quartz medium. The piezo, in turn, sets up a sound wave in the quartz crystal. The sound field within the quartz crystal produces an effect much like a diffraction grating, altering the path of the laser beam, as illustrated in Figure 3-17.

The deflection angle of the laser radiation through the AOM is proportional to the frequency of the sinusoidal signal driving the piezo, and the wavelength of the laser beam. It is inversely proportional to the speed of sound in the crystal. Efficiency of the diffraction grating is nonlinearly proportional to the amplitude of the sinusoid, and also depends upon various crystal properties.

Not all of the energy of the incident beam is diffracted into a single outgoing beam. A number of beam trajectories will simultaneously emerge, depending on the efficiency of the AOM, and the alignment of the laser beam. These beams will appear at regular angular intervals on both sides of the undeflected beam. The beams are numbered with integers, zero referring to the undeflected beam.

An objective for the SLA system is to achieve the highest possible transmission of actinic radiation on the path from the laser to the resin, while at selected intervals achieving sufficient power attenuation that polymerization does not take place. Otherwise, unwanted lines of polymerization, called "stray vectors" or "spider webs," may occur. A system with high laser energy transmission is vital, since the productivity of the entire system is directly related to the laser power at the resin surface. Unfortunately, simultaneously achieving a complete absence of laser radiation in the fully attenuated state, along with good total throughput, is not possible.

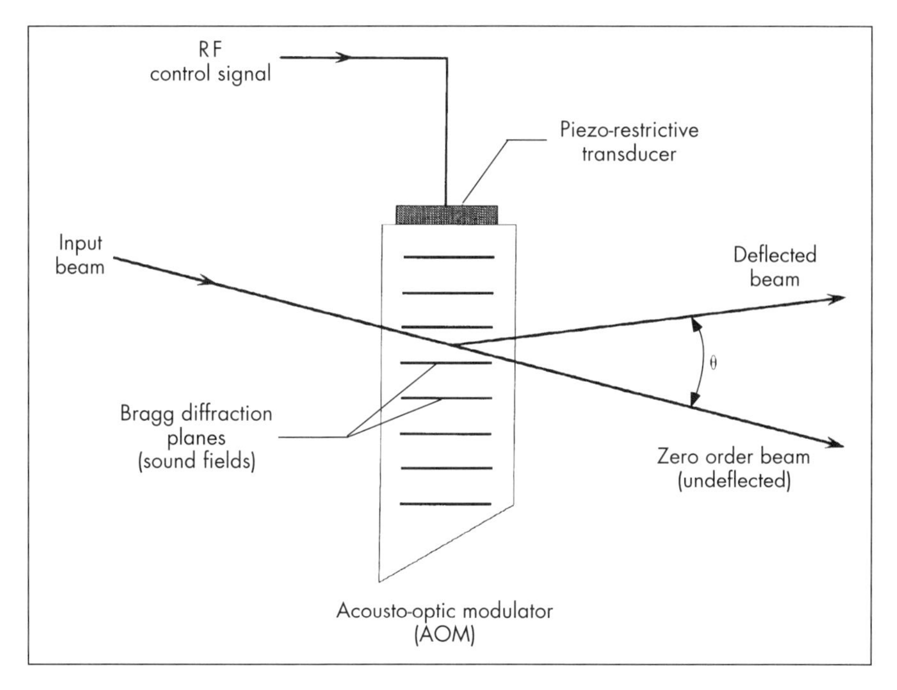

Figure 3-17. AOM alters laser beam path.

Contrast Ratio. The measure of attenuation used in SL is called contrast ratio, and is simply the ratio of the irradiance of the laser during drawing to the irradiance when attenuated. A contrast ratio of better than 20:1 allows the beam to traverse the same spot on the vat up to six times on a given layer without causing polymerization. Traversing the same spot more than six times in a single layer is very rare. In fact, a unique build file had to be developed just to test this process. Additive exposure on subsequent layers is not a consideration, since during the process of recoating, fresh resin is brought to the surface, and the previously exposed, but unpolymerized, resin is mixed with the bulk resin.

Two AOM Configurations. Two potential configurations of an AOM in an SLA optical system were considered. In both configurations, the laser beam passes through the AOM, is brought to focus, passes through a small pin hole, and is refocused back at the resin surface. The first beam and all higher order diffracted beams appear only upon introduction of the RF signal to the AOM. When the RF signal is turned off, only the *zeroth order beam* is present.

One configuration utilizes the zeroth or "undeflected" beam as the primary drawing instrument, while the second configuration employs the first order beam. In the former method, the pinhole is positioned to admit the zeroth order beam. During drawing, the AOM is deactivated, and its transmission is approximately

99%. When the AOM is activated to fully attenuate the laser beam, about 98% of the energy is diffracted out of the zeroth order and does not pass through the pinhole. Thus, the contrast ratio in this case would be 0.99 / (1-0.98) = 49.5.

In the second configuration, the pinhole is positioned to admit the first order beam. Here, the AOM is activated during drawing, and the transmission, in this state, is about 85%. Thus, the effective laser power is diminished by roughly 15%. However, when attenuation of the beam is desired, the AOM is deactivated, and practically no laser power passes through the pinhole. Consequently, the contrast ratio is effectively infinite.

One configuration affords essentially infinite contrast ratio, while the other offers a greater transmission of laser energy. Since total absence of laser radiation is not required for SL, the higher laser power option was selected for use on the SLA-500/30. This option has the advantage of increasing system throughput and productivity. However, on rare occasions, the result of finite contrast ratio and multiple scans over the same "nondrawing" location can lead to spider webs.

AOM Fast Shutter Benefits. The use of an AOM as a fast shutter has many benefits. It is clean, silent, safe, has no moving parts, and is energy efficient, requiring only about 5 watts. But the device really shines (or doesn't) in performance. The original mechanical shutter was not repeatable, taking anywhere from 600 to 1,600 microseconds (μs) to respond, depending upon factors discussed earlier. The control software could accommodate two values, one delay for opening and one for closing. Typical values were 800 μs for opening, and 1,000 μs on closing. In the case of a well-used shutter, the actual opening delay could grow dynamically during the course of part building, from the nominal 800 to more than 1,200 μs, resulting in a 400 μs error.

With the laser spot moving about 0.0001 inch (0.0025 mm) per μs at a typical drawing speed of 100 inches (254 cm) per second, a 400 μs error in shutter operation results in a 0.040 inch (1 mm) vector placement error. This directly translates into inaccuracy and poor surface finish on SLA-500 parts.

Fortunately, the AOM responds in about 1 μs, or about three orders of magnitude faster than the previous mechanical shutter. Over this time interval, the laser spot will move only 0.0001 inch (2.5 microns), or about 1% of its diameter. Based on this performance, the servo system need not concern itself with predictive control of the shutter. With the burden of shutter time-lag control relieved, the full power of the DSP is now available for servo control of the scanning mirrors.

Incorporating an AOM as the fast shutter also allows a different configuration of focusing optics, permitting an integrated mechanical design. The previous mounting requirements were driven by the position of the mechanical shutter between the two focusing lenses. The resulting open frame was troublesome to align and service, and expensive to build. The new optical mount is a one-piece structure, where optical element alignment is locked in at the time of fabrication. No adjustment is available or required. This approach has demonstrated very precise and stable positioning of the focusing optics.

The New Digital Signal Processor

The DSP used on early SLA-500 systems was an AT&T chip, the DSP32. Fortunately, DSP technology moved forward at a rapid pace while the SLA-500 was being adopted by industry. When the new Orion Imaging System™ was designed, modern DSP chips from several manufacturers were examined. The Texas Instruments TI320C30 chip was chosen for the SLA-500/30. A performance comparison table for the AT&T and TI chips is included in Chapter 4. With the TI chip, computational power is increased four fold, 10 times the memory is directly addressable, and an interrupt is provided for time critical tasks. In addition, ease of programming and the toolset available for the new chip are excellent.

The hardware configuration of the Orion Imaging System is nearly identical to early SLA-500 machines. An argon-ion laser serves as the source of UV energy, although a new higher output version is offered. The laser beam is directed via two mirrors through the new truly fast AOM shutter, into the focusing optics, onto the scanning mirrors, and finally onto the resin surface.

The keys that unlock the potential within this advanced system are the improved DSP and its new controlling software. One major advantage of digital servo control over analog control is the potential to implement more sophisticated control algorithms than would be feasible in a strictly analog system. In particular, the programs that run on the DSP can increase system bandwidth by digitally removing servo instructions occurring at frequencies that cause oscillation in the system. This is called frequency notch filtering.

Commanded Trajectory. How does an imaging system control the manner in which the actual position follows the commanded position? Figure 3-18 shows a commanded trajectory for a servo system to follow. The grid lines correspond to 0.005 inch (0.0127 mm) increments. Figure 3-19 depicts reactive servo control, without feed-forward commands, and with limited lag. As shown in this figure, the actual path significantly overshoots the command trajectory. This is because

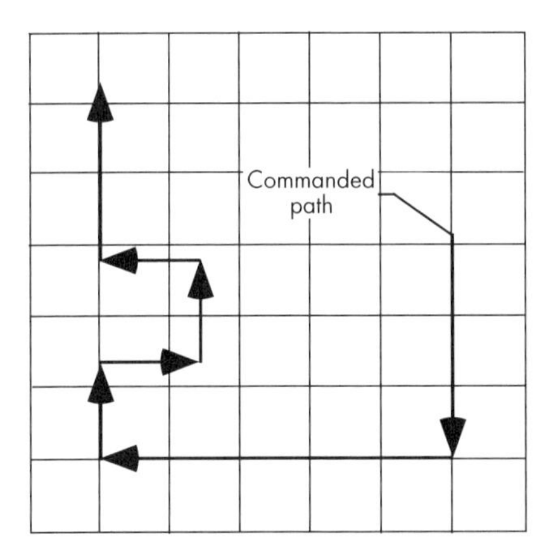

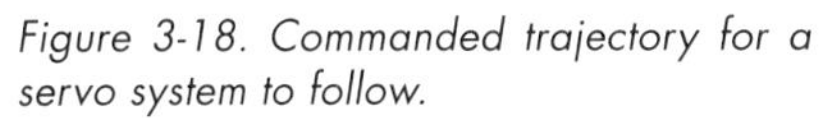
Figure 3-18. Commanded trajectory for a servo system to follow.

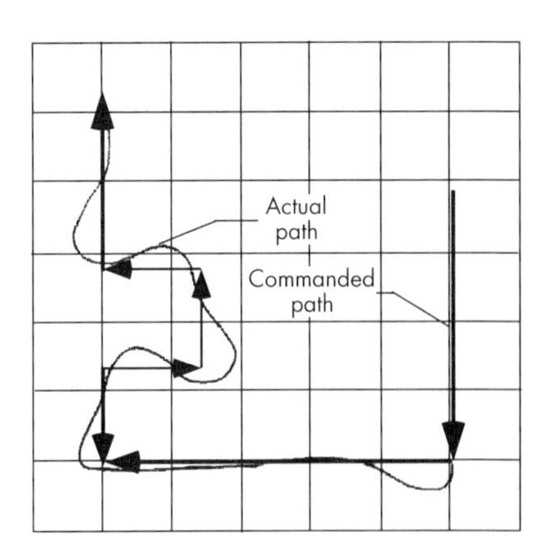

Figure 3-19. Path following with purely reactive servo.

the servo does not begin to attempt the corner turn until the actual path overshoots the command. As the actual path passes the command, the error calculated between the two becomes larger, increasing corrective action, with torque applied by the motor in order to turn the corner. The bandwidth of the servo establishes the characteristic time required to reduce the error and bring the actual path back in line with the commanded path. For a servo system with a bandwidth of one kilohertz, the error will be essentially eliminated in approximately one millisecond. The time constant of the system is described by

$$\tau \equiv \frac{1}{2\pi f} \qquad \text{where } f = \text{frequency bandwidth}$$

A system with two-kilohertz bandwidth requires about half that time to correct the error.

Servo Lag. Figure 3-7 also showed how lag can even prove useful in a reactive servo system. Here the actual path trails the commanded path by 300 μs. The error begins to increase as the commanded path traces the corner, while the actual path still lags behind. The servo has been set up to tolerate a certain error level along the straight trajectory, but an error off trajectory, as occurs when rounding a corner, starts corrective action. Thus, as the commanded path traces the corner and moves off along the X axis, the servo begins to accelerate the actual path in the X axis. If this set of moves is well balanced, the corner is cut in a smooth curve, and the actual path is back on trajectory along the X axis, with the same amount of following lag. This results in smoothly rounded features and good surface finish, even though the system is not following the true part geometry.

Figure 3-8 showed one deficiency in using significant lag as a servo control scheme. Here the commanded path has rounded a corner, but quickly makes four more turns forming a small cutaway. In this case, the actual path is too far behind the command trace, resulting in only the slightest hint of the cutaway appearing in the actual path. The lag distance, and thus the time required for smooth operation, is related to the bandwidth of the servo.

As bandwidth is increased, the lag time decreases by a similar proportion. In the example given, the reactive servo requires about one millisecond to round the corner and once again become aligned with the commanded trajectory. This time interval is roughly equal to the inverse of the bandwidth of the system. Attempts by the servo to respond to changes in the commanded path, which occur over time periods shorter than one millisecond for a 1 KHz bandwidth, will lead to poor results. The cutaway shown in Figure 3-8 represents a course change over only 300 μs. The servo system is able to make only a feeble attempt to follow the commanded path. This situation illustrates the substantial benefits that can be achieved by increasing the bandwidth of the servo system.

Mechanical implementation of the system is a limiting factor on servo system bandwidth. There are operational frequencies that will induce oscillation in the motors, encoders, and mirrors. These frequencies are referred to as system resonances. In SL, the scanners can be modeled as two masses, M_1 and M_2, connected by a spring (K_p) with associated damping (K_d). This is shown conceptually in Figure 3-20.

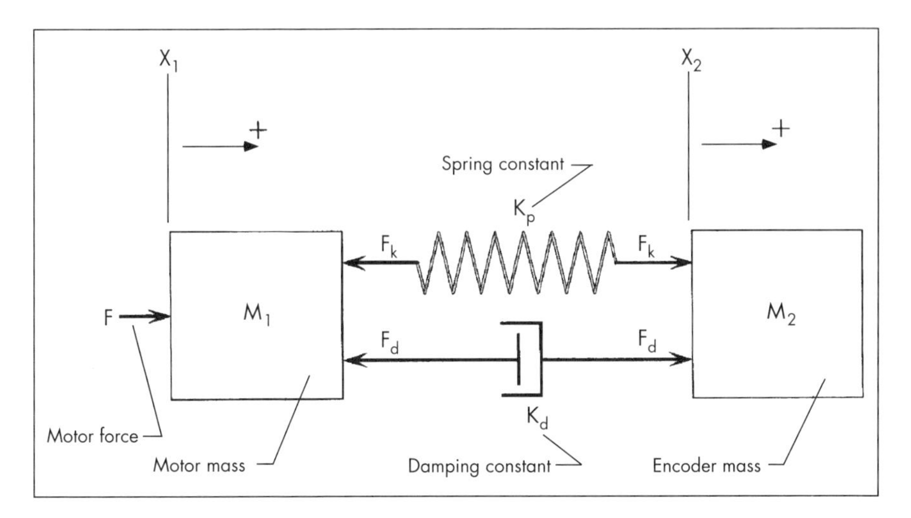

Figure 3-20. Known as system resonances, some operational frequencies induce oscillation in the motors, encoders, and mirrors.

The force is applied to the shaft through motor torque, affecting mirror motion and position encoder movement. The spring represents the stiffness of the shaft connecting the motor coils to the two masses. The behavior of this model can be described by the fundamental forced harmonic motion equation.

$$M\frac{d^2X}{dt^2} + K_d\frac{dX}{dt} + K_pX = F_0\sin(\omega t) \qquad (3\text{-}2)$$

As the frequency, ω, of the motor is increased, approaching the resonant frequency of the mass/spring system, mass M_2 begins to move out of phase with mass M_1, and this system is said to be in resonance. The resonant frequency is given by the equation:

$$F_r = \sqrt{\frac{K_p}{M_2}} \qquad (3\text{-}3)$$

In a well-balanced system, the servo must avoid applying forces at frequencies at or approaching the resonant frequencies of the mechanical parts. One method is to use a notch filter in the circuitry generating the force control. This filter prohibits the applied servo force from changing at or near the resonant frequency.

Driving Off the Road. Figure 3-21 shows a command profile being executed by the Orion Imaging System. The lines, again separated by 0.005 inch (0.127 mm), indicate the real-time association between the commanded points and the actual points along the desired path. Drawing at 100 ips, the system never strays from the commanded path by more than 0.001 inch (25 microns). The actual path is much closer to, but not exactly the same as, the commanded path. Recall that for a servo system to do its job, some error must be present. The reduced scanning errors are the result of feed-forward control algorithms decreasing the necessity of servo lag and the increased system bandwidth. Lag is intentionally used in a purely reactive servo control system to compensate for the system's tendency to overshoot its objectives. The reactive system will not begin to alter its course until an error between the commanded path and the actual path has already been measured. This is equivalent to the driver on a mountain road not starting to turn the car to follow a curve in the road until he actually begins to drive off the road. This analogy describes the result of delaying corrective action until an error is actually measured. At this point, there may be insufficient time to correct the error before it increases to such a level that the actual path no longer resembles the commanded one and the car drives off a cliff.

Servo lag will adjust the actual position to an adequate distance behind the commanded path. Thus, when the commanded path turns a corner, and error begins to grow, only then is corrective action taken by the servo to bring the actual path onto the new commanded path, as was shown in Figure 3-7. Provided the commanded path does not change faster than the bandwidth of the system, this approach will result in smooth, well connected, generally rounded corners.

By examining the subsequent commands in the queue, feed-forward allows the servo system to "see" what the commanded path trajectory will do next. The Orion Imaging System begins instituting appropriate actions, even before an error is detected. In the simple right angle turn shown previously in Figure 3-21, the feed-forward algorithm calculates the work required (applied force over distance) to alter the present path onto the new path. It also applies a correction to the actual path, so that the peak acceleration occurs at the correct location to minimize errors associated with following the requested path. The computing power of the TI DSP accommodates fairly sophisticated feed-forward algorithms, resulting in a servo system that can draw very quickly, with significantly improved accuracy.

3.6 Summary

As the understanding of all SL processes improves, goals are set for each subsystem element that, if achieved, result in a substantial improvement for the new

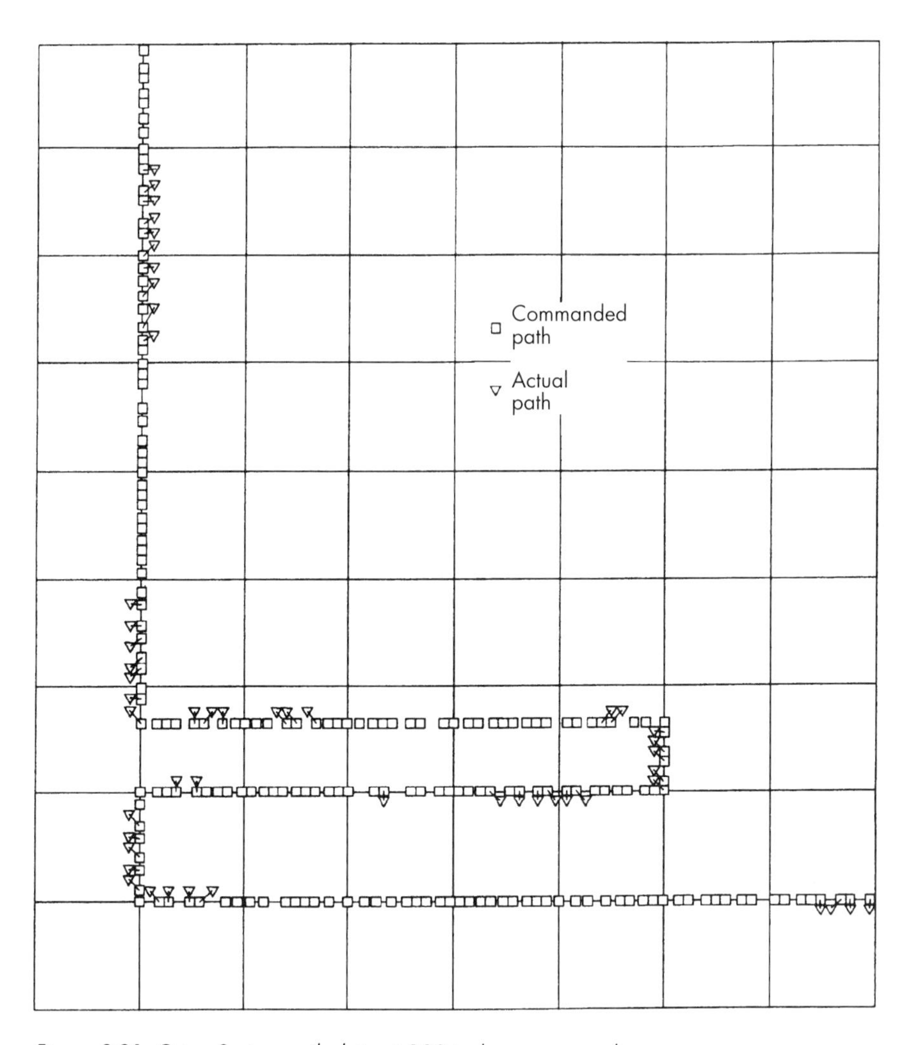

Figure 3-21. Orion System path data at 100 inches per second.

slice in time. The subsystems are continuously scrutinized for areas that can result in either productivity or performance advances.

Early in the life of the SLA-250 product, only one resin was generally used at a time, typically the latest one. As the photopolymer technology for SL matured, application-specific resins were developed. Now, one machine might be called to use several different resins in a single week. Resin interchange and system recalibration practices that formerly took considerable time became untenable. The new goal for the system was to exchange resins in only a few minutes. Adapting the hardware concepts developed for the SLA-500 allowed the SLA-250/40 to achieve this goal. Also incorporated was the new diode laser leveling hardware and software. This group of changes produced improved speed, accuracy, repeatability and reliability for the SLA-250.

For layer preparation, the ultimate goal is to quickly attain a flat, feature-free resin surface, at a controlled distance from both the model and the imaging system. The latest SLA-500/30 uses the techniques and hardware previously outlined, and is able to establish that desired result with higher resolution, repeatability, accuracy, adaptability, and speed than before.

For the imaging system, the goal is to reproduce the CAD input as closely as possible by mapping the input data onto the resin surface, and properly controlling laser exposure. By instituting new computer control architecture, and revising the software control algorithms, a two-fold increase in drawing speed and an eight-fold increase in drawing accuracy were achieved. The fault tolerance of the system was also enhanced with more robust calibration and beam profiling routines, as well as the change from the mechanical shutter to the AOM. The combination of various improvements in SLA-500 leveling and imaging resulted in a minimum throughput improvement of 30%, and depending on part geometry, a maximum productivity increase of 50%.

The changes described have all been issued either as new generation machines or as upgrades to existing equipment. As this chapter is written, a new slice of time is here, a greater understanding of SL processes is at hand, and new goals are being developed.

References

1. Mueller, T., "Introduction to Part Building," Chapter 7, Jacobs, P., *Rapid Prototyping & Manufacturing: Fundamentals of StereoLithography*, SME, Dearborn, MI, 1992.
2. Siegman, A.E., *Lasers*, University Science Books, Mill Valley, CA, 1986.
3. Jacobs, P., "Fundamental Processes," Chapter 4, Jacobs, P., op. cit.
4. Hunziker, M. and Leyden, R., "Basic Polymer Chemistry," Chapter 2, Jacobs, P., op. cit.
5. Tindel, J., Thayer, J., and Nguyen, H., "Scaling Error Due to Focal Length Differences in Low Profile Sweeping," 3D Systems Technical Report #450.

CHAPTER 4

Stereolithography Software Technology

By wisdom a house is built,
and through understanding it is established;
through knowledge its rooms are filled
with rare and beautiful treasures.

Proverbs 24:3

Stereolithography consists of many processes working together to create 3-dimensional objects. Photopolymer chemistry, laser physics, optics, viscous fluid dynamics, CAD technology, material science, DSP servo technology, motion and process control, and electrical and mechanical engineering form the "body." Software is the "brain and central nervous system" that makes the process come alive. As improvements, innovations, and new technologies are discovered in each of these areas, software's task is to integrate them into the system.

SL software can be broken in two distinct components: part preparation software and process control, or build, software. *Part preparation software* involves CAD model verification, model placement and orientation, support generation, assignment of build and recoat attributes, production of cross-sectional slices, and the convergence of all components into the appropriate build file. Then the *process control software* uses the build file produced by the preparation phase to control the entire build process on the SLA. This is shown schematically in Figure 4-1.

Advances in part preparation, slice technology, and build style techniques directed toward new applications, along with new leveling systems and a new imaging system, are all improvements integrated into the current product line during the past few years. Each of these advances affects both the preparation and process control software.

By Chris R. Manners, Manager of Software Development, 3D Systems Corporation, Valencia, California.

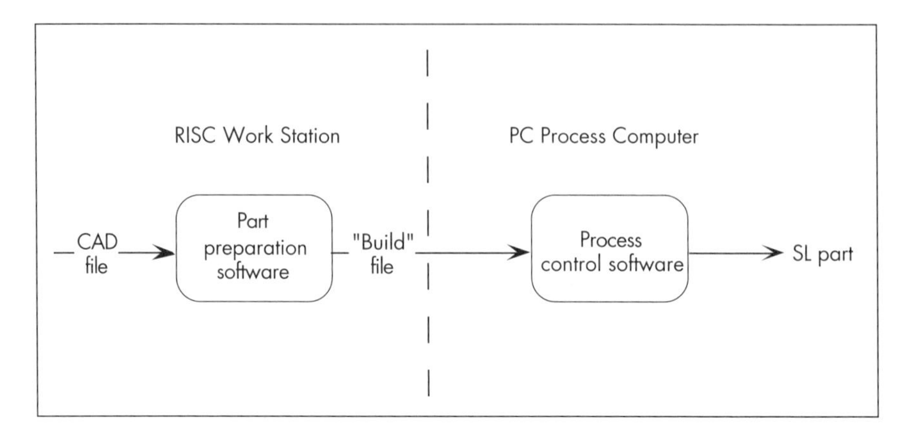

Figure 4-1. Simpified SL software flow diagram.

4.1 Part Preparation Convergence

To appreciate the significance of part preparation software convergence, it is important to understand the current SLA product mix. The SLAs can be grouped into two categories, the smaller part building envelope machines (SLA-190/250) and the larger build envelope machines (SLA-400/500). In the interest of brevity, the smaller envelope machine will be generically referred to as the SLA-250, and the larger one as the SLA-500, for the balance of this chapter. Each system builds parts using essentially the same process. Figure 4-2 diagrams the steps required in part preparation.

However, each part preparation program was developed independently and had little software commonality. Previously, a part for an SLA-250 was prepared in a completely different manner than a part for an SLA-500. This was neither logical nor efficient, since the processes are actually quite similar. SLA-250 part preparation was performed partially on a Silicon Graphics, Inc. (SGI) work station and completed on the SLA-250 process control computer itself. Part preparation for an SLA-500 has always been done entirely on the SGI work station.

Based on the similarity of the processes for both machine types, a project was undertaken to converge the part preparation software into a single package. The result of this effort is that parts for both basic product groups are now prepared in essentially the same manner, using the same software. Part preparation is currently performed entirely on the SGI work station for all 3D Systems' SLA machines. This allows the user to prepare parts for all platforms at a single location, while still employing the same programming techniques. Operators using both machine types need learn only one common part preparation process. Consequently, training time is reduced.

Figure 4-3 shows the high level data flow used in part preparation. Typically, the STL file is checked by 3Dverify for "sliceability," based on the rules set out

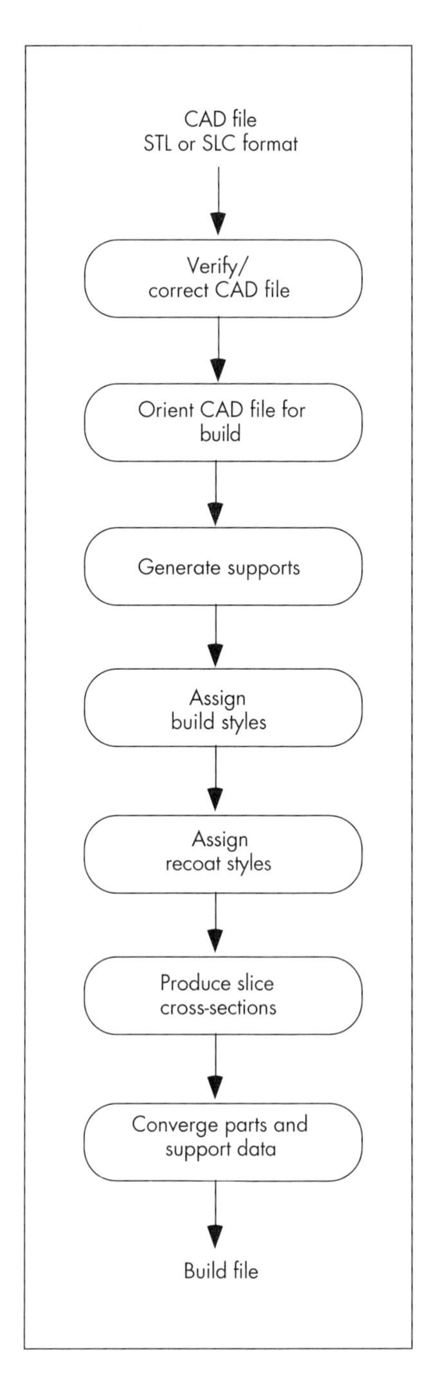

Figure 4-2. Part preparation steps.

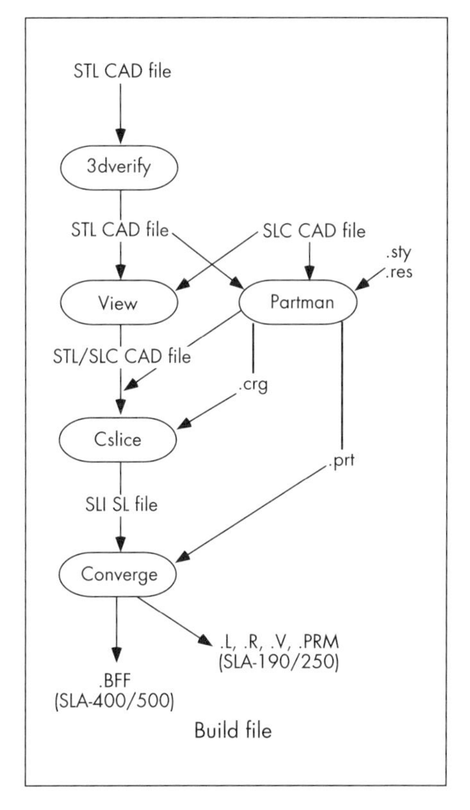

Figure 4-3. Part preparation data flow.

by the STL file specification. Then the "View" program is used to orient the part for the build process. "Partman" is responsible for the portion of part preparation where the operator inputs the desired drawing styles (.sty parameter files) and recoating attributes (.res data files). These parameters can be adjusted based on end-use application, and are the subject of extensive R&D efforts. Next, Partman produces slice argument files (.arg or .crg), that contain Z spacing, layer compensation values, slice resolution, and other ancillary slicing parameters, and the part preparation parameter file (.prt). This file saves the current programming environment, including all parameters necessary to produce complete build files. Next, each STL/SLC file is sliced into cross-sections (Slice or CSlice), producing an SLI file for each CAD input file. Finally, the process is completed by merging (*Converge*) all SLI files and build parameters into the appropriate build file(s) for each machine type.

4.2 Contour Slice

Slice is the process by which the CAD model is divided into successive cross-sectional layers. It transforms the solid CAD model into vector data, which the imaging system ultimately draws on the surface of the resin. Slicing can be a very time consuming and processor intensive process, depending on the slice thickness and the complexity of the CAD model. Therefore, it is an area worthy of continuous improvement regarding performance and accuracy. Contour Slice, or *CSlice*, is the successor to the original slicing program, Slice. CSlice has many advantages when compared with Slice. CSlice is faster, more accurate, more reliable, and accepts both the original and new input file format.

Two very different input file formats are accepted into CSlice. The data flow is shown in Figure 4-4. They are named STL and SLC. The STL file format is the *de facto* RP&M industry standard. It is a faceted solid representation of the CAD model. In contrast, the SLC format is a 2-1/2D layer representation.

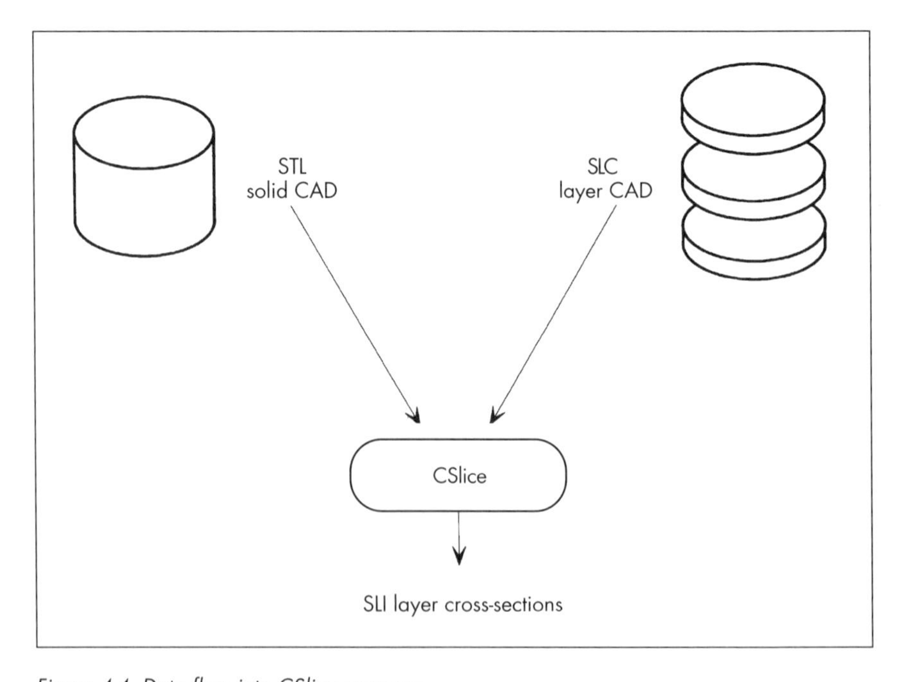

Figure 4-4. Data flow into CSlice program.

All the major CAD vendors supply translators from their CAD representation to STL. The STL file can be represented in ASCII or binary format. The binary format is preferred due to its substantially smaller file size. The SLC format is new, and translators are presently available from only a few CAD vendors.

STL Binary File Format

The STL file format interface specification was defined by 3D Systems, Inc. in 1987. It has become the *de facto* standard file format for the RP&M industry. The STL file format consists of a header, the number of triangles, and is followed by the list of triangles. The details of the STL file format structure are listed below.

STL File Data Format

Byte	8 Bits
Unsigned Integer	2 Bytes
Unsigned Long Integer	4 Bytes
Float	4 Bytes IEEE

The most significant byte is specified in the highest addressed byte. The byte ordering follows the Intel PC protocol.

Address	0	1	2	3
	Low Word	High Word		
	LSB MSB	LSB MSB		

Header		80 Bytes
Number of Triangles		Unsigned Long
For each tessellation triangle		
Normal Vector		
	I	Float
	J	Float
	K	Float
First Vertex		
	X	Float
	Y	Float
	Z	Float
Second Vertex		
	X	Float
	Y	Float
	Z	Float
Third Vertex		
	X	Float
	Y	Float
	Z	Float
Attribute		Unsigned Integer

Each triangle contains a total of 50 bytes, i.e., 12 floating point numbers plus an unsigned integer or $(12 \times 4) + 2 = 50$. Therefore, the STL file size is equal to the number of triangles times 50, plus 84 bytes for the header and the triangle count. As an example, an STL file containing 10,000 triangles would be 500,084 bytes in size.

There are two very important requirements to ensure proper processing of an STL file. First, triangle vertex ordering is used to identify interior or exterior surfaces. A clockwise vertex ordering indicates an interior surface, and a counter-clockwise ordering indicates an exterior surface. The right-hand rule states that the vertices of each facet must be ordered so that when the fingers of the right hand pass from point one through point two to point three, the thumb will point away from the solid object. This rule is illustrated in Figure 4-5.

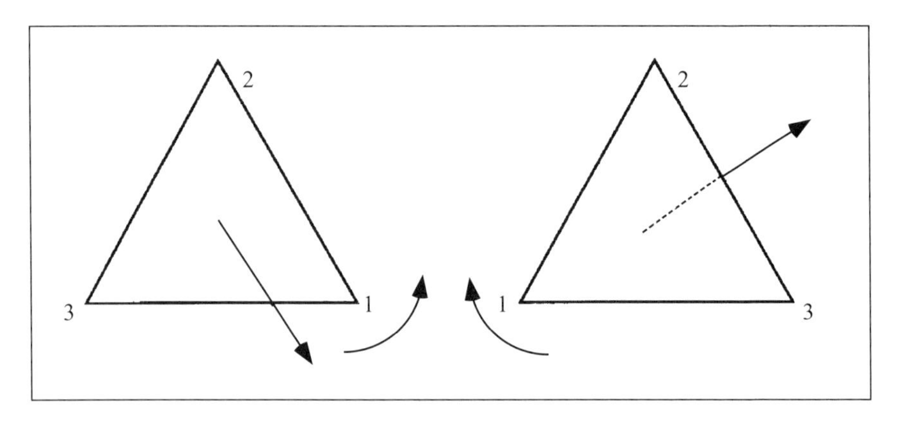

Figure 4-5. Right-hand rule.

The second rule used is the vertex-to-vertex rule. Each triangle within the STL file must meet all adjacent triangles along common edges, as shown in Figure 4-6. In other words, every triangle must share exactly two common vertices with each adjacent triangle. In no case may a triangle vertex intersect with the side of an adjacent triangle.

SLC File Format

The SLC file format was defined by 3D Systems in 1992. Unlike the tessellated solid STL representation, it is a 2-1/2D contour representation of the model's boundaries within each layer, much like an SLI file. SLC data can be generated from various sources. This may be by conversion from CAD solid or surface models, or more directly from systems that already produce data arranged in layers, such as Computed Tomography (CT) scanners. Currently, Cenit CAD/CAM GmbH offers CAT-SLICE, a product that runs under the CATIA CAD system.

The Cenit product slices a CATIA surface or solid model directly into the SLC file format.

The SLC data formats (i.e., bytes, floats, etc.) are the same as the STL data formats. The SLC file is divided into a header section, a 3D reserved section, a sample table section, and finally the contour data section.

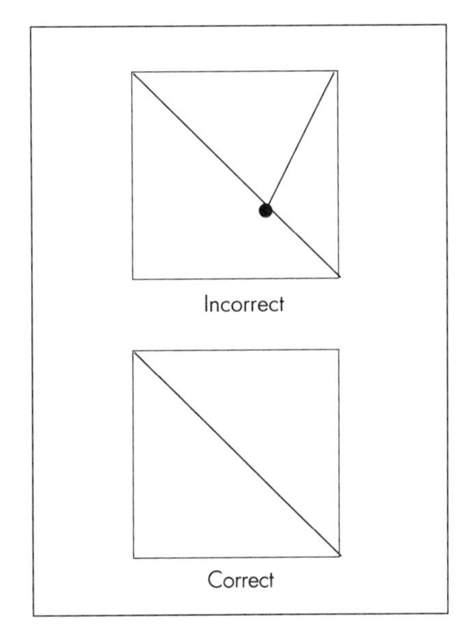

Figure 4-6. Vertex-to-vertex rule.

The header section is an ASCII character string of up to 2048 bytes. It contains global information about the part such as part units (inch or millimeter) and extents, along with the parameters used to produce the contour data. These would be chordal deviation, arc resolution, surface tolerance and gap tolerance, among others. The syntax of the header section is a keyword followed by its appropriate parameter. The header is terminated by a carriage return, line feed, and control Z string. The data in the header is currently not used by CSlice, and is only there as a convenience to the CAD vendor or user.

The sample table describes the sampling thickness (layer thickness or slice thickness) of the part. There can be a maximum of 256 different thicknesses. Each entry describes the start position in Z, the slice thickness, and what linewidth compensation is desired for that sampling range. For example, if the cross-sections were produced with a single slice thickness for the entire part, the sampling table need have only one entry.

The contour data section is a series of successive ascending Z cross-sections with the accompanying contour data. Each Z cross-section contains the number of contours, followed by the list of individual contour data. The contour data contains the number of X, Y vertex locations for that contour, the number of gaps, and finally the list of floating point vertex points. The location of a gap can be determined whenever a vertex point repeats itself.

SLC File Description

Header 2048 Maximum Bytes (including termination string)
Termination string <CR> <LF><ctrl Z>

Header Keywords:

-SLCVER <X.X> [SLC file version]
-UNIT <INCH/MM> [part building units]
-TYPE <PART/SUPPORT/WEB>

-PACKAGE <vendor specific>	[maximum of 24 bytes]
-EXTENTS <minx, maxx, miny, maxy, minz, maxz>	
-CHORDDEV <value>	[maximum chordal deviation]
-ARCRES <value in degrees>	[arc resolution]
-SURFTOL <value>	[surface tolerance]
-GAPTOL <value>	[gap tolerance]
-MAXGAPFOUND <value>	[maximum gap found]
-EXTLWC <value>	[Linewidth compensation]
3D Reserved Section	256 Bytes
Sampling Table Size	1 Byte
Sampling Table Entry	4 Floats
Minimum Z Level	1 Float
Layer Thickness	1 Float
Linewidth Compensation	1 Float
Reserved	1 Float
Contour Cross-section	
Minimum Z Level	1 Float
Number of Contours	1 Unsigned Integer (4 bytes)
Number of Vertices	1 Unsigned Integer
Number of Gaps	1 Unsigned Integer
Vertices List (X/Y)	Number of Vertices * 2 float
repeat Number of Contours - 1	
repeat Contour Cross-section until	
Maximum Z Level	1 Float
Termination Value	Unsigned Integer (0xFFFFFFFF hex value)

CSlice Processing

The flow chart shown in Figure 4-7 describes the way STL and SLC data are processed to produce the vector data necessary for the scanning system. Triangles are received, sorted, and the cross-sectional segments produced. This creates polylines, which represent the exterior and interior boundaries of the CAD model. Since SLC data is already in a boundary-type format, only data input is necessary for SLC files at this point. Note that from the point at which the polyline boundary data is produced, STL and SLC processing is identical. This readily enables future improvements that affect both input formats.

Each polyline boundary is compensated, based on the operator selected linewidth compensation value. Layer comparison is applied, producing SL-specific boundary types. These boundaries are then smoothed to eliminate redundant vertex points, and hatch/fill information is added for the SLA-250 machine type. The data is then output as an SLI file.

Three Boundary Types. The CAD boundary is divided into three generic SL boundary types: "upfacing boundaries," "downfacing boundaries," and "layer

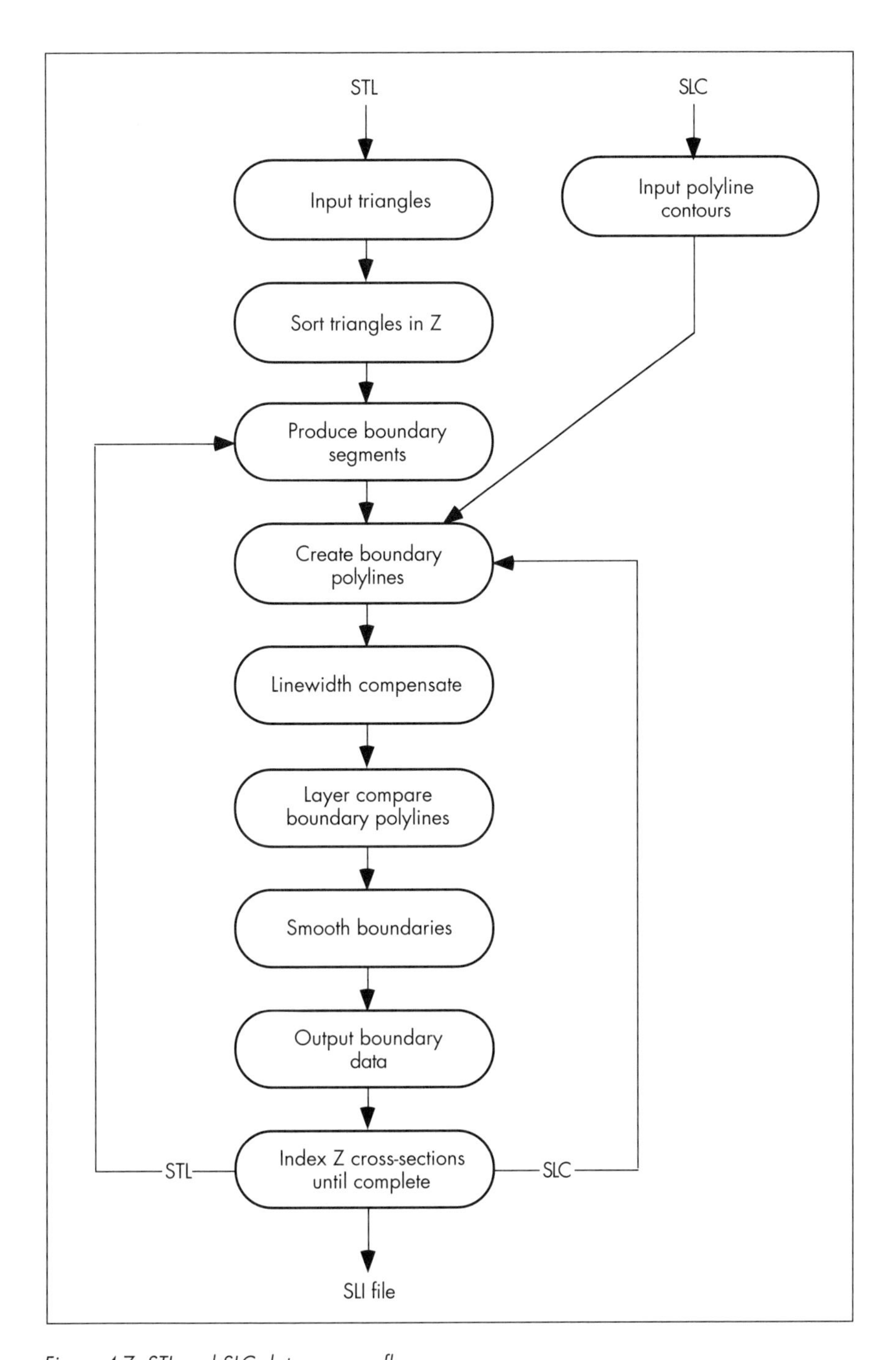

Figure 4-7. STL and SLC data process flow.

boundaries." An upfacing boundary is an exterior surface that has nothing directly above it. Conversely, a downfacing boundary is an exterior surface with nothing directly below it. A layer boundary is an interior surface that does have something both directly above and below it. Refer to Figure 4-8 for a graphical depiction of these boundaries. It is very important that the SL process is able to distinguish generic boundary types. The ability to assign different imaging attributes, such as various hatch/fill styles and cure depths, enables the SL process to achieve higher accuracy and smoother surface finishes, as well as being more adaptive to new applications and materials.

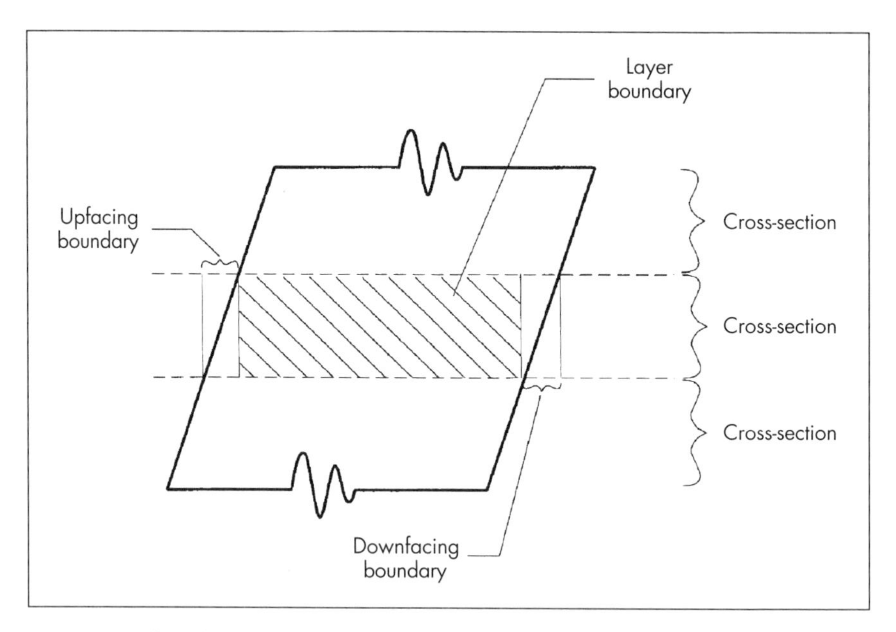

Figure 4-8. A layer boundary is an interior surface having something both directly above and below it.

Layer Comparison. CSlice uses an entirely new technique to produce the three types of part boundary data. This method is called layer comparison. In order to divide the working CAD boundary into one of these three types, CSlice must compare any contour with both the boundary above and the boundary below. Therefore, CSlice must work with three SL boundaries at any given time. We shall refer to these "working" boundaries as the "upper boundary," the "middle boundary," and the "lower boundary."

First, the downfacing boundary is created by subtracting the lower boundary from the middle boundary. Next, the upfacing boundary is created by two operations. The middle boundary is subtracted from the upper boundary, and then the downfacing boundary is subtracted from the result of the previous operation.

The layer boundary is also created by two operations. The downfacing boundary is subtracted from the middle boundary, and then the upfacing boundary is subtracted from the result of the previous operation. These subtraction operations are the heart of layer comparison slice, and are illustrated in Figure 4-9.

Once the original SL boundary has been divided into the three generic boundary types using layer comparison, each boundary is smoothed, eliminating redundant vertex points, and subsequently hatched or filled in the case of SLA-250-type machines. Note that hatch and fill operations occur in the process control software on the SLA-500 machine type. The sequential process of slice, linewidth compensation, layer comparison, smoothing, and then hatching/filling is repeated for each subsequent Z slice interval.

The output of CSlice is an SLI file, a proprietary file format developed by 3D Systems, which is essentially a 2-1/2D vector representation of the CAD solid.

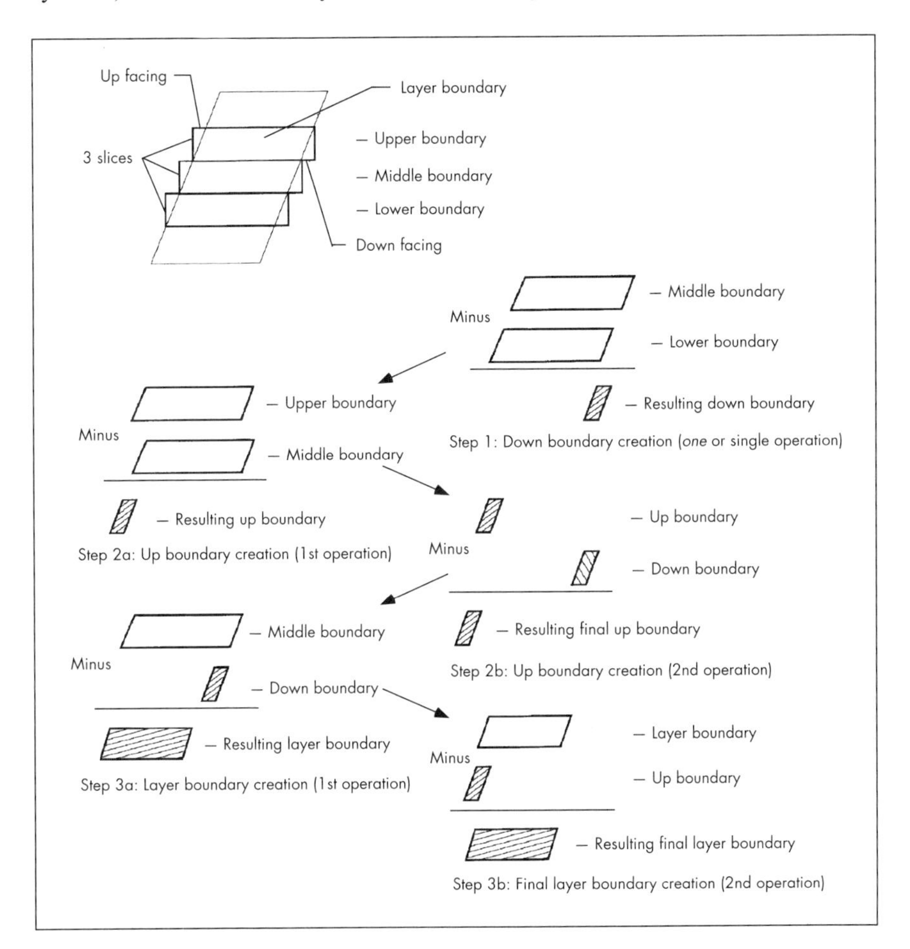

Figure 4-9. Layer boundary creation.

CSlice Performance Improvements

CSlice is significantly faster than Slice. With the improvements to linewidth compensation and hatching and filling, CSlice can be almost a factor of four faster than Slice. Since slicing can become a bottleneck in the SL build process, this advance is very significant. In addition to the development of CSlice, work station processing power has also increased more than an order of magnitude over the past five years. Initially, 3D Systems shipped work stations with 12 MHz clocks. Now, newer work stations are readily available that are equipped with 150 MHz clocks. Increased work station performance combined with CSlice reduces the part preparation time dramatically. A part formerly prepared using Slice, which took more than 20 hours on a 12 MHz work station, now takes CSlice 28 minutes on a 150 MHz equipped work station. Refer to Figure 4-10 for performance improvements using CSlice on different generations of work station platforms. The benchmark part selected to obtain this data was an STL representation of an engine block containing 163,358 triangles (8,167,984 bytes).

With the continued trend of faster and faster work stations, part preparation time will eventually be a minor portion of the total part building time.

4.3 STL versus SLC

Which is better, STL or SLC? The jury is still out on which format is better. In fact, the jury has not even been selected yet. Perhaps that is not even the correct question to ask. Each file format has its place, and each has advantages and disadvantages.

The STL file format is the *de facto* industry standard, widely used by all RP&M companies. Furthermore, all major CAD vendors have working translators. However, because the data must be translated from the internal CAD representation to a faceted representation, the STL file is always an approximation of the real surface of the object. Facets are hard to avoid on curved surfaces and are sometimes apparent in the final SL model. Unfortunately, attempts at generating triangles small enough to produce very smooth surfaces can also result in enormous data files and protracted slice times.

SLC Could Avoid Intermediary. The SLC file format, if incorporated into CAD software in a manner similar to CAT-SLICE, can virtually eliminate the problem of translating a part from the original representation to an intermediate form (such as tessellation), where some data is lost. The CAD software can produce contours directly from the original data. This results in more accurate parts with smoother surfaces. Of course, since the SL imaging system is vector based, there will always be approximations applied. However, SLC can possibly have a finer resolution using less data. In addition, the SLC file format is perfect for data that is already in a 2-1/2D representation, such as CT scan data. These data can be used directly, without any translation.

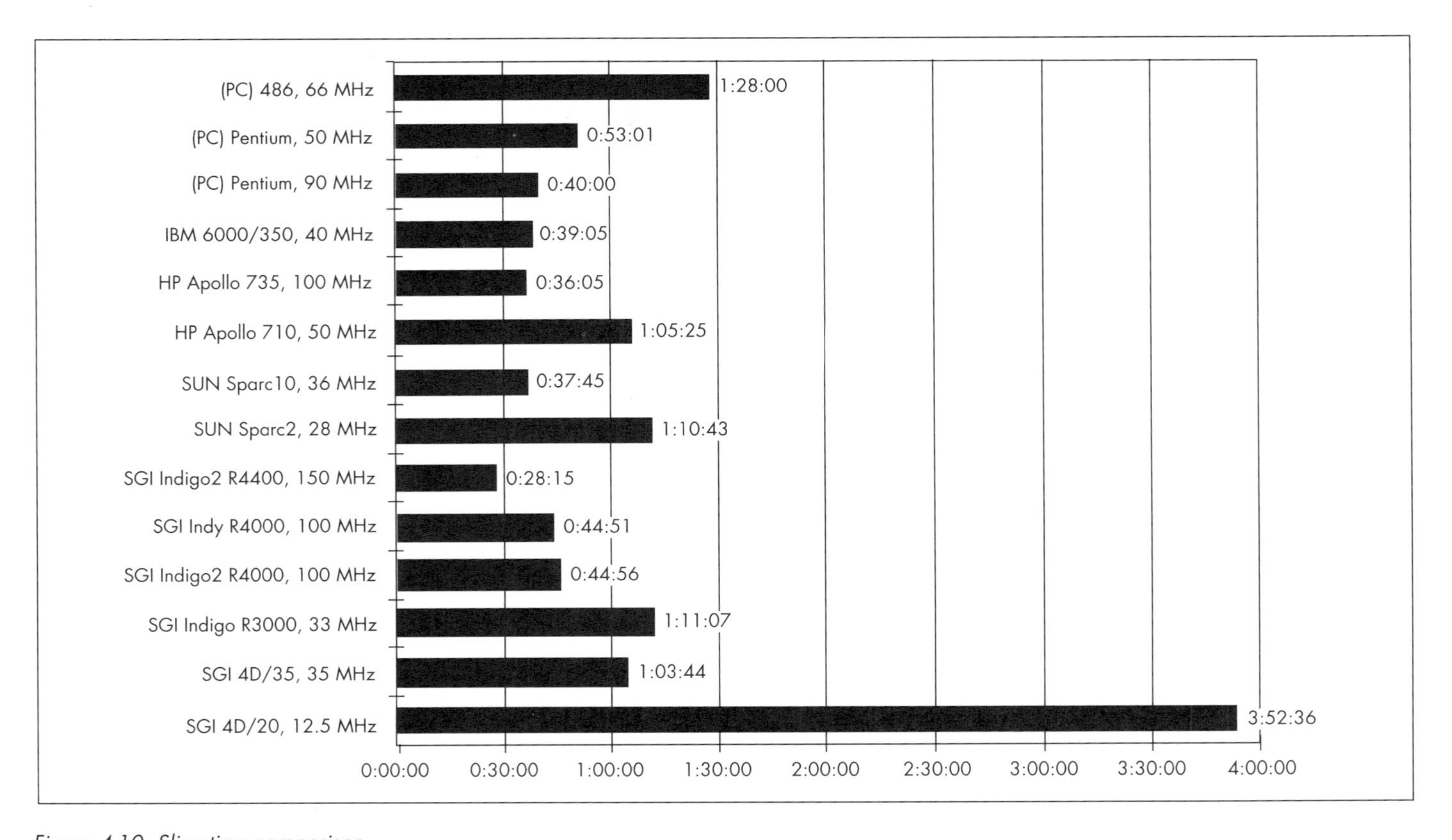

Figure 4-10. Slice time comparison.

4.4 Process Control Improvements

The process control software is responsible for operating both the hardware and the build process for the SLA machine itself. The flow block diagram illustrates the build process. Although the hardware is different in the SLA-190/250 and SLA-400/500, the build process is quite similar. Each of the leveling systems for the SLA-190/250 and SLA-400/500 have advanced, providing increased accuracy and improved performance.

SLA-250/40 Leveling System

The leveling system is responsible for adjusting resin volume before the build process, and for leveling the resin surface of each layer during part building. The SLA-250/40 introduced new leveling system hardware and software.

Previously, the leveling subsystem consisted of a HeNe laser, bi-cell sensor, and a resin displacement plunger system. The laser was pointed at the liquid surface at an acute angle, and the reflected light was incident on a position feedback bi-cell sensor. When the system read the bi-cell and detected that the liquid surface needed to be either raised or lowered, the system would lower or raise the plunger. The plunger would then displace either more or less resin, and the resin surface would be either raised or lowered. Lowering the plunger would raise the liquid surface, and likewise raising the plunger would lower the resin surface. This process worked adequately, but it was slow and suffered from inaccuracies as well as a lack of repeatability in establishing the resin surface. To address these issues, the plunger system was replaced with a vat lift system for the SLA-250/40.

Vat Lift Leveling System. The SLA-250/40 diode laser leveling system now consists of the new diode laser, bi-cell sensor and a vat lift system similar to the SLA-500. Entirely new software algorithms were developed for adjusting resin volume and leveling between layers.

The new vat lift leveling system must allow enough vat movement to compensate for the resin shrinkage due to liquid-to-solid transition, and the "expansion" resulting from lowering the part platform and support arms. The limits of vat movement necessary to provide the correct resin level while part building are computed using the maximum volumetric resin shrinkage for the largest part that fits within the build envelope, as well as the maximum resin displacement resulting from complete immersion of the platform and the platform support arms. Based upon both analysis and experiment, the limits have been set at plus 0.32 inch (8 mm) of vat movement to compensate for resin shrinkage, and minus 0.028 inch (0.7 mm) of vat movement to compensate for physical displacement.

Adjust Resin Level. In order to prepare for the build process, it is necessary to have the correct amount of resin in the vat. The *adjust resin level* software routine determines the initial resin height and manages the addition or removal of resin. There are several reasons why the system checks the necessity of adding or removing resin before starting to build. First, the platform build start position is set to a fixed location relative to the resin surface. The relative position must be maintained, otherwise the build start position would need adjustment prior to each build. Second, to ensure there is enough resin in the vat to build the maximum size part, consistent with the machine build envelope. And finally, since the bi-cell leveling feedback sensor has a fixed range relative to the rim, the resin level cannot stray too far from the anticipated limits.

The Bi-cell Sensor's Two Photocells. The bi-cell sensor contains two distinct photocells, positioned one above the other, as shown in Figure 4-11. As the laser reflects off the resin surface onto the sensor, the photocells produce electrical current. The signal forwarded to the leveling system is the difference between the electrical current of the upper and lower cells. If no light is present on either cell, the resin is either much too high or much too low, and the signal generated is zero, or bias. If the light is precisely centered between the upper and lower cells, the system is exactly level and the signal generated is also bias, since there is an equal amount of light on each cell and the difference between them is zero. As the sensor passes from the upper cell to the lower cell, the output shown in Figure 4-12 can be observed. A tolerance is established by setting upper and lower limits, assigned around bias, to adjust resin level sensitivity. The upper and lower limits are adjustable, according to the desired accuracy. Whenever the resin height is between those two limits, the system is considered level, and no fine adjustment is necessary.

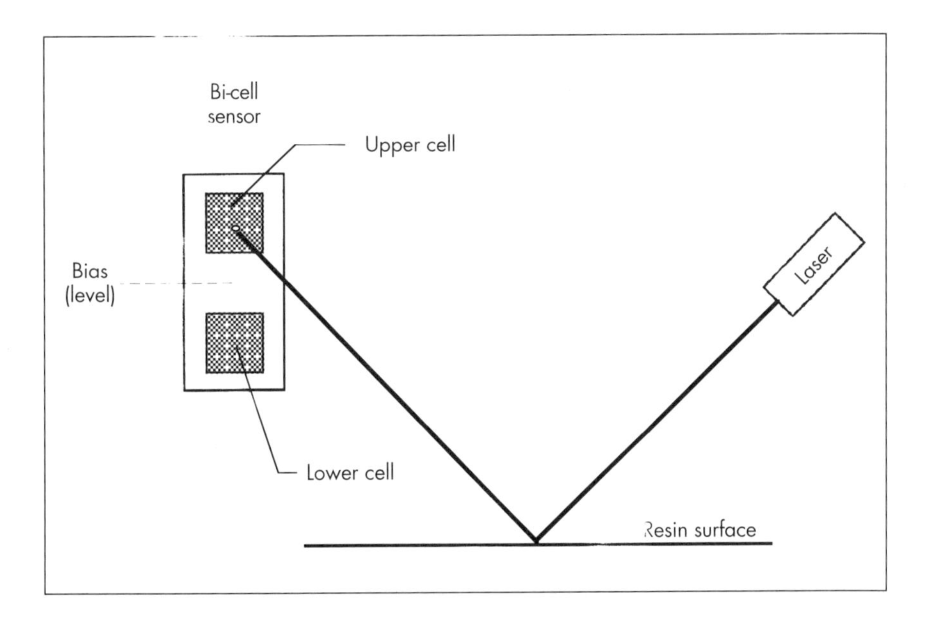

Figure 4-11. Light bounce leveling system.

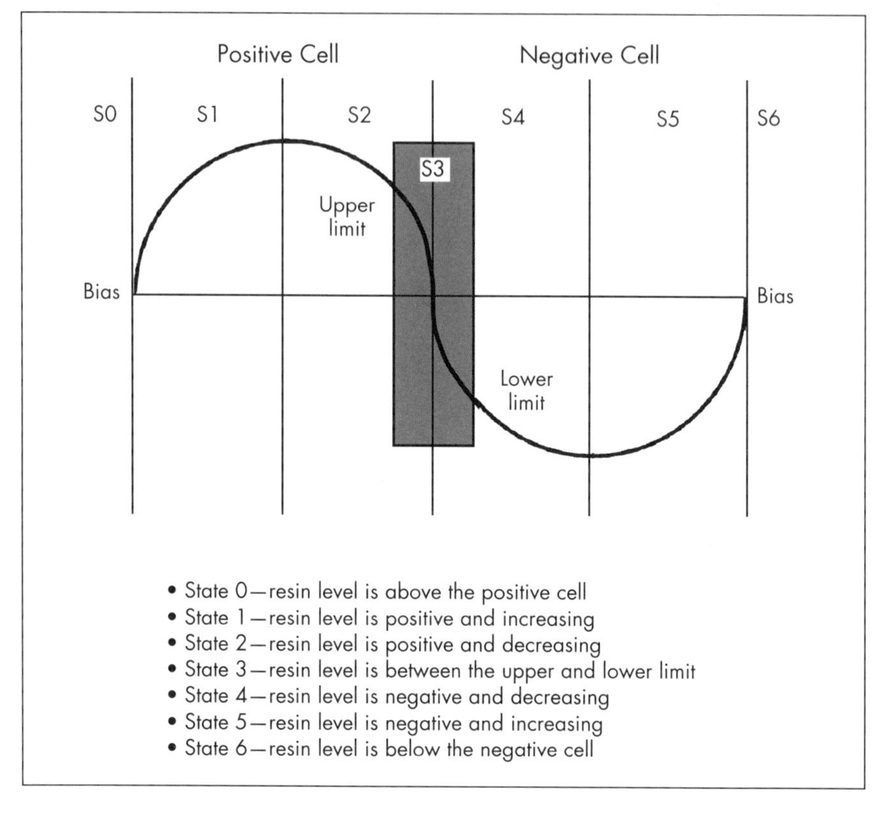

Figure 4-12. Bi-cell sensor output.

Determining Initial Resin Level. To determine the initial resin level height, the output is divided into states describing the resin surface position relative to the sensor. The following steps are used to determine initial resin level height:

- The vat is moved to the upper limit.
- The vat is lowered to the resin shrinkage height, plus the sum delta distance (0.290 inch [7.36 mm] from upper limit, or 4,350 vat steps). Both a fast and a slow speed are used to move the vat. The fast speed is used to lower the vat until 10 motor steps before the resin shrinkage height. The slow speed is used for the last 10 steps so that surface disturbance is minimized.
- A bi-cell reading is stored at the 4,350-step point.
- The vat is lowered 20 vat steps, and a bi-cell reading is stored at each step.
- The stored bi-cell readings are processed to determine the resin state.
- The vat is raised back up to 4,350.
- If the resin level state is S0 or S1, then resin must be removed.
- If the resin level is S5 or S6, resin must be added.

If the resin level is S2, S3, or S4, then there is no need to add or remove resin. The system is between the tolerance limits.

Adding or Removing Resin. Once the system determines the initial resin level state (using the routine described previously), resin is either removed or added as required. If the system requires more resin, it is added until the surface passes through the lower limit. Conversely, if the system requires resin removal, the operator removes the excess resin until the surface crosses the upper limit. Since resin is regularly removed from the vat in the form of a completed part, the situation of resin removal generally occurs only when too much resin has been added prior to a new part building session.

The fine resin level adjustment is used during the build process for every layer. If the resin level is not between the upper and lower tolerance limits, the system moves the resin surface (i.e., raises or lowers the vat), while reading the bi-cell values, until the reading is bias, indicating the resin is at the correct level. The tighter the tolerance used, the more accurate the leveling system. The tightest tolerance can be achieved when the upper limit is set to 1 vat displacement unit and the lower limit is set to -1. This will result in leveling accuracy of approximately ±0.0002 inch (5 microns) for a 10-inch (254-mm) part.

SLA-500 Leveling System

Like that of the SLA-250, the SLA-500 leveling system is responsible for adjusting resin volume before the build process, and maintaining the level of the resin surface for each layer during part building. The SLA-500 leveling system, described in Chapter 3, operates in much the same manner and uses hardware almost identical to the SLA-250. The only substantial differences are that the SLA-500 system incorporates a linear photocell detector in place of the bi-cell, and an additional sensor is used to detect when the resin level is too high.

The linear cell has two output channels. These are used effectively like the high and low sides of the SLA-250 bi-cell. The normalized difference between two outputs is determined, and this position along the slope indicates how far the system is from the desired resin level. (Refer to the leveling system section of Chapter 3 for a diagram of these two outputs.) When there is no output from the linear cell, the resin level is either too high or too low. When the incident light from the laser is below the lower end of the linear cell, the normalized output difference is zero.

"Too-high" Sensor. An additional sensor incorporated in the SLA-500 leveling system allows it to detect when the incident light from the diode laser is beyond the upper end of the linear cell. The additional sensor in the SLA-500 system is a large area photocell with binary output, such that there is a signal when the laser spot is impinging on the cell, and no signal when there is no light hitting the cell. Therefore, by adding only one "too-high" sensor, the SLA-500 system is definitive in level reporting. This prevents a circumstance that can occur in the SLA-250, where the system indicates that the user should add resin, when the resin level is already much higher than required.

To make its determination of resin level position, the software proceeds in the following manner.

- The "too-high" sensor is checked. If the system is adjusting resin level prior to building a part, the vat is moved to its baseline point. If there is a signal from this sensor, the system requests that the user remove resin. If there is no signal from this sensor, the system moves to the next step. If the system is leveling the resin during the build process, and there is a signal on the "too-high" sensor, the system moves the vat down until there is no signal on this sensor.
- The output of the linear cell is read, and a determination is made whether the resin level is above or below the desired level. If it is not at the desired resin level (either too high or too low), the system servos level by moving the vat in the indicated direction, while reading the linear cell, until level is achieved. During resin level adjustment prior to build, if the resin volume is not within the tolerance band, the operator will be asked to add or remove resin.
- If there is no output from the linear cell, and the system has already checked the "too-high" sensor, then the resin level must be too low. The operator is requested to add resin.

Previously, the SLA-500 leveling system did not adjust the resin level during approximately the first 3/8 inch (1 cm) of the part. This could result in errors in Z height, or in delamination, depending on the features of the part and the relationship between the part and its supports. The system did not level in this area because the large amount of resin displacement while the platform and support arms were descending through the resin resulted in excessively long leveling intervals. Based on the speed, reliability, and accuracy of the new system, the sensors check and/or adjust the resin level on every layer. This eliminates possible layer delamination as well as *Z* accuracy problems due to leveling system errors.

Changing the Blade Gap. Since the recoating characteristics of one resin may be more favorable using a different blade gap than that employed for another resin, the new leveling system has features that change the effective blade gap on a per-resin basis, as shown in Figure 4-12A. The blade gap used for acrylate resins is 0.025 inch (0.635 mm), and the SL systems are configured mechanically during production and installation at that 0.025 inch blade gap. However, the recommended blade gap for epoxy-based resins for the SLA-500 is 0.006 inch (0.152 mm).

By calibrating the relationship between different positions on the normalized output of the linear cell, with resin surface movement, the software is able to change the effective blade gap automatically, based on the resin type chosen by the operator. Blade gap is included as a parameter in the user-selectable resin-specific data file. Therefore, resins requiring different blade gaps can be changed by the operator, without requiring any mechanical adjustment of the system by field service personnel.

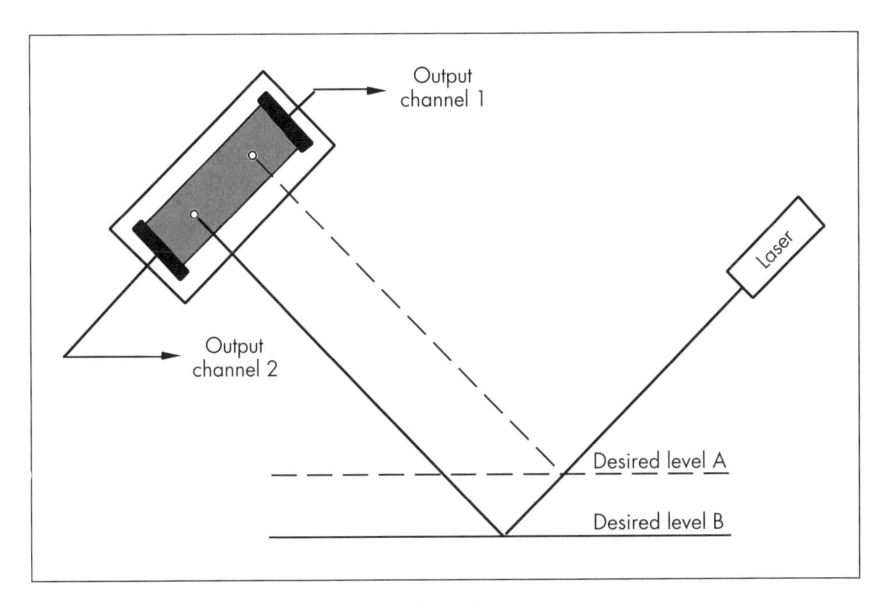

Figure 4-12A. Resin level adjustment to different levels.

4.5 Improvements in Imaging Software

The Orion Imaging Technology combines new hardware and software to drive the SLA imaging system. Using a new Digital Signal Processor (DSP), an Acousto-Optic Modulator (AOM), and a higher bandwidth servo, including a digital notch filter, Orion improves system throughput, reliability, part accuracy, and surface finish. Orion's simplified design is based on a single 33 MFlop (million floating point operations per second) DSP and an interrupt background servo control loop.

The processor executes at 16.6 MIPS, unless there are pipeline conflicts or wait states. The instruction set provides for parallel instructions, so a multiply and an add can be executed in one instruction. This is what allows a peak performance rate of 33 MFlops. The chart shown in Figure 4-13 describes the differences between the old and new DSP chips. The manufacturer of each chip is listed for clarity. However, this table is not an apples-to-apples comparison, in that it does not compare the newest offering from each manufacturer. It compares the older chip of one manufacturer to the newer chip of another, in order to assess the performance of one SLA system relative to another. The chip used in the Orion system was chosen for various reasons independent of the manufacturer.

The system software for the new DSP is more easily maintained and improved over time, because the software is written in the C programming language and a set of sophisticated diagnostic tools has been developed to help analyze system characteristics. This allows substantial performance tuning, as more information is gathered on the Orion systems in the field.

	AT&T DSP32	TI 320C30
Speed	4 MIPS	16.6 MIPS
Floating point	8 MFLOPS	33 MFLOPS
Memory	50 Kbytes	500 Kbytes
Interrupt	No	Yes

Figure 4-13. DSP chip comparison.

When the new DSP was being developed, an important objective was that its incorporation have minimal impact on the existing software architecture. Rather than rewriting all the involved process control software programs, the integration was transparent from the architectural standpoint. The same architecture is used, no matter which DSP is on board the system. This reduces the impact on software, while still allowing incorporation of all the desirable performance enhancements. The new methodology also reduces the potential for error, as well as the amount of testing required.

The control software within the imaging system performs several major functions. They are: DSP initialization, model fabrication, calibration, and programs for gathering laser beam information. The initialization program (StartDSP) establishes the scanning mirrors at their reference, or zero point. The build program (CFAB) manages the fabrication of the model, including initializing resin volume, resin leveling, laser drawing, layer recoating, and beam profiling. Calibration (CAL) corrects for the disparity between the spherical coordinates of the imaging system and the planar resin surface. The utility programs providing laser beam motion (movebeam, blbeam, beam) are used to direct the imaging system during inquiry and also for directing storage of beam shape, location, and laser power information. This information is used in the build program and to assist with system tuning by field personnel. Figure 4-14 diagrams the interface between these programs, with respect to the imaging subsystem.

The DSP architecture consists of several major components. The first component is responsible for communication between the PC and the DSP. The second performs geometric correction, and the third maintains the servo control loop. In the old system, an individual DSP chip was dedicated to performing communications and geocorrection, while another chip was dedicated to servicing the servo control loop. In the Orion system, one DSP is used to provide all these functions. This is possible due to the increased speed of the new DSP, and the availability of an interrupt. The interrupt provides the capability to run the foreground command/geocorrection loop on the same processor as the background servo control loop. This results in a tremendous simplification of the new software, since managing communication between two DSP chips is no longer necessary.

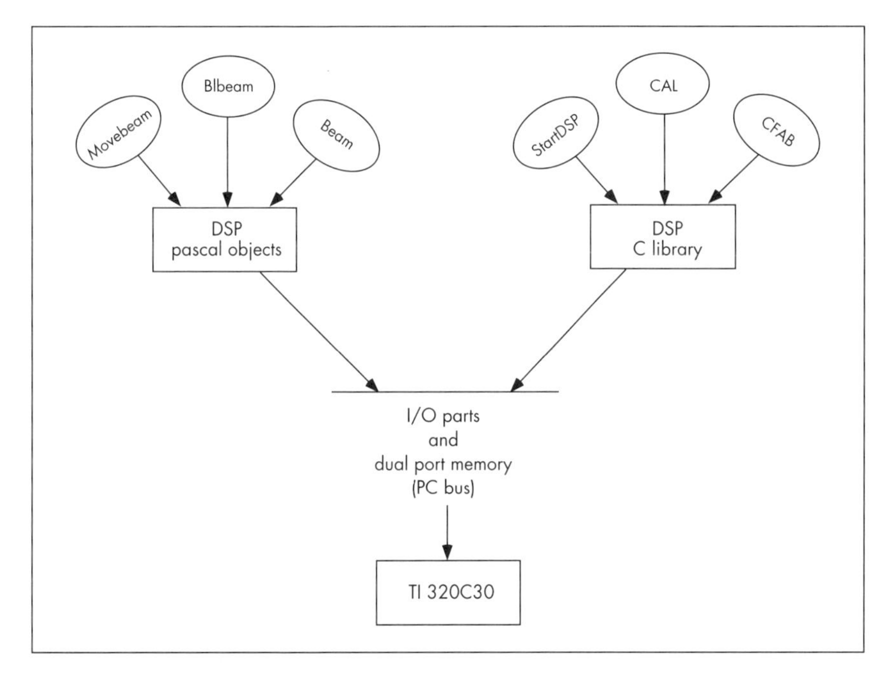

Figure 4-14. Interface between control software programs.

The servo control loop is the heart of the imaging system. It is responsible for positioning the mirrors to a commanded point. One of the most important characteristics of a servo loop is its bandwidth. The bandwidth of a system is a measure of the speed at which the system is able to respond to changes in the commanded position. If the commanded position changes at a frequency less than the system bandwidth, the system will follow the commanded position very accurately. If the commanded position changes at a frequency greater than the bandwidth of the system, the system will follow that position command rather poorly.

Imagine driving a car down the road. When you turn the wheel, the car takes some finite time to respond to that command. After the car has responded, you can now correct for any mispositioning of the car by turning the wheel again. If you turn the wheel many times in quick succession, and you have not given the car time to respond, it will become unstable and you may lose control.

In a servo system, the trick to great performance is to detect any error in real position versus commanded position very quickly, and make the required corrections, while still maintaining a stable system. The greater the bandwidth of the system, the better the performance will be in this regard. The Orion imaging system has more than doubled the bandwidth of the previous system, and adds a digital notch filter, which suppresses oscillation at the resonant frequency of the scanning mirror motor. This substantially tightens the servo control tolerances, and enables increased accuracy at much higher drawing speeds.

Imaging Software Organization

The communication function operates in the foreground and handles requests by the PC, such as move, jump, reset, etc., and communicates status back to the DSP. This communication has been simplified in the Orion design. In the previous system, the PC accessed a wide range of memory on the DSP board. It maintained a queue of commands for the foreground, by placing commands and data directly into the DSP's memory and updating appropriate pointers for the DSP. The PC also directly sets flags and parameters in the memory of the DSP.

The Orion system does not have a data queue between the PC and the DSP. It is not clear that a queue between the PC and the DSP increases performance. In fact, the overhead required to perform queue management may actually degrade performance. Commands and data are passed to the DSP through a small common memory area in the form of a message. The message consists of a single command word, which identifies the type of command, and optionally, several other parameters.

On the DSP side, the foreground task monitors the PCCommand word. When the DSP detects that it is nonzero, it decodes the command code, loads any parameters, and processes the command. It then clears the PCCommand word, which indicates to the PC that the command has been completed, and the DSP is ready for the next command. Refer to Figure 4-15, which pictorially describes the foreground command loop.

Geocorrection converts a number in CAD space, in inches, to the required encoder position. It performs three steps:

1. A two-by-three matrix transformation.
2. The geocorrection table interpolation.
3. Drift correction.

The two-by-three matrix transformation rotates, scales, and offsets the data from the CAD file to place the component at the desired location. The interpolation table corrects for the distortion caused by the projection of a spherical field onto the flat surface of the resin. Finally, drift correction compensates for any scale or offset changes in the scanning system measured by the beam profilers between each layer.

Scanning Mirror Control

The scanning mirror motors are controlled by passing the position error through a difference equation, an integrator, and outputting the result to the digital-to-analog converters (DACs). The old imaging system used a method similar to standard proportional, integral, derivative (PID) control. The new system uses a fourth order infinite impulse response (IIR) filter with a parallel integrator. The fourth order filter is implemented as two second order stages in cascade. Refer to Figure 4-16.

The four orders allow implementation of two poles and two zeroes (for lead compensation), plus a second-order notch filter to compensate for the primary

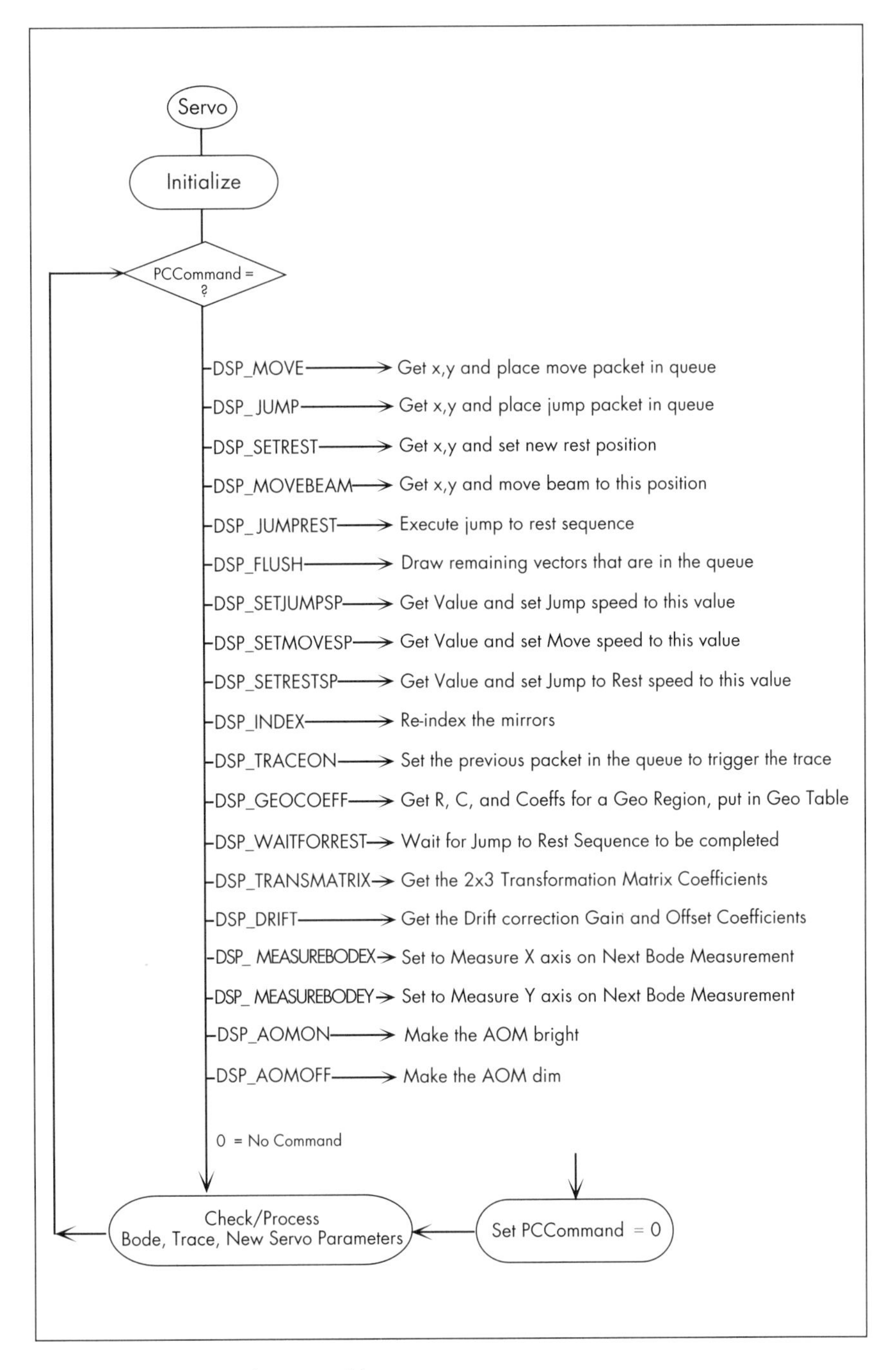

Figure 4-15. Foreground command loop.

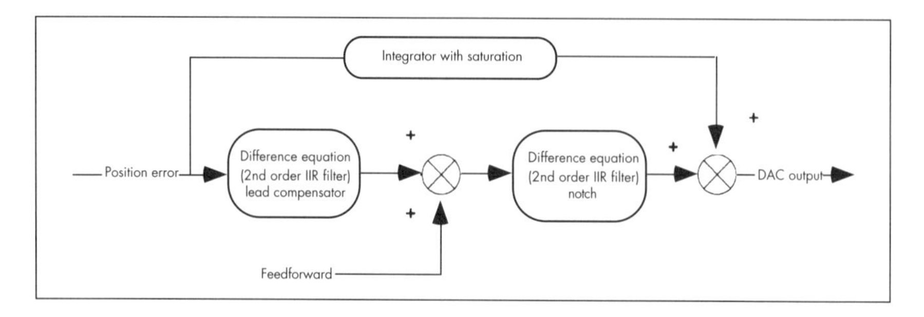

Figure 4-16. IIR filters with parallel integrator.

mechanical resonance. The filter is implemented as two second-order stages for two reasons. It avoids numerical problems that might occur in higher order polynomials and it provides a node where feed forward commands may be added, so that they pass only through the notch filter and not the lead compensator.

Scanning Mirror Flip. The integrator is implemented as a separate parallel stage for two reasons. The first is to avoid a phenomenon known as *integrator windup*. This describes a situation when a large error occurs for a protracted length of time. Imagine that the motor was manually restricted from moving for a period of time. The integrator would ramp up to a huge value, often overflowing register variables. After the disturbance is released, the servo overshoots to such an extent that the range of the counters or the maximum velocity is exceeded. This results in an effect commonly referred to as scanning mirror "flip."

To avoid this condition, the integrator is set to saturate at a maximum value equal to ±1000 DAC counts. This is far more than the integrator requires to correct for steady state errors such as friction, DAC offset, etc. Yet it is small enough to avoid the effects of integrator windup.

The other reason that the integrator is implemented as a separate parallel stage is accuracy. This may not even be an issue with the 32-bit floating point arithmetic used in the system, but the method virtually guarantees near-perfect integration and negligible steady state error. If the integrator were incorporated as another order in the difference equation, coefficient rounding might cause the integration to be inexact.

Background Data Packets

Previously, we have described foreground and background tasks. In the background, data packets are always being executed. These packets describe the vector to be drawn, including a starting location, delta, and number of clock ticks. For all the ticks where any packet is valid, the commanded position is described by:

$$\text{Command}_x = \text{Start}_x + \text{Tick}_x\,\text{Delta}_x$$
$$\text{Command}_y = \text{Start}_y + \text{Tick}_y\,\text{Delta}_y$$

The commanded position is computed in this manner, rather than performing successive addition. This method takes advantage of the fact that the DSP is able to perform floating point multiplication as fast as it is able to do anything else, and because successive addition can lead to significant errors, if it is performed enough times.

Shutter Command. There is also an AOM command, sometimes called a shutter command, in the data packet. This command is simply loaded and output to the control register on the control board during each servo tick. From the software standpoint, it is simpler and faster to do this each tick, rather than only on the first tick. It is faster, since the DSP can solve a difference equation faster than the software is able to test a flag. When computing the values that are in the data packet, several interesting issues arise. One example occurs when the number of servo ticks is other than a whole number and either the length or speed of the vector must be adjusted. Refer to Figure 4-17, which describes the format of the data packets.

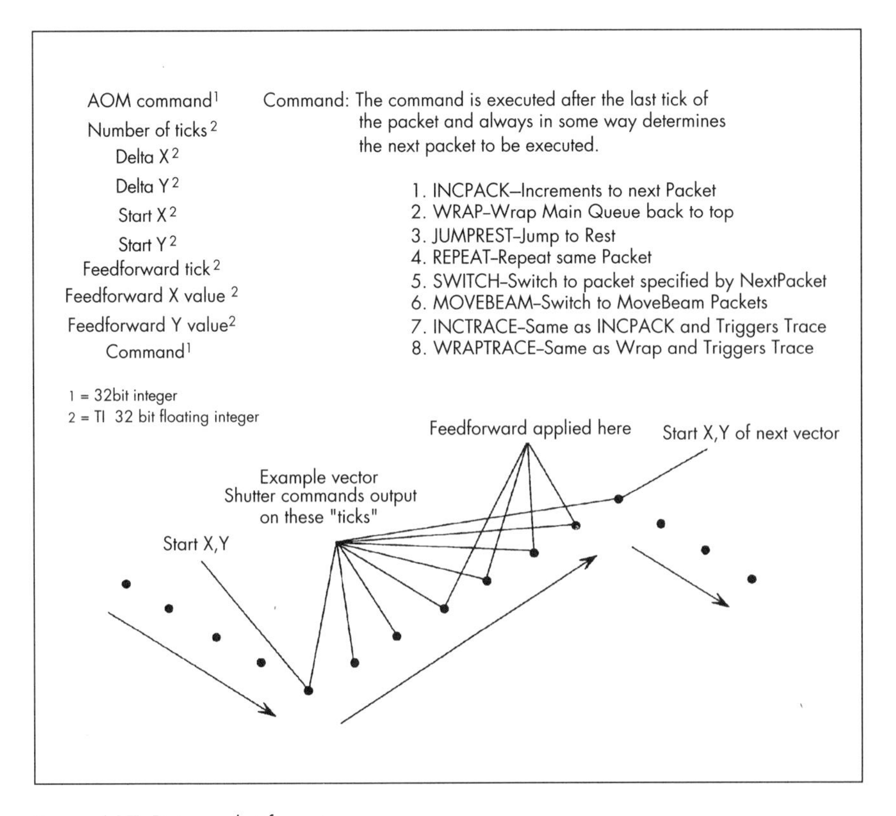

Figure 4-17. Data packet format.

Imaging Data Packets

The data packets being processed in the background reside in a queue. Refer to Figure 4-18 for a schematic description of the queue and to Figure 4-19 for information contained in an example data packet. There is a main queue, through which the background is normally implemented. In this case, each data packet contains an INCPACK command telling the background that when it is finished with one data packet, simply increment to the next packet. The main queue is circular, so that the last packet in the queue must tell the background to "wrap" back to the beginning. The foreground process places a WRAP command in this packet to cause the background to reset back to the beginning, rather than incrementing again.

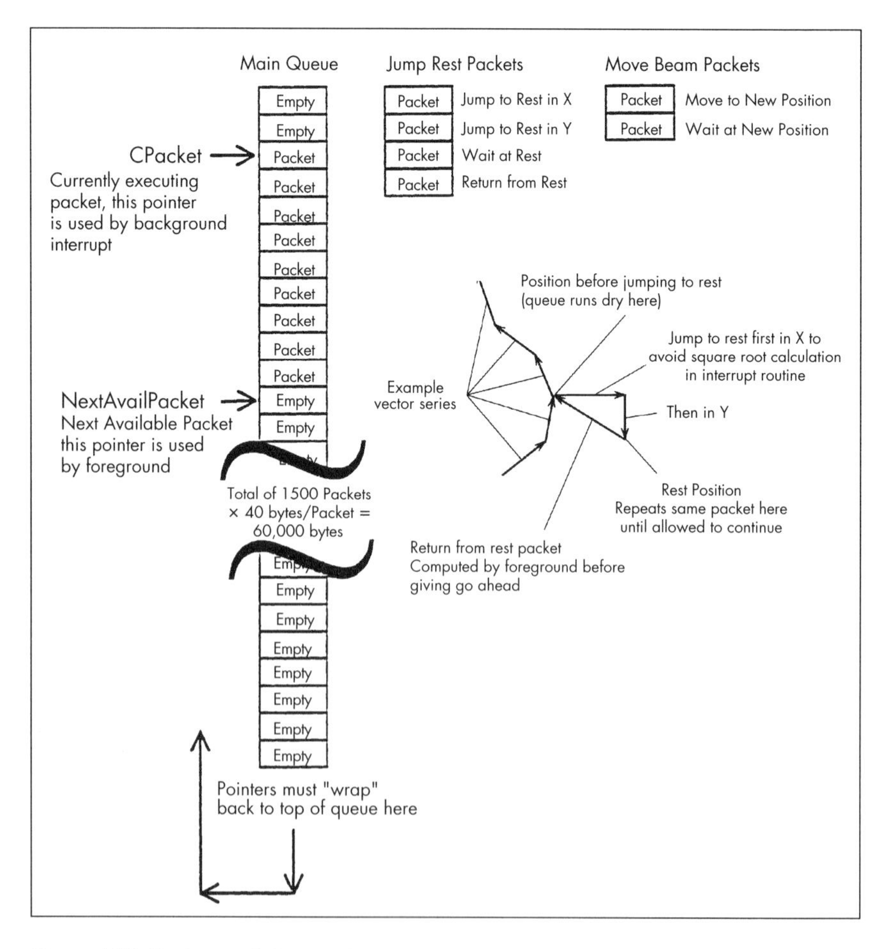

Figure 4-18. Background data packet queue.

```
/* fills in a packet given the beginning and ending points,
   and speed in mirror counts/tick                         */

Void CalcPacket(float FromX, float FromY, float ToX, float ToY, float Speed,
                                PACKET *Packet)
{
   float deltax,deltay,dist,ticks;

   deltax = (ToX-FromX);
   deltay = ToY-FromY);

   dist = sqrt(deltax * deltax +deltay * deltay);

   ticks = (float)((int)(dist/Speed+ 1.0)); /* compute the int number of ticks */

   if (ticks < ServoParams.FFTicks) ticks = ServoParams.FFTicks;

   Packet->shutter             = AOMOFF;
   Packet->NTicks              = ticks;
Packet->dx                     = deltax/ticks;
Packet->dy                     = deltay/ticks;
Packet->startx                 = FromX;
Packet->starty                 = FromY;
Packet->fftick                 = -1.0;
Packet->ffx                    = 0.0;
Packet->ffy                    = 0.0;
Packet->cmd                    = INCPACKCMD;
}
```

Figure 4-19. Example data packet.

One special case occurs when all vectors have been drawn, and the beam must be moved to its "rest" (nondrawing) position. The foreground places a JUMPREST command in the last valid data packet, which triggers the system to jump to rest. When the background executes the JUMPREST command, it fills in the special sequence of three jump rest packets and switches to them. The first packet moves the beam in the X direction, from the current position to the X coordinate of the rest position. The second packet moves in the Y direction, to the rest position in the Y direction. The third packet has zero deltas, such that the system simply remains at the rest position. It also has a REPEAT command, which continuously keeps repeating this command.

If the foreground loop wants the background to start drawing more vectors, it places additional packets into the main queue. It then computes a *return from rest* packet, which will move along a diagonal path from the rest position to the position where the first vector in the main queue begins. This packet is placed in the return from rest packet of the jump-to-rest packets. Finally, the sequence is initiated by the foreground, when the command in the jump rest packets is changed from REPEAT to INCPACK. The next packet contains a SWITCH command, which causes the software to return to the main queue.

Don't Starve the Queue. It is very important to keep the foreground ahead of the background, in order to avoid starving the queue. When the queue starves, the system must jump to rest in the middle of a part cross-section. If queue starvation is unavoidable, every attempt is made to have the jump to rest occur at the point of a planned jump command, rather than in the middle of consecutive moves, since this is possibly a part boundary. Queue starvation is unusual with the Orion system, and is usually the result of a circumstance where a tremendous number of very short (0.1 inch [2.5 mm]) vectors are being drawn.

After data is placed in the main queue, but before the background is allowed to begin executing, the last valid data packet is set to a JUMPREST command. When more space is available in the queue, the new data is placed into the queue and terminated with a second JUMPREST command. Then the previous JUMPREST command is removed, by changing it to an INCPACK or WRAP command as required. *This assures that there is always a JUMPREST command following valid data in the queue.*

The foreground monitors the background loop. If the background ever starves and jumps to rest after the queue is refilled (assuming there is enough data to actually fill the queue), it is restarted.

This queuing scheme results in the burden of queue management being placed on the foreground loop, in order to keep the background loop as simple as possible. Since the foreground is written in the C programming language, it is a simpler program than it would be otherwise.

MOVEBEAM. There is one additional sequence of two packets, which implements a function called MOVEBEAM. This function positions the beam to a fixed X,Y location, and is used by the laser beam profiling (power and shape information) and calibration programs. When the PC requests that the beam be moved to a specific location in X,Y coordinates, the foreground constructs two MOVEBEAM packets. The first packet moves the beam from the current location to the desired location. The second packet repeats that location. The sequence of events is such that the command in whatever packet is currently executing is changed to a MOVEBEAM command. This causes the background to execute the MOVEBEAM packet instructions while the beam stays at the desired location until a JUMPREST command is issued. Then the system returns to a state in which queued moves and jumps may again be issued and executed.

Resulting Benefits from New Imaging Software

Orion enabled the SLA-500/30 to briefly hold the world's record for user-part accuracy. Both the part and the measurement technique are discussed in detail in Chapter 5. The Orion-equipped SLA-500/30 had 90% of all measurements lie within 98 microns of ideal. In addition, it is now possible to develop new technologies and applications that were tremendously difficult, if not impossible, with the previous SLA-500/20 system.

4.6 Potential Software Improvements

There are several major areas where changes in software architecture would further improve machine accuracy and make the systems more user friendly. A common build file format for all machines, moving hatch/fill generation onto the process control software, for the SLA-250, and more automated part preparation software are a few examples.

Common File Format. Now that part preparation on all machines has been converged into a single user interface, the next step is to develop a *common build file format for both platform types*. Currently, the small and large platform machines use two completely different file formats. This prevents an operator from using a file prepared for an SLA-250 for part building on an SLA-500, or vice-versa, without repreparation, including reslicing. When the systems converge on a single build file format, parts could be prepared without knowledge of the platform type on which they will be built. This would be advantageous for anyone with different machine types and critical machine time scheduling issues. Imagine preparing a queue of parts, ready to execute, and building them on the next available build platform—regardless of platform type! Of course, parts larger than the designated build platform must be built on the larger machines, or else joined together during postprocessing.

The major functional difference between the two file formats is that the SLA-250 includes hatch and fill vectors from the preparation phase. The SLA-500 produces hatch and fill vectors on the process computer, while the boundary data is drawn. Therefore, eliminating hatch and fill vectors from the build file format will bring the systems closer together. This has several benefits. It reduces the slice and build file size dramatically, further decreasing slicing and file transfer time. It also simplifies modifications to the part preparation process on the small machine, since there would be no need to reslice when hatch and fill parameters are changed.

As higher power lasers become available for the 190/250, the current scanning system will reach its inherent limits, both in drawing speed and part accuracy. The next logical step would be to incorporate the Orion Imaging Technology on those machines. This would bring the two machine types closer together, both in terms of functionality and software architecture. This change, along with a common file format, opens the door to an identical process control interface on both machine types.

Simplified Part Preparation. As the technology progresses, part preparation will be simplified, to the point that a person not trained in SL will be able to build excellent parts. Currently, part preparation is somewhat an art, requiring significant training and experience to produce very accurate, high-quality finished parts. This is due to many process and machine specific parameters that must be skillfully selected by the user to produce the desired result. Advances in process techniques, along with certain hardware improvements, open the door for a much simpler, expert system. The goal of such a system would be for a CAD user, with little knowledge of SL, to be able to obtain the same part accuracy and quality as the current expert user. This has been termed "de-skilling."

From the software standpoint, automation of recoat parameters and eliminating user input of machine specific parameters would facilitate this de-skilling process to such a point that an operator would simply *pick* the desired part, *select* the resin being used, *assign* a drawing style, and press *build*. This would not, however, preclude an expert user from providing inputs to the system in order to achieve a certain unique result. The system could have variable skill levels, so that a novice user would make a few picks to obtain his part, but an expert user would have access to many parameters useful to achieve some specific set of desired part qualities.

Reference

1. Kerekes, Tom, "3D Systems Control Software," 3D Systems Technical Report, June 15, 1993.

CHAPTER 5

Advances in Part Accuracy

"I decided that I was going to learn to work just a little more accurately than any of the other apprentices around me, and when I succeeded in doing this, then I decided that I would learn to do it still more accurately than I had done before. Throughout my life that has been my one idea."

Lucian Sharpe, Co-founder, Brown and Sharpe

Rapid prototyping and manufacturing thrives on accuracy. Every time there is an improvement in accuracy, new applications are discovered that invite many new experts and innovators into the RP&M arena. These individuals provide new ideas and discoveries that further expand the horizons of RP&M. Figure 5-1 reveals the accuracy milestones achieved with stereolithography at 3D Systems. This time line represents SL's evolution from simple visualization to its present role in fabrication, or in practical terms, from rapid prototyping to rapid prototyping and manufacturing. The data also represents a consistent trend of accuracy improvements. In just five years, stereolithography RMS error has been reduced from around 9 mils to 1.8 mils. This five-fold improvement has allowed SL to be accepted into various manufacturing processes as well as many research projects aimed at significantly advancing the productivity of the manufacturing cycle.

About the time user-part accuracy reached levels that could begin to compete with the average CNC machining process, industries worldwide recognized that SL could offer far more than visualization models of their designs. In fact, QuickCast was being developed and it held the promise of metal parts actually being used for functional prototypes. After almost a year of coordinated research between 3D Systems and a few select foundries, QuickCast was commercially released together with SL 5170 in July 1993.

Although many organizations were beginning to reap benefits from the process, QuickCast was still a new technology and shrinkage factors and foundry processes were not sufficiently refined to satisfy industries that required extremely

By Bryan Bedal, Research Engineer, and Hop Nguyen, Senior Research Scientist, 3D Systems, Inc., Valencia, California.

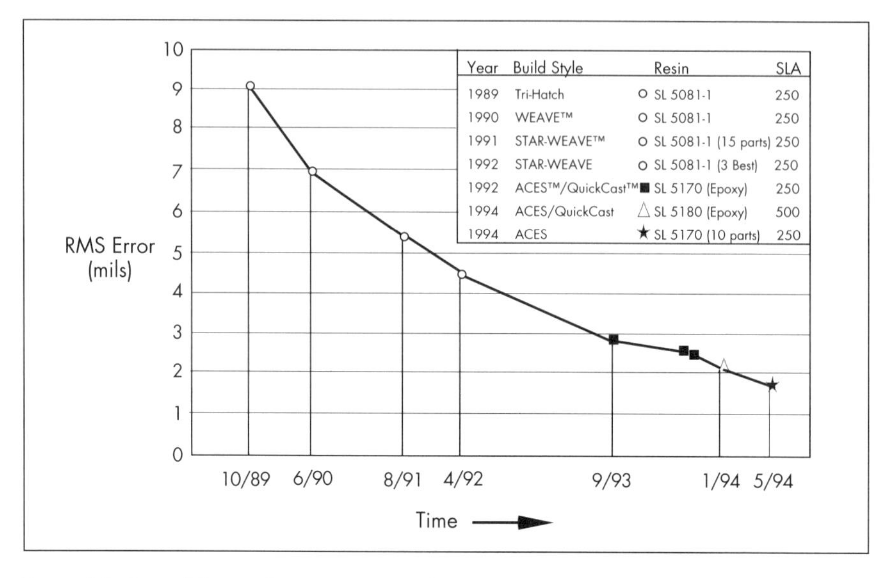

Figure 5-1. Stereolithography user-part time line.

accurate parts. However, over the course of one additional year, QuickCast pattern quality has advanced, and foundries have improved casting techniques to the point that accuracy levels and surface finishes are now sufficient for prototype tooling applications.

Solid SL parts are also currently being used and explored in a number of tooling methods. These parts are used as both models for creating tools and as tools themselves, by generating the negative of the CAD design. Fifteen consortiums have already formed to develop and optimize rapid tooling processes, and are reporting substantial progress.

With rapid tooling as a short-term goal, RP&M manufacturers are in healthy competition to produce parts sufficiently accurate to satisfy this demanding application. This chapter informs the reader of the advances in SL part accuracy and surface finish that have been achieved. Although this information pertains specifically to stereolithography, the concepts associated with determining accuracy can, and should, be applied to all RP&M systems.

5.1 Improving Accuracy

Improving the accuracy of SL requires commitment, understanding, and ingenuity. Commitment means having engineering projects devoted solely to improving accuracy while not accepting those changes that sacrifice accuracy. It requires devoting the resources necessary to discover, understand, and eliminate numerous small sources of error. This can only be accomplished through tedious and meticulous research, which is often very costly. The testing must be combined with an understanding of the physics, chemistry, and mechanics of the SL process as integrated disciplines. Finally, creativity is required to advance the process beyond current available technology. As this book demonstrates, technical ingenuity, combined with commitment and understanding, has provided numerous advances responsible for the consistent trend toward more accurate SL parts.

The technical advances that have improved part accuracy can be divided into four major areas. 1) Resin improvements that provide the desired properties of low distortion with higher strength and lower viscosity. 2) Hardware improvements that allow the laser beam to be positioned more accurately and the liquid level to be controlled with greater precision. 3) Software programs that optimize capabilities of the new laser scanning system and provide more accurate build files. 4) Process improvements using optimized build styles that take advantage of the machine, resin, and software improvements to further increase accuracy and functionality of stereolithography.

Resin Improvements

Perhaps the most exciting improvement to the stereolithography process has been a resin that exhibits the physical properties of various engineering plastics as well as extraordinarily low distortion during the SL build process.[1] The introduction of epoxy-based SL resins in 1993 nearly doubled dimensional accuracy, reducing the RMS error from 4.5 mils in SL 5081-1, to 2.8 mils in SL 5170, the SLA-250 version of the epoxy-based resin. As soon as the ACES solid build style for this resin was optimized, a series of 10 user-parts revealed a very repeatable RMS error of 1.8 mils. The strength, reduced distortion, and low viscosity of the epoxy resins also made the QuickCast build style possible, enabling accurate patterns that can be shell investment cast in a multitude of metal alloys.

Epoxy-based SL resins exhibit numerous properties critical to RP&M and especially rapid tooling applications. Chapter 2 explains the chemistry and reports the physical properties of the commercially available epoxy-based resins used in SL. Increased accuracy can be attributed primarily to the improvements in the resin physical properties that are related to distortion. Chapter 2 also discloses the results of the various diagnostic tests that are used to quantify different types of distortion, such as shrinkage, curl, swelling, and creep. These tests reflect the distortions as a function of resin, machine, and build style.

New Epoxy Resins. The predominant reason for low distortion is due to the new epoxy resins. However, new hardware and build styles, such as QuickCast and ACES, must also share credit for these advances. The data also demonstrates that epoxy resins possess the mechanical properties of several engineering plastics and exhibit low distortion both during and after the SL process.

Hardware and Software Improvements

When challenged to provide a more accurate servo controller for the SLA-500 and to design an interchangeable vat on the SLA-250, engineers implemented new designs that contributed to notable improvements in part accuracy. The SLA-500 received a new digital signal processor (DSP) and an acousto-optic-modulator (AOM) along with improved software that allows the scanning system to reproduce the requested beam path with greater fidelity. These are components of the Orion Imaging Technology.

Further contributions to improved part accuracy resulted from the new diode laser leveling systems that control the liquid resin levels to within ±0.0002 inch (±5 microns), implemented on both the SLA-250 and SLA-500. The interchangeable vat for the SLA-250 includes a dual-rail recoater and a vat hoist leveler that also help produce more accurate liquid layers.

Software Contributions. Software advances have contributed to improved part accuracy as well. In addition to the software that drives the Orion Imaging Technology, build files can now provide the machine with more accurate vector information. The SLC slice format, which uses border definitions instead of tessellation to define the surfaces of a CAD design, can provide higher resolution for curved surfaces without increasing file size. Also, a new linewidth compensation program attacks the linewidth problem in a more efficient manner. The addition of user-selectable "vents" and "drains" to the QuickCast 1.1 software eliminates delamination and uneven upfacing surfaces due to trapped volume effects, by allowing the user to place openings in horizontal upfacing and downfacing skins. These vents and drains permit the resin in the part to respond to pressure differences within the vat.

Process Improvements

Since the early days of stereolithography, resin shrinkage has been a major cause of part inaccuracy. The initial shrinkage occurs when the liquid resin is polymerized to form a green part. The term "green" refers to a part that has been laser-cured but not yet postcured in a PCA. Shrinkage factors are used to compensate for the overall change in volume, but there are other forms of distortion caused by resin shrinkage. For example, material is free to shrink in an unconstrained manner when the first layer of a part is built. However, starting from the second layer, the material shrinks while it is bonded to the previous layer. This causes a phenomenon similar to the "bi-metallic strip" effect, and is responsible for creating internal stresses within the green part. Furthermore, green parts, by definition, are not 100% polymerized. They need to be postcured to complete the cross-linking required to achieve full strength. During postcure, additional shrinkage takes place that can result in further distortion and dimensional instability. For this reason, almost all the part-building processes intended to improve part accuracy focus on minimizing the effects of internal stress.

STAR-WEAVE, QuickCast, and ACES are process techniques specifically developed to reduce distortion.

QuickCast

In addition to the ability to be investment cast, the QuickCast build style was designed to improve SL part accuracy. After realizing that the thermal expansion of solid SL patterns cracked even the most durable shell systems, it became evident that SL patterns should be built as outer borders supported by an internal structure of widely spaced hatch lines. By altering the hatch pattern to allow excess resin to be removed, "quasi-hollow" patterns are generated that can be successfully investment cast into solid metal parts.

Quasi-hollow Structure. Since cast metal parts are very likely to be used as functional components, accuracy is certainly an important factor. Fortunately, since the quasi-hollow structure of QuickCast patterns requires less material to be bonded between each successive layer, there are reduced internal stresses in the green state. Postcure distortion is also reduced because there is less material to shrink. Lower distortion results in greater dimensional stability. User-parts built in QuickCast 1.0 produced remarkable accuracy results during the diagnostic test phases of SL 5170 and SL 5180. Of course, this build style would not even be possible without the epoxy resins that combine low viscosity, high green strength, and minimal distortion during the SL process.

To create the quasi-hollow QuickCast structure, objects must be built using a large hatch spacing. Unfortunately, conventional hatch patterns form compartments, or cells, that trap uncured liquid resin. The objective of the QuickCast method is to build a part in a manner that connects these cells so that they are not isolated, allowing uncured liquid resin at any given location within the pattern to flow freely to any other location.

Mathematically, the object must be *topologically simply connected*.[2] If a drain hole is intentionally generated at any location on a topologically simply connected pattern, and a relatively small vent hole is created at a second location to relieve the partial vacuum that would otherwise be created, the uncured liquid resin can exit the pattern.

To create passages between the cells, the crosshatch pattern is repeatedly drawn for a certain Z axis height (H), then the hatch lines are translated, or "offset," by the amounts dx and dy in the X and Y axes respectively, and then drawn for another vertical height, H. The hatch pattern then returns to the original configuration and the hatch drawing style continues to alternate until the part is completed. The crosshatch pattern can be in any style, square hatch, equilateral triangle hatch, etc. Figure 5-2 shows the QuickCast 1.0 equilateral triangle hatch pattern.

Three Types of Hatch Lines. In the QuickCast equilateral triangle hatch pattern, there are three types of hatch lines: the hatch lines that are parallel to the X axis, and the lines that form 60 and 120° angles with the X axis. In each direction, the parallel lines have the same hatch spacing, h_a. These hatch lines will form equilateral triangles with h_a equal to the triangle altitude. Triangle vertices in the second section of the QuickCast pattern should land directly above the centroids of the triangles of the first section. In order to meet this condition, the required horizontal translations in the resin plane are $d_x = h_a/\sqrt{3}$ and $d_y = h_a/3$, as shown in Figure 5-3.

In the QuickCast 1.1 square hatch style, shown in Figure 5-4, the values of hatch spacing for X hatch and Y hatch are h_x and h_y respectively. For the best

Figure 5-2. QuickCast 1.0 equilateral triangle hatch pattern.

liquid resin flow rate, the offset values $d_x = h_x/2$ and $d_y = h_y/2$ are selected. It becomes evident that the cells at the first and the third vertical sections are connected via the cells of the second vertical section through passages having a height H in the Z direction. Clearly, the flow of liquid resin is also limited by H.

These examples of the QuickCast square hatch and equilateral triangle styles demonstrate how these patterns are mathematically created. These drawing styles were fully implemented in the software of both the SLA-500 and the SLA-250 systems.

Optimizing QuickCast Patterns. Although the present hatch algorithms provide functional QuickCast patterns, there is still room for improvement. In fact, there is an aggressive, ongoing effort to optimize QuickCast patterns. At the top of the list is pattern accuracy. While it is important that the pattern be designed to readily collapse during burnout, surface finish, structural integrity, and dimensional stability must not be sacrificed by making the pattern too weak.

These are just a few of the many factors that must be considered when designing new hatch patterns as well as border and skinning techniques. Although the square and equilateral triangle hatch patterns seem elementary, they were chosen for their favorable results in a comprehensive test matrix. In addition to accuracy, uncured resin drainage, pattern collapsibility, and maximum cured resin solid wall thicknesses were also considered. The following example illustrates the implications involved in determining the optimal value for hatch spacing, which is just one of many variables that must be established.

A principal goal of the QuickCast build style is to drain the excess uncured liquid resin from green parts using only gravity. The flow of the uncured liquid resin is limited by 1) size of the cells (hatch spacing), 2) height of the passages that connect the cells together (H), 3) resin viscosity, and 4) resin surface tension.

Hatch Spacing Limitations. Obviously, liquid resin will flow more freely if the hatch lines are farther apart. However, wider hatch spacing results in lower structural integrity. QuickCast parts require a minimum strength to avoid distortion in the green state (i.e., during building and support removal). Furthermore, even cured QuickCast patterns must withstand forces imposed during handling and shipment as well as in the investment casting process itself. Another limitation on increased hatch spacing appears in the quality of the upfacing skins. As hatch spacing increases, the upfacing skin tends to sag between the hatch lines and surface quality quickly diminishes.

An additional limitation to increased hatch spacing involves recoating. If QuickCast patterns use very wide hatch spacing, they are almost hollow and be-

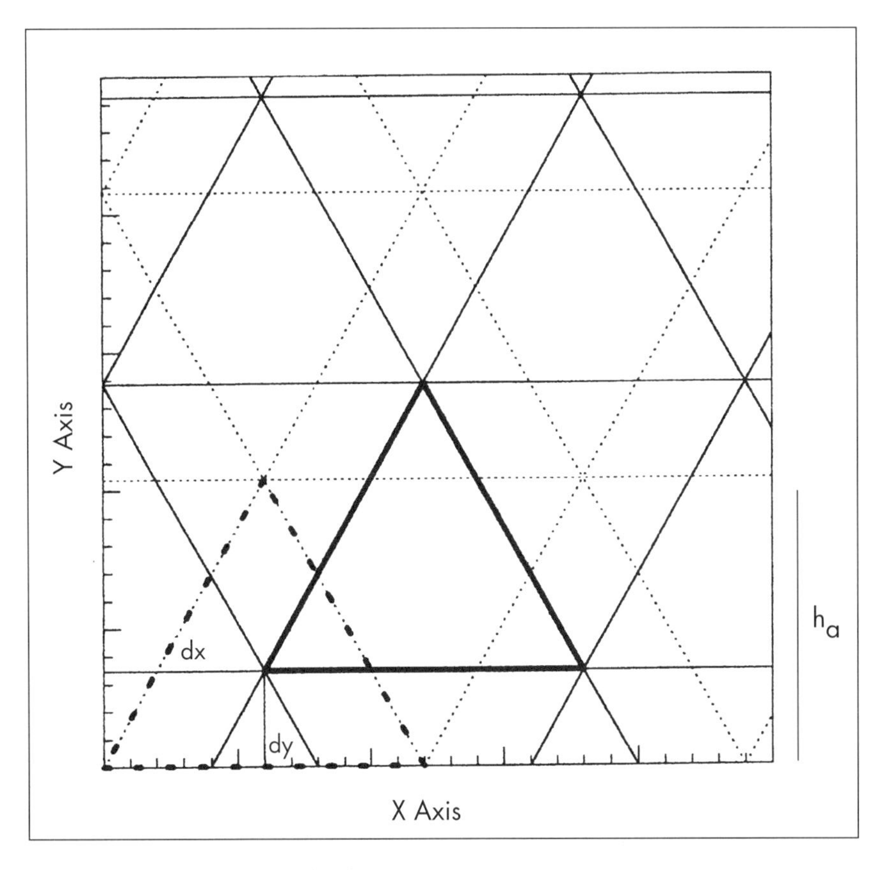

Figure 5-3. Equilateral triangle hatch pattern.

have like miniature trapped volumes. Borders can exhibit delamination and skins may experience dewetting due to uneven resin layers resulting from trapped volumes. Fortunately, this problem was overcome with the implementation of user-selectable "vents" and "drains" in the QuickCast 1.1 software. The ability to open the downfacing skins in selected areas allows the resin in the vat to communicate with the resin in the QuickCast pattern, virtually eliminating the associated trapped volume effects.

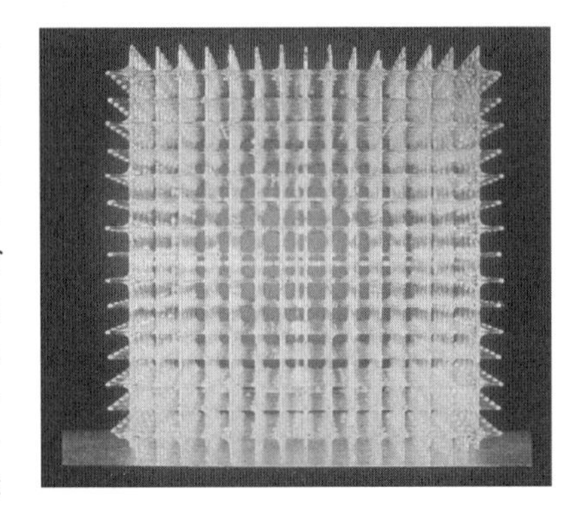

Figure 5-4. Model of QuickCast 1.1 square hatch pattern.

The implications continue. At some point, increasing the hatch spacing does not further increase the flow rate. When the hatch spacing becomes very large, the limiting factor becomes the vertical height of the passages between the cells. This size is determined by the offset height distance, H, which must be limited so that thin horizontal features are not closed off by vertical hatch sections. Of course, hatch patterns composed of dashed lines will overcome this problem and there are a few such designs currently being tested. To further complicate the matter, a foundry must be involved in this effort to provide input on structural integrity, collapsibility, and thermal expansion during the shelling and casting processes. Here, pattern drainage and structural integrity are of major concern.

This example illustrates just one of the variables that can be adjusted to optimize the QuickCast pattern design. Hatch patterns, border and skinning techniques, border and hatch exposures, and vents and drains, each having countless parameter possibilities, are constantly being explored and tested. Recommended values are carefully selected through experimentation and verification, and are designed to produce successful results for the widest possible range of part geometries. If speed is more important than accuracy, or if the geometry will allow, then deviation from the recommended parameters may provide some special advantage.

ACES

The purpose of the ACES building technique is to produce **A**ccurate, **C**lear, **E**poxy, **S**olid parts with excellent dimensional stability using the epoxy-based resins. This is accomplished by complete and uniform polymerization during the part building process, virtually eliminating postcure distortion and internal stresses. To minimize the bi-metallic strip effect, ACES uses a method of progressively curing a layer of resin so that it is almost entirely solidified *before* it is bonded to the previous layer. This is achieved by using two consecutive sequences of uniform UV exposure.

First, the surface of a liquid resin layer receives a nearly uniform dose of UV exposure, producing a planar cure depth that almost touches the previous layer. Since there is no bonding, the cured resin is free to shrink without causing distortion. Experiments have proven that the great majority of ACES shrinkage takes place when the liquid resin is solidified during this initial exposure sequence. The layer is then subjected to a second similar exposure, increasing the cure-depth by an amount sufficient to uniformly cure the remaining liquid resin, while assuring adhesion to the previous layer.

Uniform exposure is achieved through the superposition of transversely overlapped laser scans. This is accomplished by selecting a hatch spacing less than or equal to the laser Gaussian half-width (W_0). When a Gaussian laser beam is scanned on a resin surface having a certain D_p and E_c, the resulting cured "line," with its parabolic cross-section, will have a maximum cure depth (CD_0) proportional to the natural logarithm of the maximum exposure, E_{max}, located at the center of the scan line. To demonstrate the total effective cure depth achieved by overlapping the scan lines in a raster fashion, a series of graphs is presented showing the cure depth achieved with each subsequent pass of the laser. Figure 5-5 demonstrates the above case for a single pass, beginning at $y/W_0 = 2$, where the laser beam is scanned in the X direction (i.e., into the paper) and the exposure and cure depth are plotted in the Y direction, which is transversely orthogonal to the scanned line.

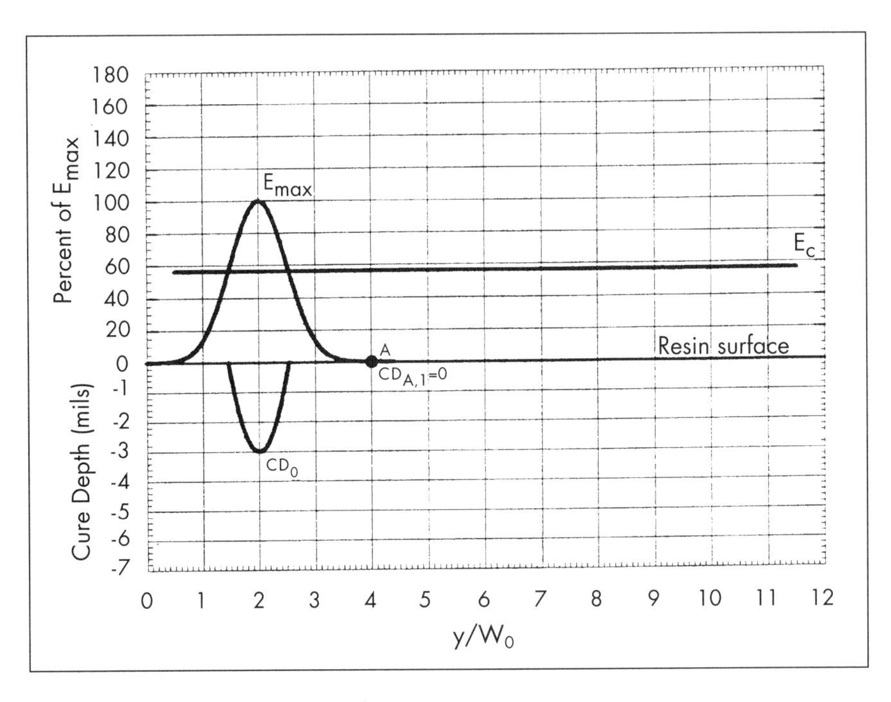

Figure 5-5. Single pass of Gaussian beam.

If a laser beam is scanned in a raster fashion, with hatch spacing, h_s, the exposure at any point will be equal to the sum of the exposure contributions from the surrounding scanned lines that affect that area. For this example, let's follow the cure depth at point A on the resin surface, where $y_A/W_0 = 4$, and use a hatch spacing, $h_s = 0.8W_0$. On the first scan, no resin is solidified at point A, thus $CD_{A,1} = 0$. Figure 5-6 represents the results of a second scan, which still does not solidify any resin at point A, hence $CD_{A,2} = 0$. In Figure 5-7, the third scan generates a combined exposure at point A that exceeds E_c, and therefore resin begins to solidify at point A. In this case $CD_{A,3} < CD_0$ and the surface of the resin is solidified slightly beyond point A. After the fourth scan, shown in Figure 5-8, the cure depth is $CD_{A,4}$, which is greater than CD_0 but slightly less than the total accumulated effective cure depth CD_I. The cure depth is considered to have reached CD_I

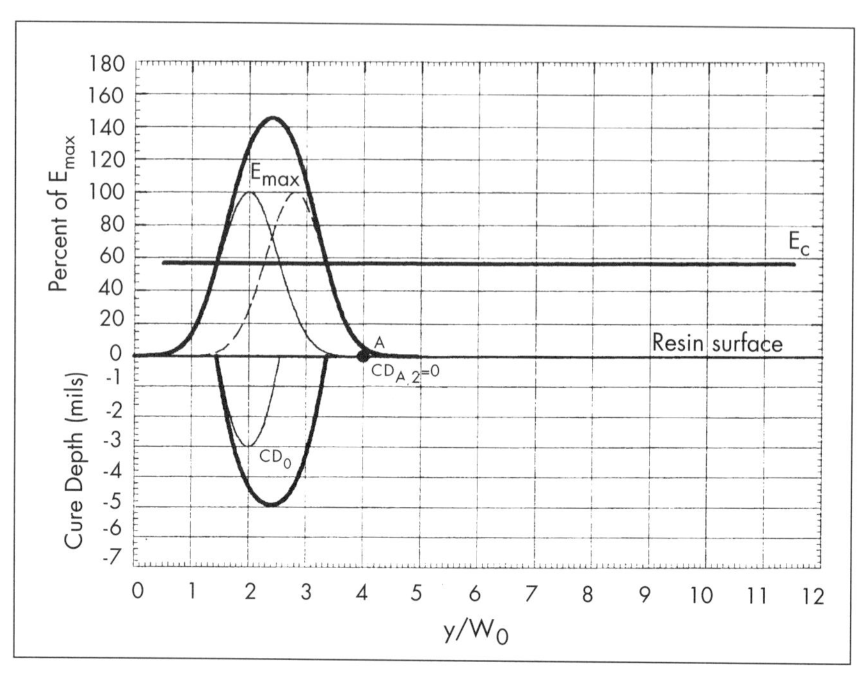

Figure 5-6. Second pass of Gaussian beam.

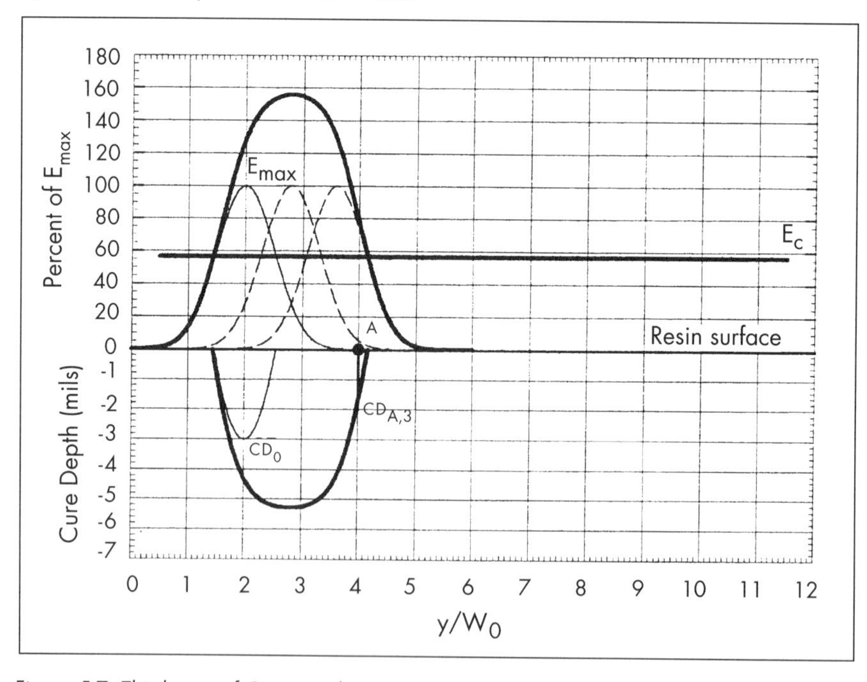

Figure 5-7. Third pass of Gaussian beam.

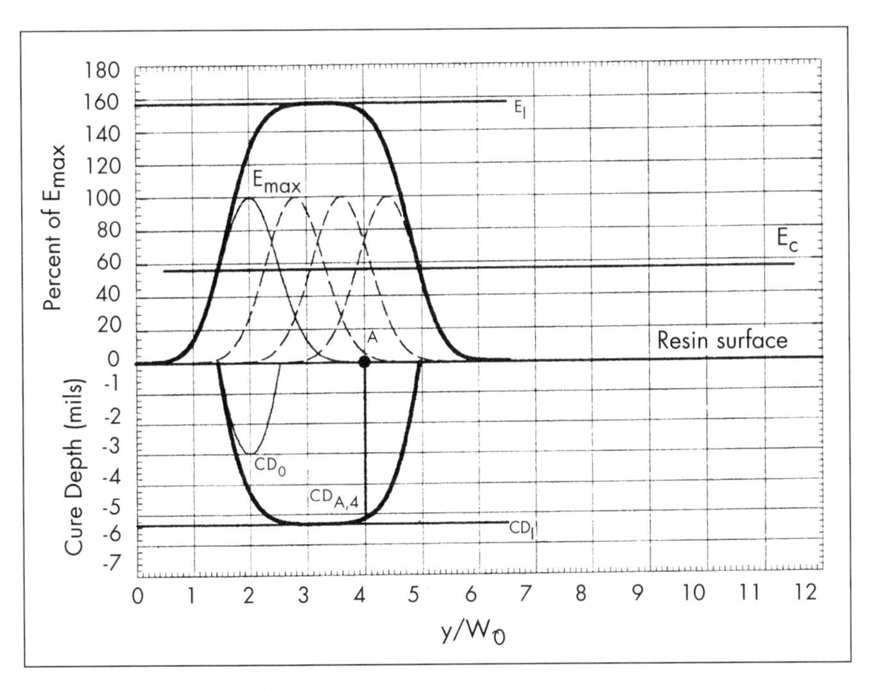

Figure 5-8. Fourth pass of Gaussian beam.

in the fifth scan, shown in Figure 5-9, and subsequent scans will no longer increase the cure depth, as seen in Figure 5-10.

Reference 3 explains the conditions necessary for "planar cure depth." Using some of the equations derived in that reference, it is possible to calculate the exposure required for each single hatch line to achieve the desired effective cure depth:

$$CD_0 = D_p \ln (E_{max}/E_c) \tag{5-1}$$

When the resin is exposed to the laser beam in a raster scan fashion, it will receive a uniform exposure, E_I, and resin is solidified with a planar cure depth CD_I. The additional cure depth attained when the scanned lines are overlapped can be calculated from the following equation:

$$CD_I - CD_0 = D_p \ln (E_I/E_{max}) \tag{5-2}$$

In the superposition of a scanned Gaussian beam, the ratio E_I/E_{max} depends on the ratio of the hatch spacing to the Gaussian half-width, h_s/W_0. Figure 5-11 is a plot of E_I/E_{max} vs. h_s/W_0. Note that E_I/E_{max} can exceed 2.0 for small hatch spacings.

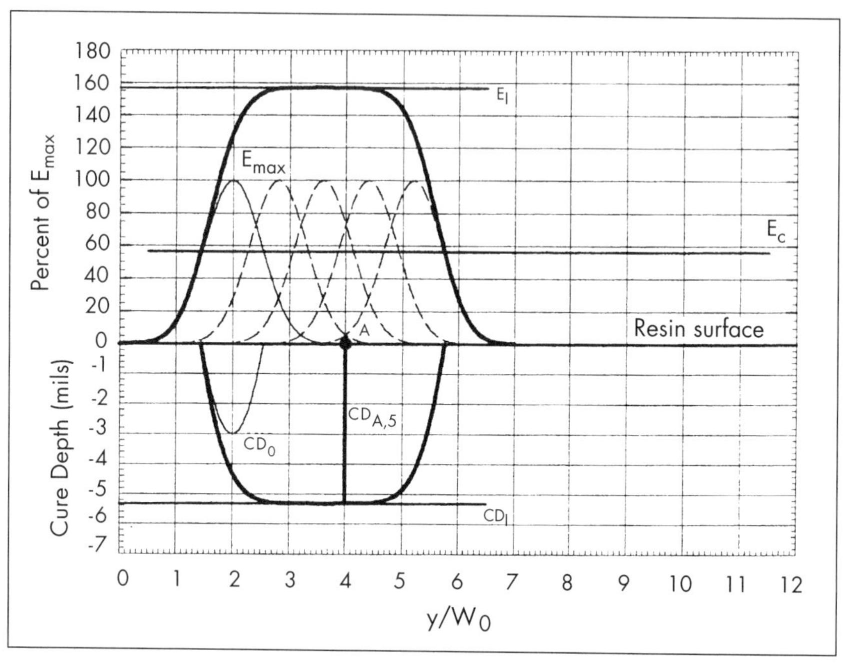

Figure 5-9. Fifth pass of Gaussian beam.

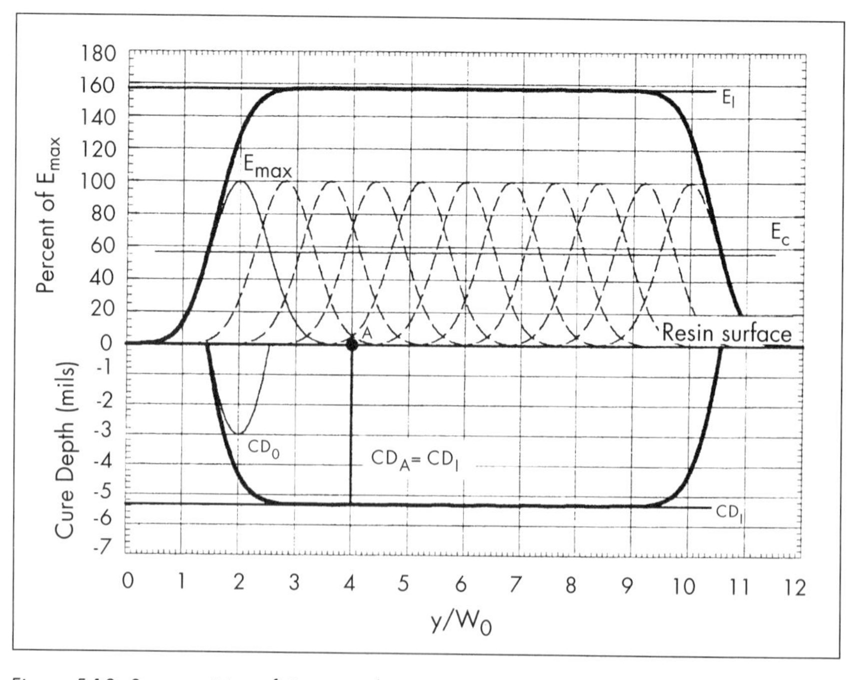

Figure 5-10. Superposition of Gaussian beams.

The ACES build style uses both X and Y hatch to achieve the two levels of cure for each layer. The first sequence (designated by roman numeral I) accomplishes the majority of the solidification, without bonding to the previous layer. The desired planar cure depth, CD_I, for this pass should be 0.001 inch (0.025 mm) less than the layer thickness, a. If all quantities are expressed in mils, where 1 mil ≡ 0.001 inch ≈ 0.025 mm = 25 microns, then:

$$CD_I = a - 1$$

For example, with a hatch spacing of 4 mils and a typical laser beam having a Gaussian half-width of 5 mils, then $h_s/W_o = 0.8$. From Figure 5-11,

$$E_I/E_{max} = 1.57$$

Therefore:

$$CD_I - CD_o = D_p \ln(1.57)$$
$$CD_o = CD_I - 0.45\ D_p$$

or

$$CD_o = (a\text{-}1) - 0.45\ D_p$$

Typically, the D_p value of the SL epoxy resins is about 5 mils. For the case a = 6 mils, we find on substituting these values in the above equation:

$$CD_o = (6 - 1) - 0.45(5)$$
$$CD_o = 2.75 \text{ mils}$$

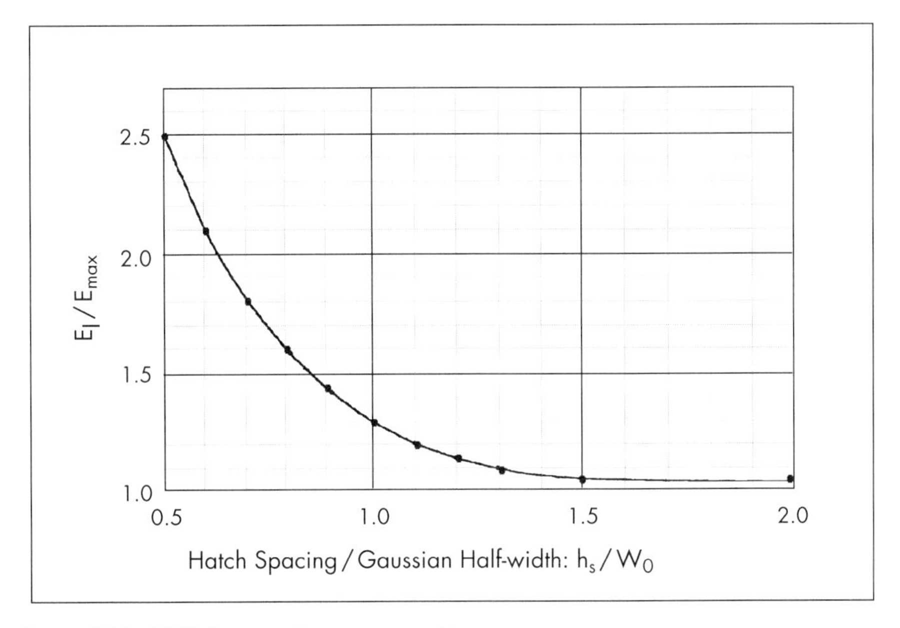

Figure 5-11. ACES Gaussian beam superposition.

For this reason, CD_0, the cure depth of a single hatch line, is specified to be 3 mils less than the layer thickness a. Or, equivalently, hatch overcure = -0.003 inches (-0.076 mm). Note that after the effects of superposition, the net cure depth on the first pass, CD_I, will indeed be 5 mils.

Since ACES involves sequential X and Y hatch patterns, the layer is scanned a second time, perpendicular to the first pass but with the same exposure. This pass provides the additional exposure necessary to bond this layer to the previous layer. When the second scan is completed, the total exposure is $E_{II} = 2\ E_I$, and the final cure depth is:

$$CD_{II} = CD_I + D_p \ln(2)$$
$$CD_{II} = (6 - 1) + 5(0.69)$$
$$CD_{II} = 6 \text{ mils} + 2.45 \text{ mils} = 8.45 \text{ mils}$$

Consequently, the layer is cured 2.45 mils into the previous layer. With the Ciba Geigy SL epoxy resins, this overcure is sufficient to provide good adhesion between layers while generating extremely uniform polymerization within each layer, which is responsible for the remarkable clarity of ACES parts. The designed overcure of -0.003 inches (-0.076 mm) must be changed if it is applied to a resin with a significantly different working curve.

Bubble Reduction

When parts are built in ACES, bubbles and micro-bubbles usually form and remain at the centers of solid build areas. These bubbles not only blemish the transparency of the SL parts, they are a source of surface roughness as well as dimensional error.

Trapping Bubbles. After each layer is solidified, the part is dipped down into the vat to provide a new layer of liquid resin. As the resin flows onto a solidified area, it converges toward a central point. At this location, a small amount of air is trapped when the resin meets, and a bubble is formed. This will occur during every recoating operation. If the part cross-sectional area is large, the recoating blade often bursts the bubble by pressing it between the part surface and the blade. However, on small solidified areas, the bubble is often simply pushed off the solidified area during the sweep. These bubbles are frequently swept back into the drawing area and are subsequently entrapped within the solidified layer.

Three Sweeps. One approach to eliminate this problem is to use a series of three special sweeps during the recoating process. The first is a conventional sweep used to separate the bubble from the solidified surface. The second sweep is done slowly with the part dipped far below the resin surface. The intent of this sweep is to move the detached bubbles well outside the build area. In the third pass, regular sweep parameters are used to provide the desired layer thickness. Most of the bubbles will be left behind because of the blade acceleration. The remaining bubbles are either burst when the blade encounters the solid part area, or are carried to the opposite side. Experience has proven that beautifully transparent, nearly bubble-free parts can be produced in ACES with this three-sweep method. Furthermore, the lack of bubbles has resulted in an additional improvement in user-part accuracy.

Summary of Build Technique Advances

Chapter 2 discusses the data that was collected during the diagnostic testing of the ACES and QuickCast build styles using the Ciba Geigy epoxy resins. Table 5-1 shows some of the recent test results, compared to past performance of the WEAVE build style using acrylate resins. This data illustrates the reduced part distortion that can be attributed to the new resins and build styles.

Table 5-1. Test Results of ACES and QuickCast Compared to WEAVE Using Acrylate Resins

Build Styles	Resin Type	Postcure Shrinkage (%)	Cantilever Curl (%)	Vertical Wall Distortion (mils)	Swelling (mils/inch/hour)	Slab6 Distortion (mils)
ACES™	Epoxy	0.12	1.90	0.00	0.00	19.3
QuickCast™	Epoxy	0.09	1.26	0.00	0.00	23.4
WEAVE™	Acrylate	0.72	11.80	1.80	0.30	65.1

At first, the ACES and QuickCast build styles were tried with the acrylate resins. The ACES build style induced catastrophic horizontal curl. Furthermore, QuickCast parts were too difficult to drain, and also demonstrated poor green strength. It was not until the development of the epoxy-based resins that these stress-reducing build styles could be implemented to produce parts with significantly improved accuracy.

5.2 Determining Accuracy

Accuracy improvement requires quantitative measurements. Does the new scanning system produce more accurate parts? What is the optimal build style for new resins? Will a more precise leveling system improve part accuracy? How does the postprocessing procedure effect part accuracy? Reliable answers to these typical questions require a practical, repeatable method of measuring accuracy. At 3D Systems, the effort to develop such a method has been extremely valuable. The task of obtaining tens of thousands of data points is rewarded with the ability to provide confident answers to important questions. Furthermore, the data from the accuracy experiments provides insights to better understand the numerous anomalies that arise in our research efforts.

Understanding Measurements. The most important aspect of obtaining accuracy data is an understanding of physical measurements. All scientific measurements have errors associated with them. It is important to understand the sources and magnitudes of these errors in order to estimate the reliability of the measurements and, more importantly, to avoid obtaining erroneous data. References 4-5 explain the fundamentals of physical measurements. This is valuable knowledge when obtaining data, as well as for those who must determine the reliability of the reported accuracy specifications.

The User-part

When quantifying the accuracy of an RP&M system, it is important to adhere to some test standards that will allow individuals to understand, compare and, if desired, check the reported values. The first step is to agree upon a standard accuracy test part. In 1990, the North American Stereolithography Users Group developed the user-part to test the capabilities of rapid prototyping machines. Since its development, 3D Systems has been employing the user-part for build style optimization, accuracy, and repeatability studies.

Figure 5-12 is an engineering drawing of the user-part and its various dimensions. Figure 5-13 is a schematic that indicates the features measured on a coordinate measuring machine (CMM). A total of 170 dimensions are taken from these features; 78 are in the X direction, 78 are in the Y direction, and 14 are in the Z direction. This geometry was chosen as a test metric because it has the following properties:

A. It is large enough in the X and Y dimensions to indicate accuracy at the edges of the SLA-250 platform as well as at the center.
B. It has a significant number of small, medium, and large dimensions.
C. It has "inside" and "outside" dimensions to verify that linewidth compensation is functioning properly.
D. It is short in the Z dimension to reduce build time.
E. It is low in mass to reduce material consumption.
F. It can easily be measured with a CMM.
G. It exhibits long thin walls, round and square holes, thin flat areas, and thick sections.

The Data Base. The user-parts' lack of complexity often raises questions about its value as an accuracy metric. Systematic errors are not readily apparent in even the simplest geometries, and experience has shown that the user-part is actually a relatively difficult part to build accurately. Although the user-part may appear to lack geometric diversity, it is interesting to consider what chance there would be for a machine to build a large, truly complex part accurately if it cannot build the

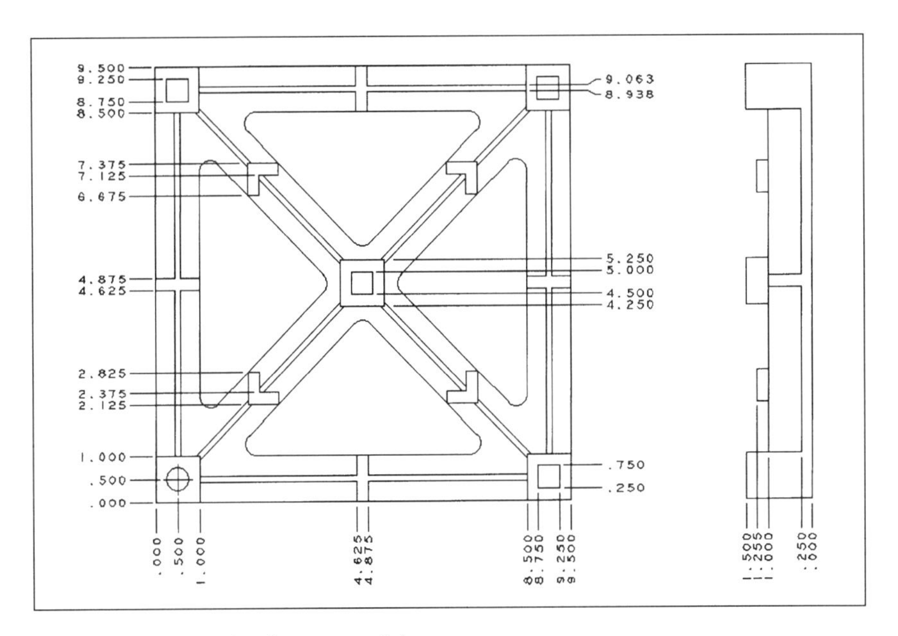

Figure 5-12. X, Y, and Z dimensions of the user part.

user-part accurately. Furthermore, after five years of measuring the user-part with the same instrument, in the same manner, a rather large data base has been established. This data base has become a valuable tool for research, as well as a metric to determine progress.

The CMM and Surface Profilometer

The CMM is an essential tool in determining the accuracy of parts built by today's advanced RP&M systems. With the number of measurements required to achieve statistically significant data, coupled with the low RMS errors currently being reported, automated measuring with high resolution is the only practical method. 3D Systems uses a Brown & Sharpe Validator, Model 8102 CMM with a Renishaw PH9A probe. The reported linear error for this configuration is +/-0.0004 inches (+/-10 microns) throughout the envelope in which the user-part is measured. A recent study revealed that the CMM measurement repeatability was within +/-0.0002 inches (+/-5 microns).

Another measure of part quality lies in the roughness of the surface finish. As the manufacturing industry embraces QuickCast and rapid tooling, surface quality has become a more important issue. A Mitutoyo Surftest 301 Surface Profilometer was purchased to determine the actual surface finish on upfacing and downfacing angled surfaces, as well as horizontal surfaces and vertical walls. The instrument is first calibrated with a NIST-certified calibration surface and the

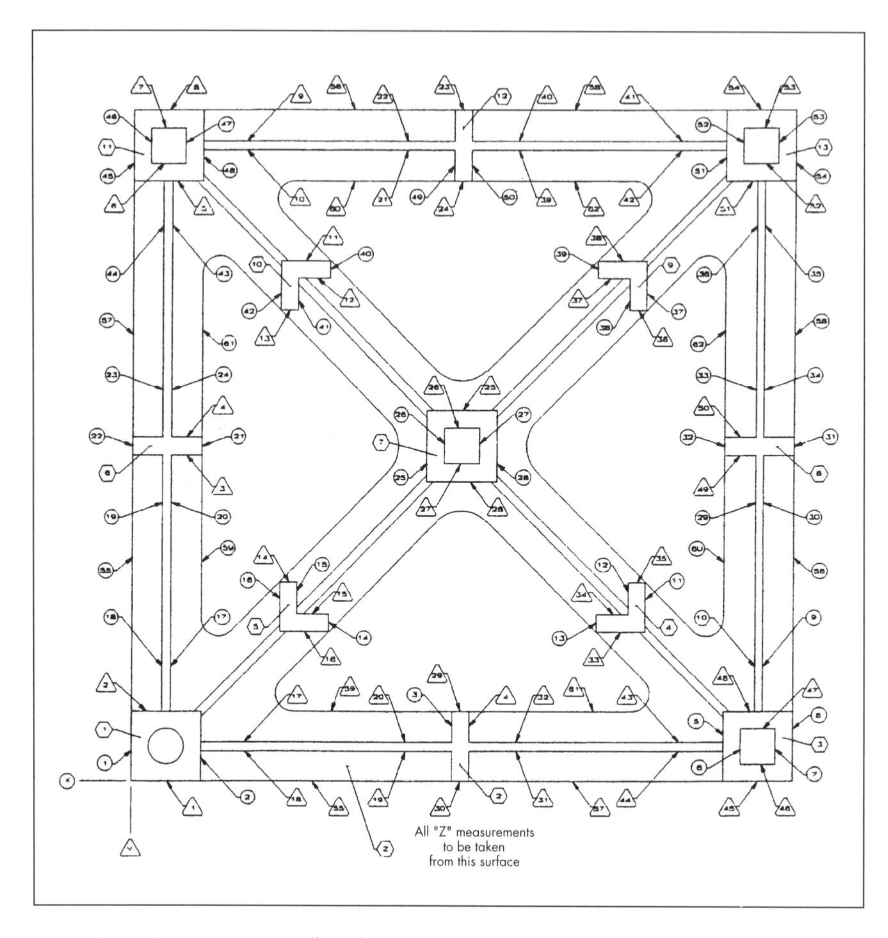

Figure 5-13. Features measured on the user-part.

tests are performed to ANSI B46.1 standards. The reported values are in terms of Ra (roughness average), the arithmetic mean of the absolute values of the profile deviation, from the centerline, within the evaluation length (L_e). Figure 5-14 is a graphical representation of Ra, an industry standard for reporting surface quality. As decreasing layer thicknesses become more prevalent, and dimensional accuracy approaches the limits of the process, surface finish will become an additional standard metric in determining RP&M accuracy, especially for rapid tooling applications.

Reporting Accuracy

A well-developed accuracy program not only aids in research and technology advancements, it has another, perhaps more important, benefit. It provides in-

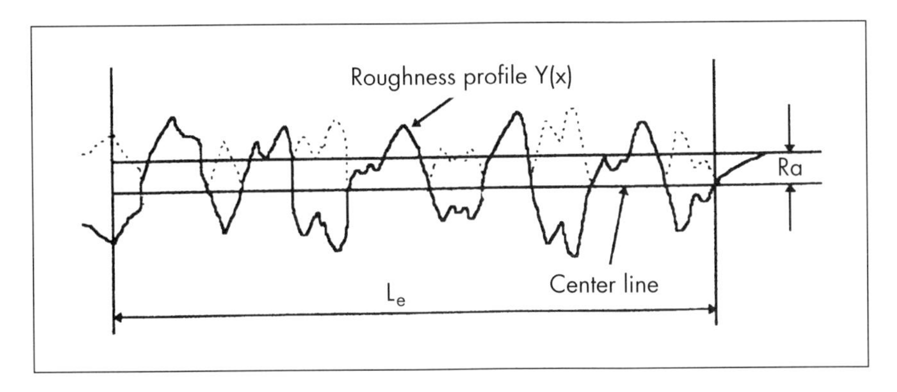

Figure 5-14. Representation of roughness average.

formation to those users and potential users who need reliable accuracy data to make either process or purchase decisions. The nature of these decisions imposes a responsibility on RP&M manufacturers to provide data that represents a realistic estimate of the tolerances that can be achieved by their system. This section describes how 3D Systems obtains and reports such numbers.

The results of an accuracy study performed in 1994 are disclosed earlier. As these results were obtained, they were shared with users and potential customers to inform them of the levels of accuracy they could now achieve with stereolithography. Most customers asked the questions that everyone should ask when given numbers presumed to represent the capabilities of a particular system. What dimensions were used to obtain this accuracy data? Why did you use the user-part? What is the accuracy of the measuring instruments? How reproducible is this result? What does Epsilon (90) represent? What is the process capability?

How Many Parts? After understanding the fundamentals of physical measurements, choosing a test part to build, and determining the capabilities of the measuring instruments, we determine how many parts must be built and measured to feel confident that results of further experiments will continue to yield numbers close to the reported value. To establish repeatability, 10 user-parts were built on the same SLA, with each test part having 170 specific dimensions measured, for a total of 1,700 measurements. This was done for both the SLA-250 and the SLA-500, resulting in an overall total of 3,400 measurements. During these tests, the only parameter that changed was literally the day the user-part was built.

Furthermore, the parts were not even built consecutively. The test was often knowingly and intentionally interrupted to build other diagnostic parts and benchmarks, as well as R&D evaluations of new software or hardware. These interruptions caused the duration of each test series to run several months. Only through extensive testing of this type can a single number convey the accuracy levels that a system can reliably achieve.

Obtaining the measurement data is not the end of the experiment. We must still answer an important and complicated question that decides the next course of action. How should we report the accuracy of an RP&M system? To answer this question, it is important to recognize the diversity of RP&M system users, as well as those executives who require accuracy information to make purchase decisions. Unfortunately, various individuals speak different languages regarding the subject of accuracy, and no single approach is likely to be fully appreciated by everyone.

The first users of RP&M technology were the visionaries, who tend to be technical and inventive. Typically, they view the new RP&M processes as leading edge technology to be further understood and improved. Their continued involvement helps to propel RP&M into the future. In order to speak their language, we must offer a statistical analysis of the data that allows them to develop conclusions about part accuracy and also determine the reliability of the reported numbers.

Nontechnical Involvement. Additionally, the results need to be expressed so that they can be understood by people who do not have a technical background. There are many people in the RP&M world that are experts in nontechnical fields, such as finance officers, corporate executives, and salesmen. They need a clear, straightforward demonstration of what the data indicates, and are usually interested in worst-case errors. Finally, since RP&M has recently evolved into an important tool for the manufacturing industry, we should also report part accuracy in language that the manufacturing community understands.[6]

To satisfy readers not involved in manufacturing, 3D Systems' R&D Department has developed a simple method to describe system accuracy using the error distribution function (EDF), the cumulative error distribution (CED), and ε(90), (read epsilon 90). These concepts are covered in Reference 4.

Describing the Accuracy

The function of the user-part is to provide a standard method of determining accuracy on RP&M systems. Therefore, it is important to have a standard method of obtaining a numerical value to describe the accuracy. The procedure adopted by 3D Systems involves graphing the 170 measurements taken from a user-part in the form of an error distribution function (EDF), which plots the frequency of the error vs. the magnitude of the error as shown in Figure 5-15. This is used to derive a cumulative error distribution (CED), which graphically displays the percentage of measurements falling within a given error versus the magnitude of the error. An ε(90) value is then calculated from the intersection of a horizontal line at 90% on the ordinate, with a smoothed version of the CED shown in Figure 5-16. ε(90) is specifically defined as that tolerance value that will encompass 90% of the measurements.

Increasing Sample Size. Each user-part returns an ε(90) value that can be used to measure part accuracy achieved by the particular RP&M system in which

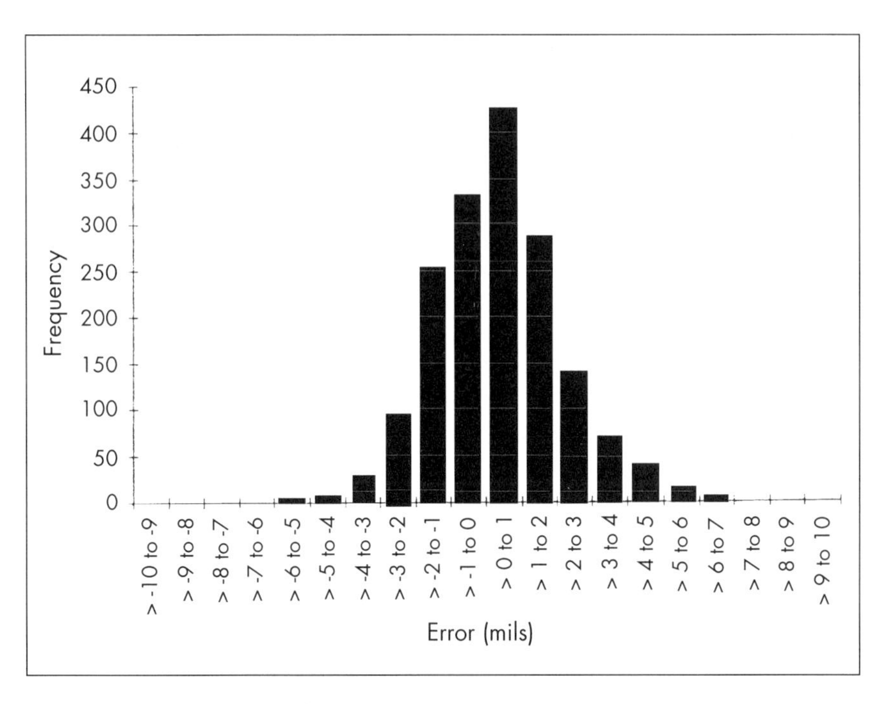

Figure 5-15. Error distribution function for 170 user-part measurements.

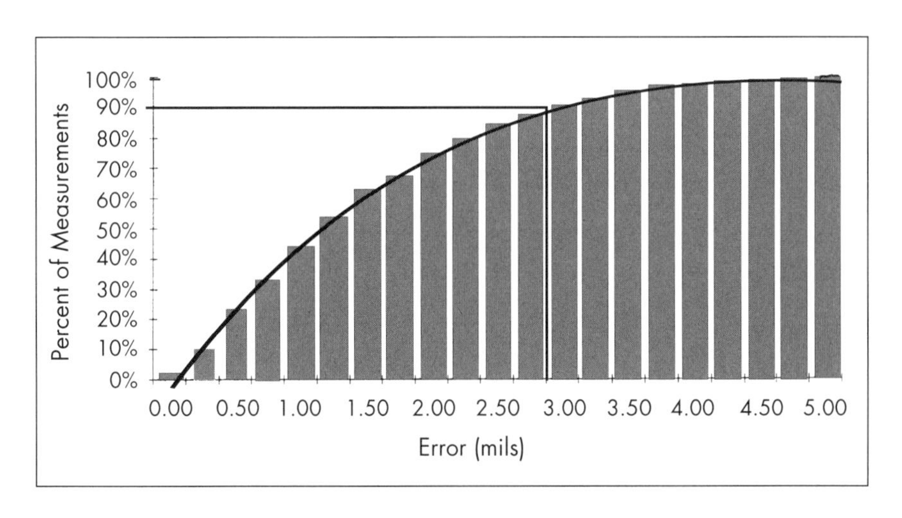

Figure 5-16. Cumulative error distribution demonstrating ε(90).

it was built. To demonstrate repeatability and also to gain confidence in the reported value of ε(90), numerous user-parts must be built and measured. Obviously, the results from 10 user-parts will be more reliable than reporting the value from one user-part. The ε(90) that we calculate, whether from one user-part or 10 user-parts, is just an estimate of some unknown, real ε(90). Our calculated value is based on only a small sample from the theoretically infinite "universe" of parts that could be built.

The question is, how good is the ε(90) sample as an estimate of the actual ε(90)? To answer this kind of question, statisticians employ a confidence interval.[7,8] That is, they describe their confidence in an estimated range. Statistically, a single point estimate of ε(90) always has a confidence of exactly zero, so we must develop a range of values centered on ε(90) to describe the reliability of the estimate. To develop a confidence interval for an estimate with a sample size less than 30, statisticians use the equation:

$$\bar{x} - c < \mu < \bar{x} + c \qquad (5\text{-}3)$$

where,

$$c = t\frac{s}{\sqrt{n}}$$

μ = the actual value for the universe
$\bar{x}$ = the mean value of the sample
s = the standard deviation of the sample
n = the number of data points in the sample
t = a value from a t distribution table, using the criteria below

The value for t is chosen based on the degrees of freedom, df = n−1, and the confidence level desired. When an interval estimate is made both above and below the sample average, i.e., Equation 5-3, a "two tail" model is used. The term "two tail" refers to the probability that the value will lie within a defined range having an upper and lower limit. When applying a confidence interval to ε(90), we are only concerned with the interval estimate greater than the sample average (i.e., a worst case) and the lower limit is ignored. Therefore, the value for t is chosen using a "one tail" model and Equation 5-3 is modified into Equation 5-4 for estimating the confidence interval for ε(90). A table of t-values can be found in appendix B of Reference 7.

$$\varepsilon(90) < \bar{x} + c \qquad (5\text{-}4)$$

As an example of how increasing the sample size gives a smaller confidence interval and thus greater reliability, we can use Equation 5-4 to calculate ε(90) using a sample of 10 user-parts versus using a sample of only 3 user-parts. It should be noted that experiments using only one user-part cannot report a value with any confidence interval, indicating extremely low reliability in that value.

The t-value for the examples in Table 5-2 was chosen using a 97.5% confidence level.

It is obvious that increasing the number of samples improves the reliability of the estimate. Another important factor is the sample standard deviation. In Equation 5-3, c is proportional to s. Therefore, a less repeatable experiment will decrease the reliability of the estimate.

Using the ε(90) results from 10 user-parts allows us to make the following statement about the accuracy of stereolithography. There is a 97.5% probability that at least 90% of the user-part dimensions will be within 3.13 mils of their intended CAD value. Although this statement may seem awkward, it allows the reader to distinguish between values that are reliable enough to form the basis of important decisions, and values that have little statistical significance. With money and careers depending on these decisions, RP&M users must demand reliable accuracy data and understand how it was derived.

Table 5-2. Increased Sample Size = Smaller Confidence Interval = Greater Reliability

n (number of samples)	$\bar{x}$ (mils)	s (mils)	t @ 97.5% confidence	$c = t\frac{s}{\sqrt{n}}$ (mils)	ε(90) $\varepsilon(90) < \bar{x} + c$ (mils)
3	2.83	0.404	4.30	1.003	ε(90) < 3.83
10	2.85	0.387	2.26	0.277	ε(90) < 3.13

Process Capability

Eliminate Systematic Errors. If stereolithography parts are going to be incorporated into a manufacturing process, they will require specifications that include tolerances. In industry, machines and processes are characterized by the tolerances that they can maintain. It is common, if not routine, for manufacturers to perform process capability studies on new equipment or processes. Given the number of measurements taken and the nature of the user-part, it is possible to extract enough significant data from the accuracy studies to give a reasonable estimate of the process capability of stereolithography. The first step in performing a process capability study is to eliminate all systematic errors so that the data is limited to random error and can therefore be considered Gaussian. Histograms of the user-part studies reveal EDF curves that allow us to make this assumption with reasonable confidence (Figure 5-15).

First we must understand the difference between the capability study using the user-part data and a typical process capability study. In the latter, a minimum of 30 parts are built to reveal any systematic errors and gain statistical confidence. Then, one measurement of a particular dimension to be studied is made on each part. These measurements are then compared to the required tolerance for that dimension to determine the capability of reproducing the part within specification. In the case of the user-part studies performed at 3D Systems, there are only 10 samples.

Nonetheless, the study of a particular dimension can provide statistically sound data because each user-part involves the intentional repetition of several measurements of similar features. For example, the 9.5 inch (241 mm) nominal dimension is measured a total of eight times on each part. This provides 80 independent measurements of that particular dimension. As an example, we will examine the process capability of stereolithography as a function of dimension length. Nominal dimensions of 0.125, 1.000, 4.750, and 9.500 inches (3.175, 25.4, 120.65, and 241.3 mm) will be evaluated for process capability.

The Capability Ratio. In order to understand the concept of process capability, let us first look at the "capability ratio." This is simply defined as six times the process standard deviation divided by the specified tolerance spread. The tolerance spread is the maximum allowable dimension minus the minimum allowable dimension.

$$Cr = \frac{6 \times \sigma}{\text{tolerance spread}} \quad (5\text{-}5)$$

The Cr ratio is a measure of how much of the tolerance range is used up by the process variation. As an example, the current SL 5170 user-part study revealed that 140 independent measurements of a 1.000-inch (25.4 mm) dimension provided a standard deviation of 0.00113 inch (0.03 mm) with a mean of 1.00038 inches (25.4 mm). If we arbitrarily choose a tolerance spread of 0.008 inch (0.2 mm) ($\pm$0.004 inch [$\pm$0.1 mm]), then Equation 5-5 yields Cr = 0.848.

Figure 5-17 is a graphic representation of this situation where the process variation falls within the specification limits. If we convert the Cr to a percentage by multiplying by 100, then we can describe the capability ratio as the percentage of tolerance "used up" by the process variation. A process is considered capable when Cr $\leq$ 100%. A Cr of 84.8% indicates that stereolithography is indeed capable of holding $\pm$0.004-inch ($\pm$0.10 mm) tolerances on 1-inch (25.4 mm) dimensions.

The manufacturing industry appears to be more familiar with the Cp ratio. This is simply the inverse of the Cr ratio.

$$Cp = \frac{\text{tolerance spread}}{6 \times \sigma} \quad (5\text{-}6)$$

Here, the numbers get larger as the process becomes more capable. The minimum requirement for a Cp ratio is 1.00, which equates to 99.7% probability ($\pm 3\sigma$)

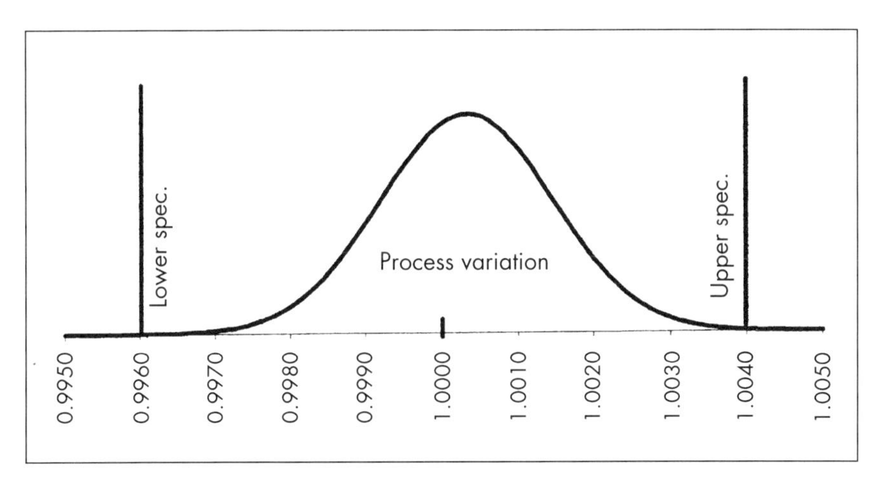

Figure 5-17. Process variation within specification limits.

that the measurement will fall within the specified tolerances. For the case just cited, Cp = 1.18. A Cp ratio of 1.33 corresponds to a failure rate of 1 measurement per million.

The Cr and Cp ratios only tell if the process is capable of holding tolerances. They do not tell whether the system is actually producing parts *within* the specified tolerances. Figure 5-18 represents a situation where the Cp ratio is equal to 1.179. This is considered a capable process, yet almost all of the measurements are out of tolerance.

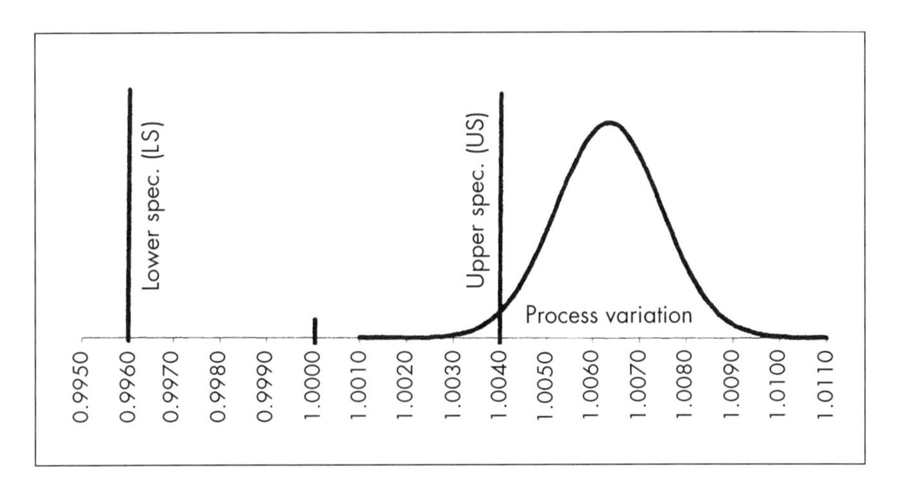

Figure 5-18. Process variation out-of-specification limits.

The ideal process capability index would indicate if the process is capable of holding tolerances, and whether the process variation is centered on or near the target value. The manufacturing industry uses the Cpk index to provide this information. The Cpk index is determined through the use of Equation 5-7. Here one calculates the difference between the mean and the closer of either the upper or lower specification (US or LS), depending on which side of the target value the mean happens to be located.

$$\text{Cpk} = \text{the minimum of} \left[\frac{\text{US} - \bar{x}}{3\sigma}, \frac{\bar{x} - \text{LS}}{3\sigma} \right] \quad (5\text{-}7)$$

The same industry standards used for acceptable Cp ratios apply to the Cpk index. A value greater than or equal to 1.000 indicates that the process is capable of holding the specified tolerances. A Cpk index that is less than 1.000 indicates an incapable process by industry standards. When applying the Cpk index to the user-part study, we do not have a pre-established tolerance range. Therefore, the Cpk indices were calculated for each of the reported dimensions at tolerance ranges of ±0.002 to ±0.009 inches (±0.05 to ±0.23 mm) (Table 5-3).

Using this table, we are able to make reliable statements about the process capability of stereolithography in the X and Y axes. For each specific dimension we can locate the lowest tolerance range where the Cpk is greater than 1.000. In this range, the Cpk index establishes that the stereolithography process is capable of maintaining the stated tolerance for those dimensions on the user-part.

For example, 40 measurements were made on a 4.750-inch (120.65-mm) dimension that resulted in a Cpk equal to 1.100 with tolerances of ±0.004 inches (±0.1016 mm). This equates to greater than 99.7% probability that the measurement of a 4.750-inch dimension will be within ±0.004 inches. Although the tolerances for the smaller dimensions may seem higher than expected, it is important to note that they were studied in the context of a 9.500-inch (241.3-mm) part.

Reporting the accuracy of an RP&M system requires a responsibility to provide values that represent the capabilities of the system. These values must be supported with a significant amount of "real" data. It is also important to agree upon a standard test part for both determining and reporting accuracy values. Finally, since RP&M is becoming a tool for the manufacturing industry, reporting accuracy using the Cpk index provides statistical process control personnel with data that they can readily comprehend.

However, the greatest responsibility lies with the person who uses the accuracy data for making decisions. Asking to see the experimental data, understanding the method used to obtain accuracy values, and requiring confidence intervals to determine the reliability of the information are necessary steps that must be taken to avoid serious misconceptions about the accuracy of a particular system.

Table 5-3.
Cpk Indices Calculated for Reported Dimensions at Tolerance Ranges of ±0.002 to ±0.009 inches

Nominal (inches)	n	$\bar{x}$	σ	Cpk ±0.002	Cpk ±0.003	Cpk ±0.004	Cpk ±0.005	Cpk ±0.006	Cpk ±0.007	Cpk ±0.008	Cpk ±0.009
0.125	160	0.1261	0.0008	0.375	0.792	1.208	1.625	2.042	2.458	2.875	3.292
1.000	140	0.0003	0.0011	0.515	0.818	1.121	1.424	1.727	2.030	2.333	2.636
4.750	40	4.7493	0.0010	0.433	0.767	1.100	1.433	1.767	2.100	2.433	2.767
9.500	80	9.5011	0.0018	0.167	0.352	0.537	0.722	0.907	1.093	1.278	1.463

3D Accuracy Study Results

Results of the accuracy study described herein involved thousands of man hours and hundreds of thousands of dollars in equipment and resin. Twenty user-parts were built and measured, as well as a matrix of surface finish diagnostic parts. The raw data is available for examination and the methods used to obtain it have been disclosed. All of the numbers have confidence intervals that declare their statistical reliability and demonstrate their level of reproducibility.

Furthermore, since nothing unusual, other than attention to detail, careful calibration of the SLA system used for the measurements, understanding the fundamental characteristics of the SL process, patience, and persistence were used on this test program, every SL user should be able to attain the results that are disclosed in this section.

It is also important to recognize that these numbers represent user-part accuracy levels achieved during a test program conducted in 1994. Another look at the SL accuracy progress chart, Figure 5-1, indicates a trend fueled by a sustained commitment to build more accurate parts. This effort has been rewarded with the continuing infusion of SL into new applications as discussed in Chapters 10-12. It is also important to recognize that this expansion provides a growing intellectual pool of talented, creative, and diverse people responsible for many of the advances that have been discussed.

Epsilon (90)

Table 5-4 reports the results of the 1994 epoxy resin user-part accuracy study intended to establish $\varepsilon(90)$ with a 97.5% confidence interval.

Table 5-4. Results of 1994 Epoxy Resin User-part Accuracy Study

Machine Type	Sample Size	Experiment Duration	Confidence Level	User-part Accuracy $\varepsilon(90)$ (mils)
SLA-250	10	2/5-5/15/94	97.5%	$\varepsilon(90) < 3.13$
SLA-500	10	3/25-5/25/94	97.5%	$\varepsilon(90) < 4.05$

Cpk Index

Table 5-5 lists the tolerances where the stereolithography process is capable. These values correspond to those cases where Cpk ≥ 1.000, for the indicated nominal dimensions. The numbers listed in Table 5-5 were derived from the data obtained in the user-part accuracy studies discussed previously. It is important to note that the nominal dimensions reported were studied collectively, NOT indi-

Table 5-5. Tolerence where Stereolithography is Capable

Nominal (inches)		0.125	1.000	4.750	9.500
Sample Size		160	140	40	80
SLA-250	Tolerance (inches) Cpk	±0.004 1.138	±0.004 1.084	±0.004 1.042	±0.007 1.083
SLA-500	Tolerance (inches) Cpk	±0.004 1.000	±0.005 1.083	±0.006 1.176	±0.009 1.067

vidually. These numbers represent the results obtained when examining the process over the full range of measurements.

Surface Finish

Initial measurements involving surface roughness on a series of ramped stereolithography parts revealed many variables significantly affecting the surface quality as well as the readings obtained with a surface profilometer. Before the surface quality of SL parts can be quantified, these factors must be studied and understood. Furthermore, surface quality can be improved with reduced layer thickness, adjusted recoating parameters, meniscus smoothing, and special cleaning methods. For example, QuickCast patterns are wiped clean, leaving a film of liquid resin that fills in the "stair-stepping" to various degrees. Fortunately, this reduces the effects of stair-stepping. Unfortunately, the reduction is difficult to quantify because variable amounts of the liquid resin may be removed.

One measurement, however, does produce consistent and reliable results. The surface roughness values for flat upfacing surfaces on parts produced in QuickCast and ACES on both SLA-250s and SLA-500s repeatedly measure Ra $\leq$15 microinches.

Future Advances in Accuracy

The current ability to achieve user-part RMS errors less than 2 mils is the result of 1) producing accurate build files that compensate for the cured linewidth and resin shrinkage, 2) correctly positioning the laser beam on a precisely leveled liquid surface, and 3) building with a low distortion build material using an optimized build style. These technological advances have brought stereolithography to the point where we must now address four remaining sources known to cause small, but systematic, errors. They are known as stair-stepping, quantization error, overcure error and print-through. "Meniscus smoothing" uses a combination of recoating and part drawing techniques to reduce the effects of stair-stepping on the upfacing sloped surfaces. The other three sources of error are unique to the Z axis and are addressed in the technique entitled "simultaneous multiple layer curing" (SMLC) (patent pending). Definitions of these errors and explanations of their proposed solutions will be discussed in this section.

Meniscus Smoothing

An obvious source of RP&M inaccuracy is "stair-stepping", which results in the rough texture apparent on sloped and curved surfaces due to building with finite layer thicknesses. All layer-by-layer building processes exhibit this characteristic, which is similar to the effect of cutting a sloped wall with a milling machine.

In May 1993, U.S. patent no. 5,209,878 was granted for "Discontinuity Reduction."[9] This procedure involves laser curing the liquid meniscus, which is naturally formed in the individual stair steps during the build process. The method is known as "active meniscus smoothing," as illustrated in Figure 5-19. First a layer, n, is drawn. After the platform is ultimately lowered one layer thickness, the system draws the next layer, n+1. If there is sufficient offset that skin fills are required, the machine performs step 3. This step entails raising the n+1 layer above the free resin level and then drawing the border and skin fill appropriate for the previous layer, n. This cures the naturally formed liquid meniscus present in the internal corners. This method was used to build the surface roughness test part on an SLA-250 with SL 5170 at 0.006-inch (0.15-mm) layer thickness. A comparison with a similar part built without meniscus smoothing demonstrated a roughly 90% reduction in surface roughness for a 45° upfacing surface.

Due to the trend toward building with smaller layer thicknesses and the additional recoating time required to perform this operation, "active meniscus smoothing" has not yet been implemented into the present software. In the meantime, techniques such as "passive meniscus smoothing," conservative part building parameters, as well as material removal processes, can greatly improve surface quality.

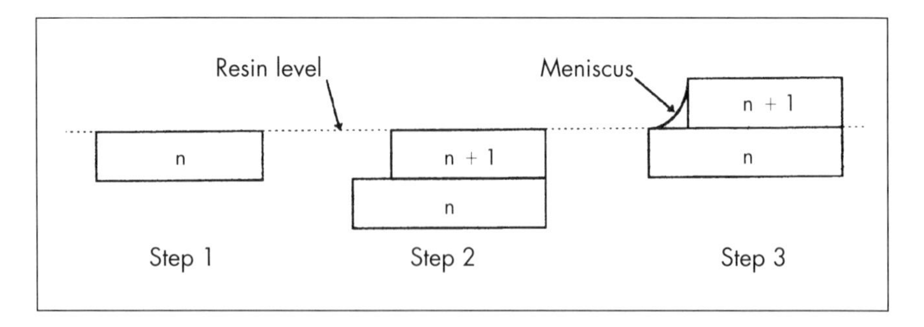

Figure 5-19. Meniscus smoothing diagram.

Simultaneous Multiple Layer Curing

SMLC is a set of building techniques involving patented software algorithms that address the known SL errors that are currently present in Z axis dimensions.[10] The technique used involves modifying the slice data in order to produce build files that will compensate for "quantization error," "overcure error," and "print through." In order to convey the concept of SMLC, it is important to understand the causes of the errors that it corrects.

Quantization Error. Simply stated, there will be error in the Z axis if any Z dimension is not evenly divisible by the layer thickness. Imagine attempting to construct a 1.002-inch (25.451-mm) vertical dimension with 0.005-inch (0.127-mm) layers. The part will be either 1.000 or 1.005 inches (25.4 or 25.53 mm) high depending on the software algorithm used to slice the part. A possible 0.003 inch (0.076 mm) error can be reduced to less than 0.001 inch (0.0254 mm) by slicing the part in 0.001-inch (0.0254-mm) layers, and using the SMLC software algorithms to decide when resin should be cured on those levels that are not evenly divisible by the build layer thickness.

Minimum Recoating Depth. Unfortunately, the process is not as simple as just curing resin where it is needed. Due to the surface tension properties of SL resins, and the associated viscous fluid dynamics, there is a limitation on the minimum uniform layer of liquid resin that can be formed on a substrate of cured resin. This is known as the minimum recoating depth (MRD). At present, the MRD is 0.006 inch (0.15 mm) for ACES and 0.004 inch (.10 mm) for QuickCast when using the epoxy resins.

If a part has a nominal Z axis dimension of 1.002 inches (25.451 mm) in a particular region, then building the part in five mil layers to the 1.000-inch (25.4-mm) level would require recoating that region with a final 0.002-inch (0.051-mm) layer of liquid resin. At present, this cannot be done, since 0.002 inch (0.051 mm) <MRD. However, SMLC employs an algorithm that intentionally avoids curing the portions of the previous layer, or layers, which lie below this particular area, in order to increase the recoating thickness at that location to a value greater than the MRD. As an example, such a feature could be built with SL5170 in QuickCast by utilizing 199 layers at 0.005 inch (0.127 mm), and one final 0.007-inch (0.178-mm) layer, since both 0.005-inch (0.127-mm) and 0.007-inch (0.178-mm) layer thicknesses are greater than the MRD value.

Overcure Error and Print Through

Overcure error and "print through" are the result of curing downfacing features to a depth greater than one layer thickness. Imagine drawing the first 0.005-inch (0.127-mm) layer of a part with a skin fill cure depth CD_f = 0.010 inch (0.254 mm), a hatch overcure of 0.006 inch (0.152 mm) (CD_h = 0.011 inch [0.28 mm]) and a border overcure of 0.009 inch (0.229 mm) (CD_b = 0.014 inch [0.356 mm]). Overcure errors are those errors associated with cure depths greater than the layer thickness. In this case, an excess of 0.005 inch (0.127

mm) in skin fill areas, 0.006 inch (0.152 mm) in hatch areas, and 0.009 inch (0.228 mm) at the borders.

"Print through" is the result of additional exposure that a downfacing feature may receive from layers lying above that feature. Layers that are drawn subsequent to the downfacing layer add exposure, which causes the liquid resin below the downfacing layer, previously partially exposed, to reach or exceed E_c. The expected 0.009 inch (0.229 mm) border overcure error is subsequently further increased by some finite amount, usually less than 0.003 inch (0.076 mm), due to the optical transmission characteristics of the downfacing skin. The solution to minimizing these errors also lies in the SMLC algorithms. Again, modified SLICE data is used which delays the curing of selected portions of some layers. The layers are then drawn after the MRD has been attained. The exposure is subsequently adjusted to achieve the required cure depth.

Eliminating systematic errors is a continuous effort at 3D Systems. The current system is now capable of achieving RMS errors on the user-part of less than 0.002 inch (0.051 mm) in the X and Y dimensions, which is sufficient for the majority of the current practical applications. Engineers understand that holding tight tolerances costs money and they generally make an effort to design components that do not require overly exacting specifications. Additionally, many RP&M applications simply involve form studies or concept models. Furthermore, the Z errors can often be compensated in the postprocessing stage with files and sandpaper. As with all manufacturing processes, there will always be some degree of error involved. However, we believe a continuous effort to minimize part errors will open additional manufacturing applications.

Summary

As efforts to increase the accuracy capabilities of stereolithography continue, careful balance must be maintained with other machine characteristics in order to advance the overall RP&M market. Material properties, throughput of the machine, build envelope, cost of the machine, reliability, and ease of use must also be considered. However, as RP&M evolves into a necessary step in the manufacturing cycle, there will be markets that demand still more accurate parts. For instance, the concept of rapid tooling, which will turn CAD designs into production injection-molded plastic parts in a matter of weeks, relies on tight tolerances and smooth surface finishes. Photopolymer laser printing most likely has an ultimate resolution limit at the sub-micron level.[11] Ingenuity, science, and dedication will help us approach that level. For this journey, the ability to determine part accuracy will tell us, and the manufacturing world, where we are at any given time.

References

1. Pang, T.H., "StereoLithography Epoxy Resins SL 5170 and SL 5180: Accuracy, Dimensional Stability, and Mechanical Properties." Proceedings of

the Solid Freeform Fabrication Symposium, University of Texas, Austin, TX, August 8-10, 1994.

2. Blake, P., Baumgardner, O., Haburay, L., and Jacobs, P.F., "Creating Complex Precision Metal Parts Using QuickCast." Proceedings of the SME Rapid Prototyping and Manufacturing Conference, Dearborn, MI, April 26-28, 1994.
3. Nguyen, H., Richter, J., and Jacobs, P., Chapter 10, *Rapid Prototyping and Manufacturing: Fundamentals of StereoLithography*. SME, Dearborn, MI, 1992, pp. 249-285.
4. Richter, J. and Jacobs, P., Chapter 11, *Rapid Prototyping and Manufacturing: Fundamentals of StereoLithography*. SME, Dearborn, MI, 1992, pp. 287-315.
5. Taylor, J.R., *An Introduction to Error Analysis: The Study of Uncertainties in Physical Measurements*. University Science Books, Mill Valley, CA, 1982.
6. Lockard, M., 3D Systems, Inc., private communication, May-July, 1994.
7. Clements, R.B., *Handbook of Statistical Methods in Manufacturing*. Prentice Hall, New Jersey, 1991, pp. 29-68.
8. Juran, J.M., (ed.), *Quality Control Handbook*, 4th ed. New York, McGraw-Hill Book Co., 1988, pp. 16.14-16.47.
9. Smalley, D., et. al., "Discontinuity Reduction." U.S. Patent no. 5,209,878, May 11, 1993.
10. Smalley, D., et.al., "Simultaneous Multiple Layer Curing." U.S. Patent no. 5,192,469, March 9, 1993.
11. Hull, C.W., "Rapid Prototyping in the 1990's." Second International Conference on Rapid Prototyping, University of Dayton, Dayton, OH, Conference Proceedings, June 23- 26, 1991, pp. 211-218.

CHAPTER 6

The Development of QuickCast

Every man's work shall be made manifest:
for the great day shall show it,
because it shall be revealed by fire;
and the fire shall try every man's work.

1 Corinthians 2:13

Functional metal prototypes are often required in numerous industrial applications. These components are typically needed in the early stages of a project to determine form, fit, and function. Until 1993, RP&M techniques such as SL could assist with the first two items, but the latter was accomplished only with considerable difficulty.[1,2]

If the component in question was ultimately to be made of metal and only one part was required, it would typically be CNC machined. However, if four or five prototypes were needed for destructive and/or nondestructive testing, then the fabrication of tooling to injection mold wax patterns for investment casting probably would be necessary, or patterns and core boxes for sand casting might suffice if the tolerances were less stringent. In all three situations, the time and expense required could be considerable.

Under these circumstances it is difficult to amortize the tooling cost over such a limited number of parts. And these difficulties could be compounded if proper operation of the final component has not yet been established. In such situations redesign is commonplace and, unfortunately, tooling rework is more the rule than the exception.

Nonetheless, this continues to be a fact of life on a great many industrial projects. Budgets are generally tight and contractual deadlines are absolutely critical. In

By Paul F. Jacobs, Ph.D., Director of Research and Development, 3D Systems Corporation, Valencia, California.

this scenario there is little room for problems, iteration, and rework. The margin for error is slim, and a serious difficulty with even a single component in a complex system can lead to significant time delays, budget overruns, and in extreme cases, project termination.

However, a new technique is now available to all designers and engineers faced with this dilemma. Where small quantities of functional metal prototypes are needed, but the time and money for tooling is difficult to justify, a new method called QuickCast is now available to generate quasi-hollow patterns. No tooling whatsoever is needed. The QuickCast pattern is invested with conventional slurry to form a ceramic shell coating. The pattern is then eliminated by means of flash firing, and molten metal is then cast in the ceramic shell. The result is an accurate, metallurgically sound, functional prototype rapidly available in a wide array of alloys.

With this new method, neither expensive nor time consuming CNC machining or tooling is required. As a consequence, at the time of this printing, over 5,000 QuickCast patterns have been successfully converted into precision shell investment castings. Simply stated, one can now proceed from a CAD model, to a QuickCast pattern, to a functional metal prototype ready for test and evaluation, within three weeks. In numerous cases involving various SL users in the aerospace, automotive, medical, and consumer product fields, it has been documented[3-6] that these organizations were able to successfully accomplish form, fit, and functional testing for critical components within one month.

If the resulting investment cast metal part is qualified successfully, substantial time and money are saved. However, even where the part fails due to some unforeseen functional or structural shortcoming, the CAD design can be modified, a new QuickCast pattern generated, and a second iteration metal part subsequently investment cast. What is truly remarkable is that even in this otherwise problematic case, a second iteration functional metal component can be available more rapidly using QuickCast than the first flawed part can be obtained with conventional methods.

However, by using RP&M, the designer or engineer has already been made aware of a potentially critical problem, and can take specific action to rectify the design deficiency. While nobody likes to admit they make mistakes, it is far less expensive to detect an error early in a design, as opposed to catching it after significant funds have been spent for tooling. The advantages of the new method with respect to saving time, saving money, and improving final component quality can be very significant.

For these advantages to be fully realized, each of the following key items is necessary:

1. The entire QuickCast process must be reliable and repeatable.
2. The patterns must be sufficiently accurate to accommodate normal investment casting tolerances.
3. The process must achieve sufficiently high foundry yield that the potential economic advantages can truly be realized.

4. The resulting castings must be metallurgically sound.
5. All the various steps required, including CAD modeling, pattern building, resin draining, part cleaning, UV postcuring, testing for and sealing all holes, shipping the pattern to the foundry, wax gating, shell investing, pattern burn-out, ash removal, metal pouring, casting cool down, ceramic shell and gating removal, inspection testing, and shipment back to the user must be sufficiently rapid that the net overall time savings are real.

6.1 Fundamentals of Resin Pattern Thermal Expansion

Early in the development of RP&M technology, it was recognized that numerous users wanted the ability to proceed directly from a model to a functional metal prototype. While various secondary techniques are possible, including precision sand casting and plaster solid mold casting, the method of choice for the generation of a limited number of accurate, geometrically complex, metallurgically superior, functional parts is definitely shell investment casting.[4]

Melting the Wax

However, the shell investment casting process requires complete elimination of the pattern without damaging the shell. Traditionally, this was accomplished by melting wax patterns out of some type of ceramic mold material. The nature of the process accounts for the well-known "lost wax" title given to this method of metal casting used by the Egyptians to create jewelry over 4,000 years ago. For many centuries the wax patterns were simply melted out of the mold with heat from a fire.

Autoclave Process. Partially due to increased environmental concerns, as well as the ready availability of steam boilers, the use of saturated steam in autoclaves was widely adopted about 35 to 40 years ago. An additional advantage of using high-temperature steam is that it melts the wax pattern from the outside inward. This helps reduce stresses on the relatively thin ceramic investment shell. However, disadvantages of the autoclave process are:

1. Moisture from the condensing steam alters the composition of the melted wax, which must then be collected and forwarded to a special processing facility to enable reuse by the foundry.
2. For a large foundry using many tons of wax per year, reclaiming and processing the wax can be rather expensive.
3. There are definite and very real personnel safety hazards associated with the use of saturated steam at roughly 300 to 350° F (150 to 175° C) and pressures of 4 to 9 atmospheres. The autoclaves must be periodically inspected and certified to guard against serious injury.

Even with almost four decades of experience, the industry-wide yield of high quality, micro-fissure-free ceramic investment shells subsequent to autoclave elimination of the wax pattern and gating system is still only about 85%. Of course, this varies somewhat from foundry to foundry, and is dependent on the type of shell system used (raw silica, fused silica, alumina, zirconia, or various hybrid combinations thereof). While autoclave is, by far, the most common means for pattern removal, it is not without its own problems, and is hardly beloved by foundrymen.

Ceramic Shell Stresses. During the period from 1990 through 1992, a number of SL users recognized the potential economic and schedule benefits that might be achieved by proceeding directly from an RP&M pattern to a functional metal casting. Consequently, they initiated various experiments with solid WEAVE or STAR-WEAVE patterns. The results were not encouraging. Time after time, the ceramic investment shell failed catastrophically during attempted autoclave wax gating removal. The fundamental problem with using a solid SL object as a pattern for direct shell investment casting lies with the stresses generated in the ceramic shell. These stresses result from the thermal expansion of the pattern during the autoclave heating process used to remove the associated wax gating system.

The following is a simplified mathematical model of the physical processes at work inside the ceramic shell. Consider a section of a solid cylindrical SL specimen having a radius R, surrounded by an annular ceramic investment shell with a uniform thickness a. Figure 6-1 shows a schematic cross-section where the neutral axis of both the specimen and shell is at N-N, and the shell/pattern interface is at S-S. The linear coefficient of thermal expansion (CTE) of the resin used to build the SL pattern is designated α_r, having the units of mm/mm -°C. The pattern and shell are both simultaneously raised from ambient temperature, T_a, to an elevated temperature, T, where $\Delta T \equiv T - T_a$. The resin pattern will expand about N-N, with the displacement at S-S, toward the left, being given by

$$\Delta R = \alpha_r R \Delta T \tag{6-1}$$

However, at the same time, the shell material is also expanding based upon its CTE, α_s. The annular shell undergoes thermal expansion about N-N. The motion at S-S is also toward the left, since the inner diameter of a cylinder increases when heated. Shell expansion is calculated as if the entire cylinder were made of ceramic material, thus

$$\Delta a = \alpha_s R \Delta T \tag{6-2}$$

The net shell strain, ε_s, at section S-S, due to the combined effects of pattern and shell expansion, is given by subtracting equation 6-2 from equation 6-1, and then dividing by the shell thickness, a.

$$\varepsilon_s = [\Delta R - \Delta a] / a = (\alpha_r - \alpha_s) R \Delta T / a \tag{6-3}$$

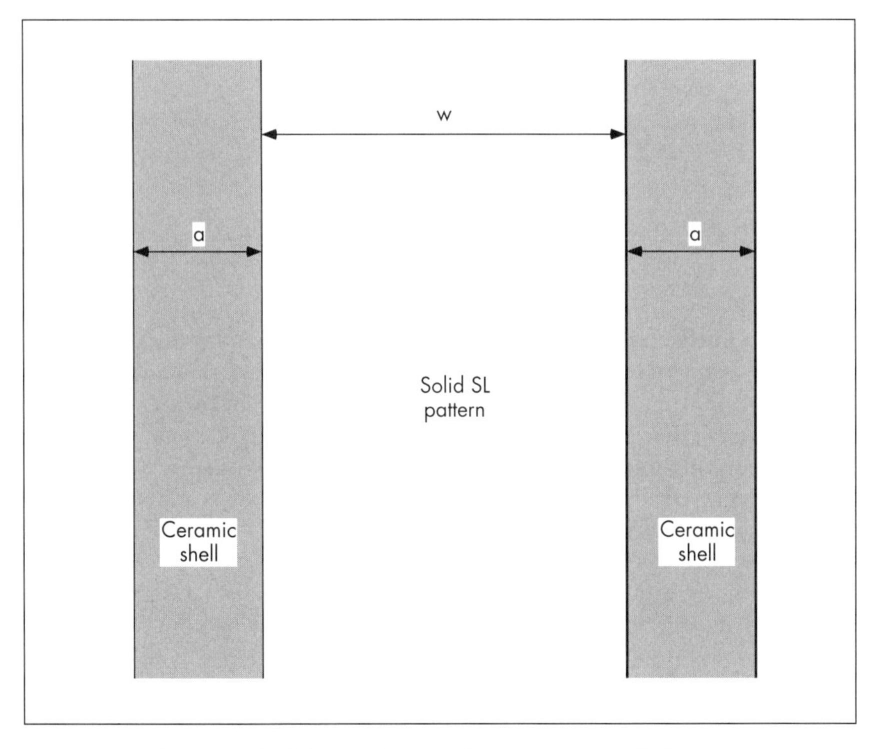

Figure 6-1. Schematic of solid stereolithography resin pattern shell investment casting geometry.

Assuming the ceramic material obeys Hooke's Law, then the shell stress is given by:

$$S_s = Y_s \varepsilon_s = Y_s (\alpha_r - \alpha_s) R \ \Delta T/a \qquad (6\text{-}4)$$

where Y_s is the Young's modulus of the shell material. Note that if the CTE of the polymerized resin exactly matches the CTE of the ceramic shell, then the stress on the shell vanishes. This is reasonable since the pattern and the shell would then be expanding together as if they were effectively the same material.

If the CTE of the resin exceeds the CTE of the ceramic material, then finite shell stress will exist for $R > 0$, $\Delta T > 0$ and $Y_s > 0$, unless the shell thickness, a , approaches infinity. Interestingly, the latter limit, far from being absurd, is the practical situation for the case of solid mold, or flask investment casting. In fact, flask casting was the original method used by Texas Instruments[2,7] to convert solid SL patterns to metal.

Provided that $S_s < S_{crit}$, the critical breaking stress of the ceramic material, the shell will remain intact. However, whenever $S_s > S_{crit}$, shell failure will occur. The critical case for imminent shell cracking occurs whenever the thickness of the solid resin section is equal to the "critical section thickness", W_{crit}. Thus, whenever $R = W_{crit}$ then $S_s = S_{crit}$.

Substituting into equation 6-4 and rearranging, we obtain the following important relationship, in dimensionless form:

$$\frac{W_{crit}}{a} = \frac{S_{crit}}{Y_S(\alpha_r - \alpha_S)\,\Delta T} \qquad (6\text{-}5)$$

Thicker Shells. Equation 6-5 indicates that the critical section thickness, W_{crit}, of a solid SL pattern is proportional to: the shell thickness and ratio of the critical stress divided by the Young's modulus of the shell. It is inversely proportional to the CTE mismatch between the resin pattern material and the ceramic shell material. Consequently, thicker shells will enable thicker solid pattern sections. Unfortunately, thicker shells require additional investment coatings, which increases total shelling time, increases cost, and may adversely affect the final metallurgy of the casting by retarding the cooling rate.

Shell systems with greater critical stress levels will enable larger W_{crit} values for solid resin patterns. More importantly, the maximum allowable thickness of any undrained and subsequently postcured QuickCast section will also increase. This explains why alumina based shell systems are more forgiving of incompletely drained QuickCast patterns than fused silica shell systems, while raw silica systems are the least forgiving.

Finally, since $\alpha_r >> \alpha_s$ for all SL photopolymer resins and every known shell system, it is clear that some finite value of critical section thickness, W_{crit}, will occur in all cases. To estimate this value, let us consider the case of a pattern made from epoxy resin SL 5170, invested in a raw silica based shell system, and undergoing wax gating removal in an autoclave. Note that we are only removing the wax gating, and have not yet even begun to eliminate the QuickCast pattern. Figure 6-2 shows test data for the CTE of resin SL 5170. Since autoclaving takes the resin from about 75° F to about 330° F (25° C to about 165° C), the average value of $\alpha_r \approx 90 \times 10^{-6}$ mm/mm -°C. The relevant values are tabulated below.

α_r	$\approx 90 \times 10^{-6}$ mm/mm -°C	(data from Figure 6-2)
α_s	$\approx$ 7 × 10-6 mm/mm -°C	(typical of raw silica)
ΔT	= 165° C - 25° C = 140° C	(typical of autoclave wax removal)
S_{crit}	$\approx$ 30 MPa	(3D test data for a raw silica shell)
Y_s	$\approx$ 8200 MPa	(3D test data for a raw silica shell)
a	= 6 mm	(typical shell thickness)

Substituting these values we obtain the result:

$$W_{crit} = 6 \text{ mm } [30/8200] / (83 \times 10^{-6} \times 140) = 1.89 \text{ mm}$$

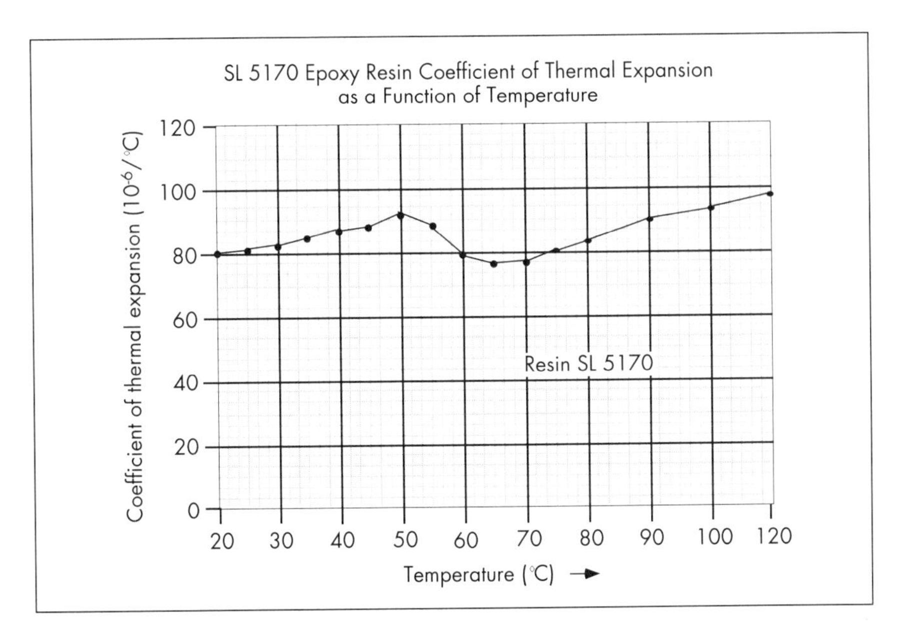

Figure 6-2. Test data for the CTE of resin SL 5170.

Thus, equation 6-5 predicts that for the case of a raw silica shell system 0.24 inches (6 mm) thick, undergoing autoclave wax gating removal at 330° F (165° C), the critical section thickness for any solid pattern built using SL 5170 resin is about 0.075 inch (1.91 mm). Therefore, any pattern having solid, postcured, residual resin sections thicker than 0.075 inch (1.91 mm) will probably lead to raw silica investment shell cracking. Critical section thicknesses for other shell systems, such as those based upon fused silica, alumina, or zirconia, are only slightly larger, in the range of 0.090 to 0.120 inch (2.29 to 3.05 mm).

Furthermore, the quantity $W_{crit}/a \approx 1/3$ or equivalently, $a \approx 3\ W_{crit}$, so increasing shell thickness to avoid ceramic shell cracking when using solid SL patterns is futile, since one would have to increase it to three times the thickness of the largest solid resin section. As an example, if a solid pattern happened to have a maximum section thickness of 2 inches (50 mm), the shell would have to be about 6-inches (150-mm) thick to avoid cracking. Clearly, this is no longer a shell, but rather essentially a solid mold. This result probably explains the cause of numerous early QuickCast failures where shell cracking occurred in autoclaving. Shell failure was almost certainly due to incomplete resin drainage, which left fully cured solid resin sections thicker than 0.08 inch (2 mm) after postcure. These solid resin sections led to excessive shell stress during autoclave heating and, thus, shell fracture.

Solid Pattern Incompatibility. Clearly, solid resin patterns of reasonable size are not compatible with standard shell investment casting procedures. This was confirmed in Reference 8, where solid test coupons produced on an SLA-250 using resin SL 5170, and having thickness values of 1, 2, 3, 4, and 5 mm (± 0.15 mm), were invested in a raw silica shell system about 0.25 inch (6.4 mm thick). The samples were then exposed to standard autoclave conditions. The 0.04-inch (1-mm) thick coupon exhibited no visible signs of shell cracking. The 0.08-inch (2-mm) thick coupons caused an occasional extremely fine micro-fissure to appear in the shell. The 0.12-inch (3-mm) thick coupons caused small but distinct shell cracks, the 0.16-inch (4 mm) coupons caused multiple major shell cracks, and the 0.2-inch (5-mm) thick coupons literally caused the shell to explode. While equation (6-5) is only a simple phenomenological model, it does seem to predict the critical section thickness for raw silica shell systems reasonably well, based upon comparison with available, albeit limited, experimental data.

It is evident from equation 6-5, as well as foundry test results, that any technique having as its goal the direct shell investment casting of SL patterns of arbitrary size cannot utilize solid resin patterns. This was further confirmed by the early foundry experience at Solidiform, Inc., Fort Worth, Texas; Cercast Group, Montreal, Quebec, Canada; and Precision Castparts Corporation, Portland, Oregon. During initial shell investment casting tests with solid SL patterns at all three foundries, shell cracking was observed in every pattern involving solid sections thicker than 0.12 inch (3 mm), while section thicknesses greater than 0.2 inch (5 mm) caused catastrophic shell fracture during autoclave removal of the wax gating.

6.2 The Development of QuickCast

Recognizing that solid SL patterns would not work with conventional shell investment casting methods, tests of quasi-hollow patterns began in June 1992. It was reasoned that if the structure were partially hollow, perhaps the patterns might collapse inward, rather than cracking the ceramic shell by expanding outward. Building quasi-hollow parts with SL is not intrinsically difficult. The challenge is to develop an internal geometric structure that will simultaneously accomplish all of the following:

- retain pattern accuracy;
- remain sufficiently rugged at ambient temperatures to avoid breakage in normal handling as well as shell investing;
- allow uncured liquid resin to drain effectively from patterns of arbitrary shape; and
- provide a means for the pattern to collapse inward at elevated temperatures, thereby avoiding shell cracking.

To assure drainage of uncured liquid resin from within the quasi-hollow interior, the pattern structure must be, in mathematical terms, topologically simply connected. Or, phrased differently, any liquid resin at a given location within the pattern must be able, in principle, to flow to any other location without encountering a dead end. If the second location is an intentionally generated drain hole, and a vent hole is also provided to allow air to enter the pattern (thereby avoiding flow retardation due to the creation of a partial vacuum), the uncured liquid resin can then exit the pattern within the limitations of gravity, surface tension, and viscous drag forces.

Draining the Resin. QuickCast is based on the concept of draining the uncured liquid resin from the interior of the pattern before postprocessing and ultimate investment casting. However, as we have seen from previous analysis, it is absolutely fundamental that the QuickCast pattern be sufficiently well drained, so that no solid section having a thickness greater than W_{crit} shall remain anywhere within the pattern.

The initial QuickCast tests, in the summer and fall of 1992, used patterns generated from acrylate resins SL 5149 and SL 5154 having viscosity values around 2,000 centipoise. Effective drainage was extremely difficult as the uncured liquid resin flow rates were excruciatingly slow. Simple geometries would take 12 to 24 hours to drain. Complex geometries, such as curved airfoil trailing edge sections, would take 4 to 7 days. And even after these intervals, complete drainage was still not achieved. Subsequent to postcure and exposure to autoclaving, the solidified sections would often crack and sometimes literally explode the ceramic shells.

Fortunately, just enough patterns were sufficiently well drained that their shells did not crack, and successful metal castings were obtained. While the foundry yield was initially very poor, the fact that some QuickCast patterns were occasionally successful suggested that at least the approach was on the right track.

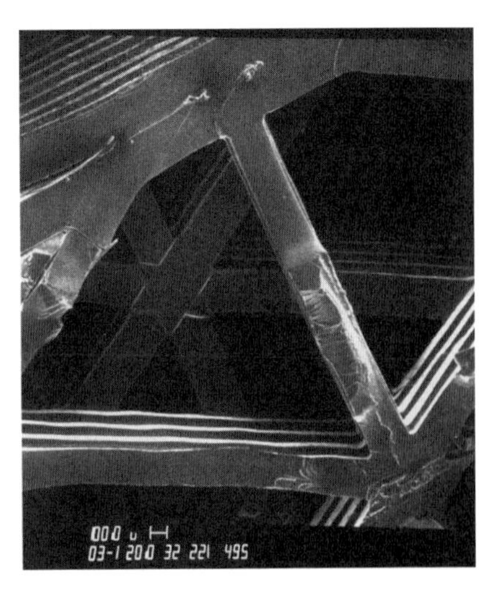

Figure 6-3. Scanning electron microscope image shows a portion of a QuickCast pattern intentionally broken open to reveal the interior structure.

Figure 6-3 is a scanning electron microscope image taken at Allied Signal, Garrett Engine Division, Phoenix, Arizona, showing a portion of a QuickCast pattern intentionally broken open to reveal the interior structure. Figure 6-4 is a photomicrograph of the top portion of a QuickCast test sample with the upfacing skin intentionally omitted. The QuickCast 1.0 geometry involves numerous small equilateral triangle hatch structures or cells, in the X - Y plane. This cellular structure is repeated in the Z direction for a prescribed number of layers, N, and then offset for an equal number of layers. The hatch pattern subsequently returns to the original configuration after 2N layers, and continues

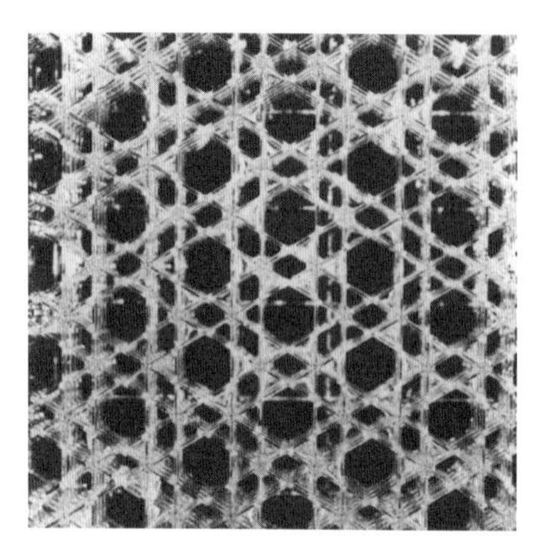

Figure 6-4. Photomicrograph of the top portion of a QuickCast test sample with the upfacing skin intentionally omitted.

to alternate offsets every N layers until the part is completed, with the topmost layer being an upfacing skin. These layer offsets create openings or passages through the interior hatch walls, satisfying the "topologically simply connected" requirement.

Impact of Epoxy Resins. The steps necessary to accomplish the successful generation of a QuickCast pattern and the resulting fabrication of a precision shell investment casting with high foundry yield have been understood for some time. However, a dramatic advance in the reality of successful shell investment casting directly from SL patterns had to wait until the development of epoxy resins. They have enabled levels of pattern accuracy previously unattained.

The multiple reasons for the remarkable success of these photopolymers as the source of QuickCast patterns are described below.

1. In order to achieve excellent pattern accuracy, a resin with very low shrinkage was required. As described in Chapter 2, the effective linear shrinkage of the epoxy resins SL 5170 and SL 5180 is about 0.06 ±0.04%, compared with the earlier acrylate resins such as SL 5149 and SL 5154, which exhibit effective linear shrinkage values of about 0.80 ±0.20%. Thus, the new epoxy resins show effective linear shrinkage values over an order of magnitude smaller than the previous acrylate resins.
2. In addition to their initial accuracy, it was important that the patterns remain dimensionally stable. In September 1992, an early attempt to build a QuickCast pattern of a large jet engine front frame for the Garrett Engine Division of Allied Signal Aerospace Company required the assembly of multiple annular segments. The patterns were built in QuickCast using SL 5149 resin. Although the segments were generated with reasonable initial accuracy, the higher levels of internal stress resulting from the much greater effective linear shrinkage values typical of the acrylate resins led to considerable creep.

   As a result, the various annular QuickCast segments displayed significant dimensional instability. Consequently, despite 3D Systems' best efforts to build and postprocess each one identically, Cercast Group discovered that none of the annular segments fit together. This happened even with the use of a special jig and fixture available at Cercast for assembling multiple segments.

Shortly thereafter, in October 1992, Ciba-Geigy formulated a new epoxy based resin that ultimately proved to be the precursor to SL 5170. The front frame segments were again built in QuickCast, but this time using the new epoxy resin. Now, not only was each segment initially far more dimensionally accurate than before, but creep was almost nonexistent, with each of the annular segments fitting together like a glove.

3. In order to achieve good drainage from parts of arbitrary geometry, especially those involving relatively long, curved, thin sections, it was fundamentally imperative that the resin viscosity be as low as possible. The prior acrylate SL resins had viscosity values typically between 2,000 and 2,500 centipoise at 86° F (30° C). With the development of the new epoxy resin that ultimately became SL 5170, a viscosity of 180 centipoise was measured at 86° F (30° C). The viscosity of SL 5180, which was later developed and released for use with the SLA-500, is also about 200 centipoise at 86° F (30° C). In these cases the resin viscosity was not just incrementally lowered, it was reduced by more than an order of magnitude. The significantly decreased viscosity of the epoxy resins dramatically enhances QuickCast pattern drainage.
4. The new epoxy resins also possess much higher green strength than the previous acrylate resins. Higher green strength enables the use of larger equilateral triangle hatch. In turn, the larger triangles significantly speed the drainage of uncured liquid resin from within the pattern. Even with the epoxy resins, drainage flow velocities are still typically rather small, lying in the range of about 0.1 - 1 mm / sec. Calculation of the Reynolds number, N_{Re}, appropriate for the drainage of liquid resin from the QuickCast interior cells, leads to values of $0.002 < N_{Re} < 0.02$. Thus, the drainage process will definitely involve laminar, as opposed to turbulent flow.

   This is important, as we may therefore utilize the results of the Hagen-Poiseuille equation for laminar flow in small channels. From Reference 9, it can be shown that the drainage rate from a QuickCast pattern scales as the fourth power of the cell size. Thus, increasing the altitude of each equilateral triangle by only 20% will more than double the drainage rate. Unfortunately, from Reference 10, if the triangles are made larger, the distortion or sag of a uniformly loaded upfacing skin is also proportional to the fourth power of the cell size, and inversely proportional to the modulus of elasticity of the green resin.

   The green flexural modulus of the epoxy resins is about four times that of the acrylate resins. Thus, one could increase the epoxy cell size by about 40% and still achieve the same distortion previously found with acrylates. Or, one could increase the cell size by 20% with epoxy resin, which will double the drainage rate and yet produce only half the upfacing skin sag. This was the compromise selected for the initial release of QuickCast 1.0.

5. From Reference 11, the cured linewidth of an SL-generated hatch vector depends on the laser beam diameter at the resin surface, the cure depth of the hatch vector, and the penetration depth of the resin. However, none of these parameters are varied as the QuickCast triangles are simply made larger. Let us define the void ratio, R_v, as the fraction of the pattern's volume that is ultimately filled with air, subsequent to drainage of the uncured liquid resin. If the triangle cell size is increased, but the cured linewidth remains constant, then the void ratio must also increase. Figure 6-5 shows the important relationship between pattern void ratio and the foundry yield of investment castings at quality Grade C or better, based upon experience gathered during the development of QuickCast.

   As we had determined earlier, equation 6-5 indicates that solid patterns having section thickness values greater than W_{crit} will result in shell cracking. Since the value of W_{crit} for typical shell systems is in the range from about 0.060 to 0.080 inch (1.5 to 2 mm), and since the vast majority of typical investment casting patterns have at least one section thicker than 0.12 inch (3 mm), it should not be surprising that the yield of successful castings for $R_v = 0$ (i.e., a solid pattern) was essentially zero. This result is shown in Figure 6-5 as a point at the origin. Although flask casting can produce metal parts from solid SL patterns, prior results indicated that almost all such castings were Grade D (Reference 7). Thus, in accord with our definition of yield, involving Grade C or better radiographic quality castings, a foundry yield of essentially zero for $R_v = 0$ is correct.

   Conversely, at the other limit, namely $R_v = 1$, the pattern would consist of an infinitely thin skin with nothing but air inside. In short, the pattern would be much like a soap bubble. Clearly, this type of pattern would collapse inward with incredible ease, would generate almost no shell stress or residual ash, and autoclaving/casting yield would be essentially 100%, provided the pattern could survive shipment, handling, and shell investment.

   The data of Figure 6-5 shows that as the void ratio increases, the probability of successfully eliminating the pattern without damage to the shell also increases. Due to the superior green strength of the epoxy resins relative to the earlier acrylates, larger cell sizes are now possible. The increased cell size results in significantly higher values of R_v.

6. The epoxy resins SL 5170 for the SLA-250 and SL 5180 for the SLA-500 also exhibit significant softening at elevated temperatures. Figure 6-6 shows the tensile modulus of SL 5170 resin versus temperature over the range from 32 to 320° F (0 to 160° C). While the resin is quite rigid at typical build temperatures (28 ±2° C, or 82 ±4° F), it begins to weaken by about 140° F (60° C), has become much less rigid at 194° F (90° C), and has effectively lost all strength above 248° F (120° C). Combined with the quasi-hollow interior structure of QuickCast patterns, this reduced stiffness of the epoxy resins at the elevated temperatures typical of either autoclaving or flashfire

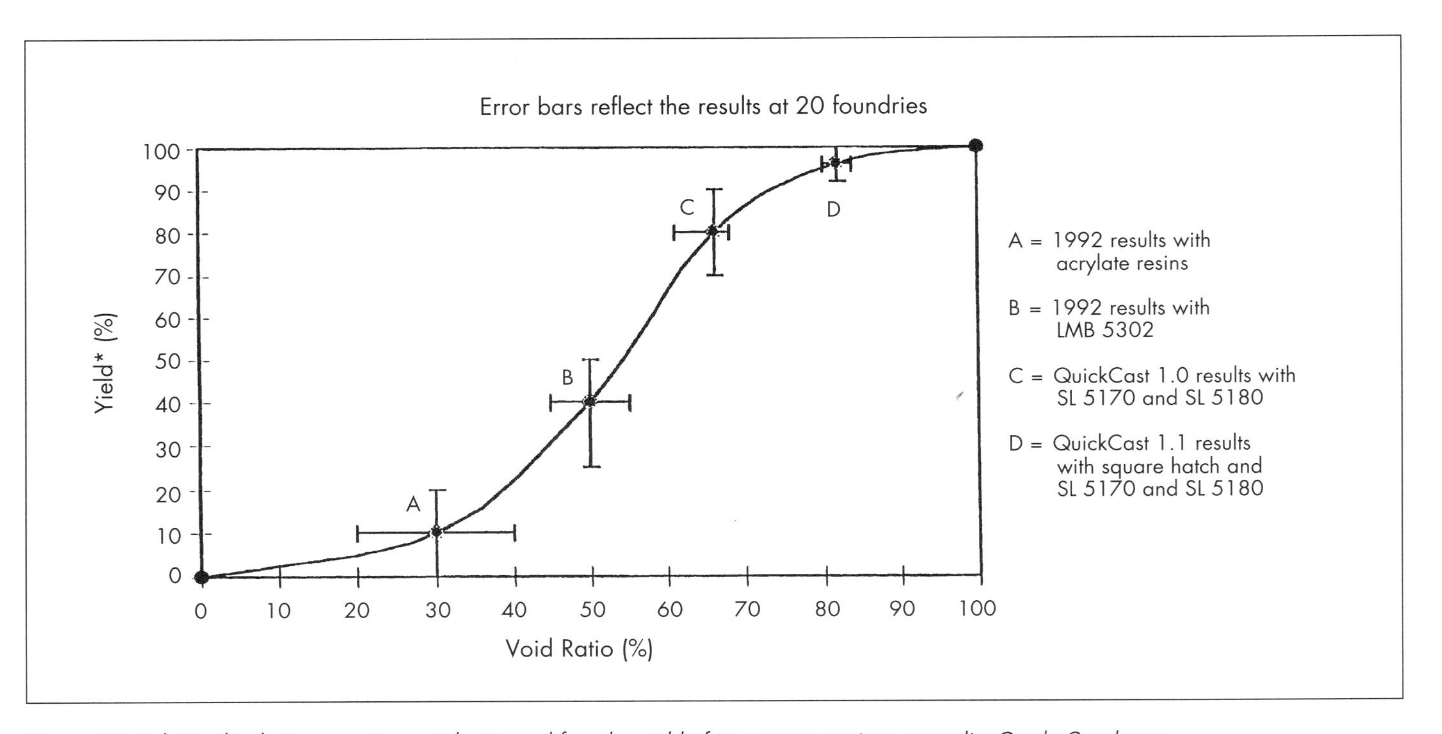

Figure 6-5. Relationship between pattern void ratio and foundry yield of investment castings at quality Grade C or better.

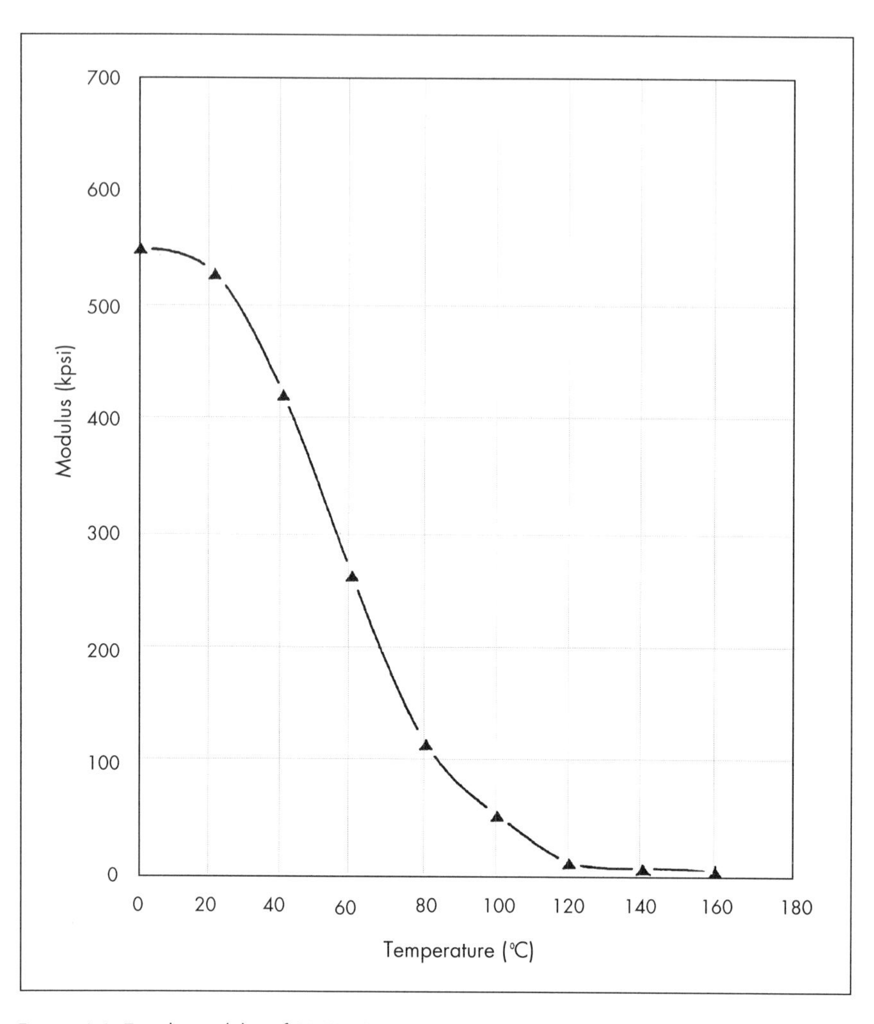

Figure 6-6. Tensile modulus of SL 5170 versus temperature.

further decreases the stresses induced in the ceramic shell during pattern elimination.

7. The epoxy resins have provided very clean burnout during oxygen-rich flashfire at temperatures in the range of 1,650 to 2,000° F (900 to 1,100° C). Resins SL 5170 and SL 5180 generate little ash provided the burnout is performed with abundant free oxygen (i.e., 8 to 12%). Remember, cross-linked photopolymers do NOT melt at elevated temperatures. Rather, the hydrocarbon based resin undergoes molecular breakdown above ≈750° F or about 400° C.

At temperatures above ≈1,500° F or roughly 800° C, in the presence of sufficient oxygen, the carbon is chemically converted to carbon dioxide and the hydrogen is converted to water vapor. Measurements of the total ash content have been in the range of 0.01 to 0.05% of the mass of the original pattern. Chemical analysis indicates that over 95% of this small amount of residual ash involves unburned carbon, with trace amounts of hydrogen, nitrogen, fluorine, sulfur, and antimony. Since QuickCast patterns are mostly hollow and have relatively low mass to begin with, the total amount of residual ash is minimal. This is important since ash trapped within the evacuated ceramic shell can lead to inclusions in the final metal casting. Excessive inclusions are cause for downgrading or even rejecting investment castings.

QuickCast Findings. After approximately 33 months of test, evaluation, and commercial shell investment casting at more than 40 foundries, some important information has been compiled. Much of the data that has been gathered is due to the outstanding work of Larry André and La Mar Daniels at Solidiform, Inc., Fort Worth, Texas; Steve Kennerknecht and Bassam Sarkis at Cercast Group, Montreal, Quebec, Canada; Jeff Smith at Precision Castparts Corporation (PCC), Portland, Oregon; and John Cordes, an investment casting foundry consultant to 3D Systems, Valencia, California.

1. When executed properly, the QuickCast process is capable of excellent results. Through the skill, innovation, experience and persistence of La Mar Daniels, Solidiform has successfully converted more than 1,200 QuickCast patterns in over 230 different geometries into aluminum castings, with an astounding 100% foundry yield of parts meeting either Grade C and/or Grade B for military and commercial nondestructive testing standards. Cercast has also achieved a yield in excess of 90% on over 800 parts in both aluminum and copper alloys. PCC has been successful with a number of large, complex geometries in inconel, stainless steel, and titanium.
2. It is extremely important to check for, detect, fill, and positively confirm the sealing of any and all pinholes. By definition, pinholes are small holes or tears in downfacing QuickCast pattern surfaces, resulting from overzealous support removal. If one happens to exert too much force while removing supports, it is possible to damage the pattern surface where the support was previously attached. Typically, a small hole, roughly 0.02 to 0.04 inch (0.5 to 1 mm) in diameter, may result.

   If this hole is not detected and filled with investment grade wax or UV cured resin, the face coat slurry will invade the interior of the pattern (remember, a QuickCast pattern is mostly hollow). This will ultimately result in an inclusion and possibly a defective casting.
3. To assure that all pinholes (as well as the intentionally generated vent holes and drain holes) have been found and sealed with investment casting wax, a number of techniques have been developed. Probably the most successful method to date involves checking with a combination of low positive inter-

nal air pressure (up to about 0.3 atmosphere), as well as low negative pressure or vacuum testing. If the pressure gage shows a slow air leak, then at least one hole must still be present.

The process should be repeated until all pinholes, vents, and drains have been filled and a constant pressure reading indicates that no further leaks are present. Under NO circumstances should the pattern be immersed in water, alcohol, or other liquids to check for pinholes. Experience has shown that any fluid immersion procedure will result in permanent softening and distortion of the pattern, with the attendant compromise of pattern accuracy.

4. Also, one must check for portholes. By definition, a porthole is a small hole in a vertical or steeply ramped surface. These occurred in early QuickCast patterns whenever resin dewetting happened on successive identical border segments, ultimately leaving a cured resin gap in the outer boundary of the part. Even if dewetting occurs on a given border segment of layer N, the QuickCast build style provides sufficient process latitude that the problem is normally solved on the following layer (N+1).

   However, if somewhat diabolically, the very same border segment on layer (N+1) also suffers from dewetting, the system overcure on layer (N+2) may not be sufficient to cure through three layers to patch the missing boundary. The result is then a porthole, which would manifest itself as a casting inclusion in the same manner as a pinhole. Fortunately, with the release of SL 5170 and SL 5180, resin dewetting at the recommended layer thickness values is greatly reduced and portholes are now rare.

5. Because autoclaves are commonly used by investment casting foundries for removing wax, it is quite natural that they also are used to remove the wax gating attached to QuickCast patterns. Most of the foundries went through a classic trial and error process in an attempt to optimize their parameters. Under careful control, autoclaving can be successful in this role. Cercast has achieved yields in excess of 90%, using autoclaving to remove the wax gating attached to QuickCast 1.0 patterns. Other foundries have been less successful. The overall average foundry yield of quality metal castings has only been about 70% for those cases where QuickCast 1.0 patterns have been exposed to autoclaving.

   Although a foundry yield of 70% is not bad, 3D Systems, SL users, and the foundries themselves were pushing for even better performance. With QuickCast, the user has indeed saved considerable time and expense by avoiding the fabrication of tooling. Nonetheless, since QuickCast patterns represent a nontrivial cost, and the shell investment process may take 2 to 4 weeks, shell failures need to be eliminated, or at least greatly reduced.

6. Fortunately, a process known as FlashFire Dewax™ has undergone test and evaluation. The FlashFire Dewax procedure, developed by Glenn Foster at Pacific Kiln and Insulation Co., Rancho Cucamonga, California, provided some very important results. Basically, QuickCast patterns are assembled

with standard wax gating into conventional trees. The trees are then positioned upside down so that the pouring cup and sprue are directly over drain holes located on the surface of a moveable bed. The chamber temperature is elevated to ≈1,560° F or about 850° C and the bed is then moved into the furnace.

Within 5 to 7 minutes the wax gating is completely melted out of the shell, and has run down into a catch basin protected by an inert gas blanket (so that the wax will not burn). After it's cooled, the wax can be reused for subsequent gating applications without any further processing. This is an important advantage for foundries since the wax has not come into contact with condensed steam from an autoclave.

After about 10 minutes at roughly 1,560° F (850° C), to assure that all the wax gating has been removed, the furnace temperature is ramped up to about 1,900 to 2,000° F (1,040 to 1,100° C) over a period of 20 minutes, and then maintained at this level for an additional 30 minutes. Finally, to avoid thermal shock to the shell, the temperature is ramped back down to 1,560° F (850° C) over an interval of another 20 minutes. The shell is then removed from the furnace. The entire procedure takes less than an hour and a half.

Although the stoichiometric ratio for air/natural gas combustion is 10:1, the FlashFire Dewax System uses a ratio of 20:1. This is done so that the free oxygen level in the chamber remains at about 10%. As a result, resin pattern burnout using flashfire dewaxing is extremely effective. Shells containing QuickCast patterns that were built using both SL 5170 and SL 5180 have been tested at the Pacific Kiln facility. The resulting ceramic shells were exceptionally clean, showing little or no evidence of microfissures as well as minimal ash or other deposits. Certainly, additional testing is required for a range of different shell systems, temperature/time programs, oxygen concentrations, and various QuickCast pattern geometries. These efforts are underway. So far, the results for the elimination of QuickCast patterns and associated wax gating using FlashFire Dewax have been extremely positive, with foundry yields in the 85 to 90% range.

7. To date, the following metals have been successfully shell investment cast directly from QuickCast patterns:
   - A 356 Aluminum
   - A 357 Aluminum
   - Copper
   - C-20 Beryllium-Copper
   - 4140 Carbon Steel
   - Nodular Iron
   - Chrome-Cobalt Steel
   - 718 Inconel
   - A2 Tool Steel
   - P20 Tool Steel
   - Ti 6Al-4V Titanium
   - Everdur Silicon Bronze
   - 304 L Stainless Steel
   - 17-4 Stainless Steel
8. QuickCast has been successfully used to shell investment cast objects having maximum dimensions as small as 0.48 inch (12 mm) to components as large as 27.4 inches (695 mm). Section thicknesses have been as large as

5.9 inches (150 mm) and as small as 0.020 inch (0.5 mm). Some patterns have utilized fillets and corner radii, while others have not. However, pattern fillets and radii are strongly recommended, wherever possible. The use of fillets on interior corners is always good practice to avoid stress concentrations in the final castings. Also, edge radii are especially useful to avoid shell thinning near sharp exterior corners. If severe, this thinning of the ceramic slurry at exterior corners can result in shell cracking during burnout of the QuickCast pattern.

6.3 QuickCast Pattern Considerations

During the development of QuickCast, a number of techniques have emerged that enhance pattern quality as well as subsequent metal casting grade and foundry yield. The following are offered as recommendations based upon experience to date.

Preparing the CAD File

1. Interior corners of a pattern should definitely have fillets added to reduce stress concentrations in the final castings. Similarly, exterior corners need either radii or chamfers for better metal flow, and to avoid shell thinning. If the user's CAD system cannot readily perform these functions, the outside sharp corners of the QuickCast pattern should be broken with a file or sandpaper. Investment casting wax may also be applied as fillets on inside corners.
2. In general, although QuickCast patterns can generate walls as thin as 0.016 inch (0.4 mm), the minimum wall thickness for reliable investment casting is about 0.060 inch (1.5 mm). When attempting to investment cast walls thinner than 0.060 inch (1.5 mm), consultation with the intended foundry is strongly recommended.
3. Parts should be oriented to maximize feature definition in the X-Y plane. This will minimize the effects of stair-stepping on ramped surfaces, and overcure, print-through, and quantization errors on any downfacing surfaces in the Z direction.
4. The CAD modeling process must account for the potential of nonunioned solids that would create independent sections in the STL file. Thus, STL files should consist of a single interconnected closed shell to avoid seams that create internal pressure pockets. For similar reasons, it is preferable that patterns fit within an SLA in a single piece. If the final QuickCast patterns must be assembled from multiple QuickCast parts, it is very important that overlapping vent holes are provided along common borders to relieve any internal pressure differences that may occur during burnout. If these steps are not taken, pressure buildup during pattern elimination can cause shell fracture.

Preparing the STL File

1. For thin wall sections, increasing the hatch spacing up to 0.32 in (8 mm) and decreasing the layer offset height, H, to half the value of the smallest Z-axis feature will facilitate improved pattern drainage by ensuring the presence of interior passages.
2. Coordination with the intended foundry will allow the user to include appropriate metal shrinkage factors during the part preparation process. Of course, conventional resin shrinkage factors must also be incorporated.
3. Supports that attach to the part in short sections followed by unattached segments, such as the castellated techniques used by Bridgeworks 3.0 and MAGICS 2.2, are strongly recommended when building QuickCast patterns with epoxy resin.

Pattern Draining

1. Exposing green parts or liquid resin to many hours of low-level fluorescent or UV radiation, especially while liquid resin is trapped in a QuickCast part, will cause resin on the exterior and interior of the part to partially cure. This may cause tacky surfaces and inadequate drainage. Thus, good practice involves draining the parts inside the SLA, wiping them clean with a paper towel, and postcuring them as soon as possible.
2. If QuickCast patterns contact water or solvents containing water, they will soften and distort. Due to their large surface-to-volume ratio, exposing green QuickCast patterns to high humidity for only a few hours will also cause pattern softening and distortion. The use of dehumidifiers and desiccants in both the SLA facility and the part finishing area is, therefore, also strongly recommended.
3. The best results have been achieved when patterns were drained and postcured in a low-humidity environment, within eight hours of build completion.
4. There is a high probability that undrained wall thicknesses greater than 0.080 inch (2 mm) will exert expansion stresses that may cause shell cracking. Thus, to be safe, it is strongly recommended that no solid regions thicker than 0.040 inch (1 mm) should exist in a QuickCast pattern prior to burnout. To the extent that such regions do exist, the probability of successful pattern burnout, without any shell cracking or residual ash, is reduced.
5. Depending on part geometry, drain holes of approximately 0.25 to 0.4 inch (6 to 10 mm) diameter, spaced about 2 inches (50 mm) apart, may be necessary to promote thorough drainage. Vent holes can be as small as 0.040 to 0.080 inch (1 to 2 mm) in diameter.
6. To shorten drainage time, position the longest dimension of the part vertically. This increases the gravitational pressure head and speeds resin drainage.
7. Ideal draining temperature for either of the epoxy resins SL 5170 or SL 5180 is between 82 and 95° F (28 and 35° C). While resin viscosity drops as

temperature increases, above 104° F (40° C) the resin modulus also decreases with increased temperature, which can lead to pattern distortion.

8. A low-speed centrifuge device, or a vacuum pump, may accelerate drainage. Never use high-water content compressed air to assist with the drainage of QuickCast parts. The water content in the compressed air may soften and distort the pattern.

Support Removal

1. Support removal may be facilitated with a hot knife.
2. Be careful not to rupture pattern surfaces when removing supports.

Cleaning Pattern Exteriors

1. As noted earlier, if QuickCast patterns contact water or solvents containing water, they will soften and distort.
2. Exterior pattern surfaces should simply be wiped with absorbent paper towels. Reagent grade isopropyl alcohol or RG-IPA, containing less than one percent water content, may be used to wipe QuickCast patterns, provided that the solvent is never allowed to enter the pattern's interior. To prevent distortion, one should eliminate any excess liquid RG-IPA from the part's exterior surface by immediately using dry nitrogen or dry compressed air to speed the evaporation process.
3. Alcohol and other solvents inhibit the photocuring of liquid resin. Therefore, the pattern's exterior surfaces should be cleaned with RG-IPA, and then fully dried before being placed in the PCA for final postcure.

Postcuring

1. After cleaning, postcure patterns as quickly as possible. High humidity can inhibit surface curing, resulting in a tacky surface. Extended exposure to humidity levels above 50% can cause part softening and, worst of all, pattern distortion.
2. Postcure time is dependent on part geometry. However, QuickCast patterns generally postcure about 25% faster than solid parts.
3. No changes to the standard bulb configuration in the PCA-250 or PCA-500 are necessary for postcuring epoxy parts.

Pattern Preparation

1. If pattern sections are joined together, overlapping vent holes must exist between the mating faces.
2. If necessary, surfaces may be sanded, provided that one is careful not to sand through the pattern's exterior.
3. All holes in the pattern surface must be sealed, either with investment casting wax or epoxy resin thickened with finely ground, solidified, epoxy resin.
4. QuickCast patterns should never be immersed in liquid to check for leaks.

5. Low pressure, dry compressed air may be used to locate pinholes in pattern surfaces. Using a maximum pressure of 0.3 atmosphere, sweep an automotive stethoscope over all surfaces while listening for leaks.
6. An additional method for locating holes utilizes a leak down tester that can be purchased at an automotive store. This hand-operated vacuum pump can confirm that there are leaks present in the pattern if the pressure indicated on the gauge varies when pumping is stopped. Once all holes have been properly sealed with wax, the gauge reading should remain constant for at least one to two minutes.

6.4 QuickCast Improvements

QuickCast has made remarkable progress since June 1992. In this regard, QuickCast is very much like RP&M itself, which has advanced considerably in the seven years between 1988 and 1995. However, there is still room for further improvement. Remember, investment casting has been used for over 5,000 years, and in its modern form for about 40 years. Thus, it is remarkable that within only 33 months, QuickCast began being used by hundreds of companies on a regular basis. Nonetheless, the QuickCast process still needs to achieve better foundry yield of high-quality, Grade B, metal castings.

Improvements to QuickCast are being developed in the following areas:

1. Elimination of "pinholes." Pinholes can occur on downfacing surfaces during support removal from QuickCast patterns. Three solutions have been identified.

   First, the use of a so-called hot knife has been quite successful for epoxy resin support removal. The intent is to soften and burn through the supports, rather than attempting to cut them off with a blade. This method essentially eliminates shear or bending stresses on the pattern surface, and is therefore less likely to create pinholes. The use of a pinpoint butane flame has also produced excellent results.

   Second is the use of supports generated in a castellated fashion, as created by various automatic support generation programs. When using epoxy resins, the area of the supports in contact with downfacing surfaces of the part can be significantly reduced. This is due to the minimal curl forces induced by epoxy resins relative to the previous acrylate resins. Castellated supports have been found to work very well with epoxy resins. They are easier to remove, improve resin drainage, diminish stresses on the pattern surface during support removal, and significantly reduce the incidence of pinholes.

   Third, 3D Systems has recently released QuickCast 1.1, an advanced version that utilizes both triple downfacing and triple upfacing skin layers rather than the single skins used with QuickCast 1.0. From Reference 10, the deflection of a thin slab in bending is inversely proportional to the moment of inertia, which varies as the cube of the slab thickness. By using triple layer skins on upfacing and downfacing surfaces, one can increase their stiffness by a factor of $3^3 = 27$. It is therefore much more likely that the support (rather than the skin) will break during support removal. The use of a hot knife in conjunction with castellated supports has significantly reduced the incidence of pinholes. The use of triple downfacing skins during beta testing of QuickCast 1.1 has effectively eliminated pinholes.

2. Improve uncured liquid resin drainage from QuickCast patterns.

   Three methods by which this may be accomplished are clearly evident. First, one could use a resin with reduced viscosity and surface tension. This was a major reason why the low viscosity epoxy resins were essential for QuickCast. Obviously, still newer resins, with even lower viscosity or surface tension values, are possible. Unfortunately, the formulation, parametric, diagnostic, alpha, and beta testing sequence for new resins is a lengthy one. Advances in chemistry take time.

   The second approach involves increasing the driving force for pattern drainage. At present, the great majority of QuickCast patterns are drained solely by gravity. Techniques using air pressure, vacuum methods, or centrifugal force are certainly possible. However, they may be problematic for patterns that include delicate, thin wall features. Pattern damage due to excessive force can be a potential difficulty. To date, numerous patterns have been drained with a commercially available low-speed centrifuge originally designed to extract honey from beehives. The device is definitely helpful for some geometries, but more data is needed for a wide range of parts.

   The third and most significant approach involves increasing the size of the internal hatch structure. QuickCast 1.0 uses an equilateral triangle hatch with an altitude of 0.150 inch (3.8 mm). Smaller triangles retard resin drainage. Larger triangles improve resin drainage, but cause sag on upfacing surfaces, resulting in poor surface quality. The 0.150-inch (3.8 mm) triangle size was chosen as the best compromise when using a single upfacing epoxy skin. However, by employing triple upfacing skins, QuickCast 1.1 was free to explore the use of still larger triangles.

   As noted earlier, the laminar flow rate of a viscous fluid in a small channel scales as the fourth power of the channel diameter. Thus, increasing the triangle size from 0.150 inch (3.8 mm) to 0.250 inch (6.4 mm) will increase the drainage rate by a factor of almost eight. Improved resin drainage assists in eliminating solid sections of resin. Parametric, alpha, and beta test results indicate that increasing the triangle size has proved to be very valuable. Using QuickCast 1.1 software, triangular hatch patterns that previously took all day to drain are now well drained in a few hours.

   QuickCast 1.1 also utilizes square hatch patterns. Figure 6-7 is a photomicrograph of a QuickCast 1.1 pattern utilizing 0.250 inch (6.4 mm) offset squares. The result is a very uniform structure consisting solely of half-scale squares, which drains more quickly and completely than the complex pattern of hexagons, tiny triangles and small rhombuses resulting from an offset equilateral triangle hatch. Also, since a square hatch pattern has interior angles of 90°, compared to 60° for an equilateral triangle, the effects of surface tension are further reduced. The decreased meniscus in the corners of each cell has resulted in better uncured liquid resin drainage from QuickCast 1.1 patterns built using square hatch. With a void ratio of 83%, only one sixth as much resin is needed to build a square hatch QuickCast 1.1 part relative to that required to build a solid part. The remaining liquid resin can be drained back into the vat and used again.

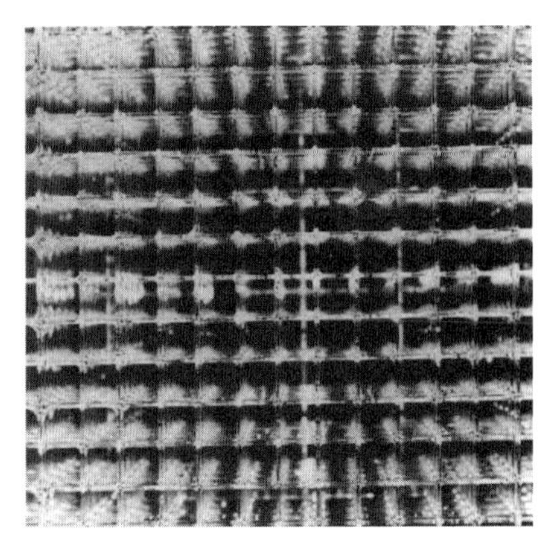

Figure 6-7. Photomicrograph of QuickCast 1.1 pattern utilizing 0.250 inch (6.5 mm) offset squares.

Finally, of special importance to the foundries, squares are intrinsically less rigid than equilateral triangles. During recent burnout tests at Solidiform and Cercast, square cell QuickCast 1.1 patterns apparently generated lower stresses on the ceramic shells. This was indicated by the complete absence of microfissures even with autoclave wax gating removal.

3. Improve the quality of upfacing surfaces.

   As previously noted, the size of the equilateral triangle hatch structure in QuickCast 1.0 was a compromise between resin drainage and upfacing surface quality. The single skin on upfacing surfaces of a QuickCast 1.0 pattern is supported only by the underlying triangular hatch structure. As a result, the skin sags ever so slightly as it spans each triangle. The fact that the epoxy resins possess high green strength minimizes, but does not eliminate this effect.

   However, with QuickCast 1.1, the first skin layer acts as a support for the next, and so on. The resulting triple skinned upfacing surfaces show outstanding smoothness. Measurements with a surface profilometer indicate that patterns built using QuickCast 1.1 had RMS roughness values of 10 to 18 microinches (0.25 to 0.45 micrometers), on upfacing surfaces compared with values from 100 to 160 microinches (2.5 to 4 micrometers), for QuickCast 1.0. An order of magnitude improvement in upfacing pattern surface quality has already been achieved using QuickCast 1.1.

4. Improve the quality of downfacing surfaces.

   The primary causes of surface roughness on QuickCast 1.0 downfacing surfaces are support remnants and print through. Remedies for the former were previously discussed (castellated supports, and a hot knife for their removal). Print through, however, is due to triangular hatch vectors being drawn on a layer just above a downfacing skin. The laser exposure required for these hatch vectors also increases the exposure just below the downfacing skin, per the Beer-Lambert law.[11] This "surplus" exposure produces an increment of additional cure depth on the downfacing skin. Since triangles are drawn on the second layer, the net result is a "ghost" of the equilateral triangle hatch pattern formed on the bottom of the downfacing surface, evident on all QuickCast 1.0 patterns.

However, with QuickCast 1.1, three layers of downfacing skin are used. Thus, when hatch vectors are drawn on the fourth layer, the "print through" all the way to the bottom of the first downfacing skin is very much less, since the exposure is attenuated exponentially through the additional two layer thicknesses of photopolymer resin.

As a result, "print through" is decreased and downfacing surface quality is improved. Further, with triple downfacing surfaces, it is now possible to use very fine sandpaper to remove support remnants, with far less danger of "breaking through" the skin of the pattern. This is important because it reduces the amount of manual labor required to complete each pattern.

5. Increase the void ratio of QuickCast patterns.

As seen in Figure 6-5, the higher the value of the void ratio, R_v, the higher the probability of successful pattern burnout. QuickCast 1.1 permits the use of larger triangles. Since the cured linewidth of the triangular hatch vectors remains constant, larger triangles result in higher void ratios. Tests involving different sized triangles have been performed to determine their corresponding values of R_v. The maximum void ratio for QuickCast 1.0 is about 66%.

However, when using QuickCast 1.1, with 0.250 inch (6.4 mm) equilateral triangle hatch, void ratios as high as 76% have been measured. This decreases shell stress, and improves QuickCast foundry yield. Increased void ratio also reduces the already low residual ash content, and requires less resin to build a given pattern.

Furthermore, QuickCast 1.1 has also recently been released using an offset square hatch pattern. This build style results in excellent pattern drainage, induces still lower shell stresses, leads to less residual ash, and consumes very little resin. Foundry tests show that square hatch patterns do indeed collapse inward more readily than equilateral triangle hatch patterns. Coupled with void ratios as great as 83%, the beta test results involving numerous patterns built using the QuickCast 1.1 square hatch style indicate foundry yields in excess of 95%. Also, a number of these patterns were processed using standard autoclave parameters for wax gating removal, with negligible or zero shell cracking occurring. While certainly very positive, additional testing will continue in order to statistically confirm these results.

The foregoing discussion addressed five specific areas of improvement for QuickCast. These are not the only advances possible. Items such as continued improvements in pattern accuracy and dimensional stability remain ongoing activities for all aspects of SL. Part building in thinner layers for improved resolution of ramped surfaces (i.e. minimizing the effects of "stair-stepping") is also being studied. Finally, improved machine throughput, lower cost materials, and simpler, more user-friendly software are perennial goals of RP&M in general, and SL in particular. Efforts continue in all of these areas.

References

1. Jacobs, P.F., Chapter 1, *Rapid Prototyping & Manufacturing: Fundamentals of StereoLithography*, SME, Dearborn, MI, 1992, pp. 1-23.
2. Blake, P., "The Applications and Benefits of Rapid Prototyping at Texas Instruments Inc.," Proceedings of the First European Conference on Rapid Prototyping, University of Nottingham, Nottingham, England, July 6-7, 1992, pp. 267-288.
3. Jacobs, P.F., "Recommended Foundry Procedure for Shell Investment Casting using QuickCast StereoLithography Patterns," April, 1993, pp. 4-7, *QuickCast™ Foundry Reports*, 3D Systems, June 1993.
4. Kennerknecht, S. "StereoLithography Based Rapid Prototyping for Aluminum Investment Castings using QuickCast™," April, 1993, pp. 8-12, *QuickCast™ Foundry Reports*, 3D Systems, June 1993.
5. Smith, J. and Hanslits, M., "Allied Signal Impeller Shroud: A Rapid Prototype Pattern Evaluation for Investment Casting," May, 1993, pp. 13-17, *QuickCast™ Foundry Reports*, 3D Systems, June 1993.
6. André, L., "QuickCast™, an Adventure", March, 1993, pp. 18-19, *QuickCast™ Foundry Reports*, 3D Systems, June 1993.
7. Blake, P. and Baumgardner, O., "Solid Modeling and StereoLithography as a Solid Freeform Fabrication Technique at Texas Instruments Inc., Proceedings of the Solid Freeform Fabrication Symposium, University of Texas, Austin, Texas, August 1990, pp. 170.
8. Schulthess, A., "Eighth Report on Formulations for Investment Casting," Ciba-Geigy research report No. 185,035, Marly, Switzerland, November 1992.
9. Schlichting, H., Chapter 1, *Boundary Layer Theory*, McGraw-Hill, New York, NY, 1960, pp. 10-11.
10. Marin, J. and Sauer, J., *Strength of Materials*, The Macmillan Co., New York, NY, Second Edition, 1954, pp. 144 and 481.
11. Jacobs, P.F., Chapter 4, *Rapid Prototyping & Manufacturing: Fundamentals of StereoLithography*, SME, Dearborn, MI, 1992, pp. 91-94.

CHAPTER 7

QuickCast Foundry Experience

"All things were made through Him, and without Him nothing was made that was ever made."

John 1:3

7.1 A Look at Investment Casting

Historians tell us that the "lost wax" investment casting process had its origins in Mesopotamia around 3500 B.C. In Egypt the process was used to make jewelry over 4,000 years ago, and was later employed in China. It was also used in the Greek classical period and later during the Renaissance in Italy. In Colombia, South America, the Indians were casting gold by 200 A.D. In the United States, dentists utilized investment casting starting in the early 1900s, followed by jewelers in the 1930s. World War II saw the process turned into a highly technical means of manufacturing super-charger buckets for aircraft reciprocating engines and, later, for jet engine parts.[1] These examples are typical of applications that drove the development of investment casting processes needed to serve numerous markets. The process capabilities for both ferrous and nonferrous investment casting have expanded and matured to a relatively sophisticated state over the past 50 years. The size and complexity of the configurations that can be produced by investment casting illustrate the advances in materials and procedures achieved in all areas of foundry practice. Nevertheless, in spite of all that has been accomplished, there is a growing need for both prototype and limited production castings at significantly reduced lead times. These requirements also address maintaining or exceeding current quality levels. Of course, it would also be very beneficial to do this at lower cost. Those in the foundry industry who have lis-

By Larry E. André, Vice President, and LaMar Daniels, Director of Research & Development, Solidiform, Inc., Ft. Worth, Texas; Steve Kennerknecht, Vice President of Technology, and Bassam E. Sarkis, Rapid Prototype Engineer, Cercast Group, Montreal, Quebec, Canada.

tened closely to this request from their customers saw that it was not just the same old plea for suppliers to do it better, faster, and cheaper. This new agenda is based on customers' urgent need to find creative ways to become more competitive in an increasingly global marketplace.

Time Reduction

In many cases, the common denominator in achieving a more competitive position is TIME. A significant time compression involving numerous prototyping, engineering, and manufacturing tasks during the product development cycle often results in a company with a substantial competitive advantage. If you have the ability to participate in a manufacturing process that can make your customers more competitive in their markets through time compression, you have probably made your company more competitive in your market.

QuickCast has been one of the primary techniques enabling investment casting foundries to participate in this market segment by giving them the capability to dramatically reduce casting lead times. Nonetheless, the ability of a foundry to provide a substantial time compression is only partially dependent on the properties of the QuickCast pattern. All aspects of the evolving manufacturing scheme, from the receipt and manipulation of engineering data for quotation purposes, to the transport of the completed item to the customer, are open for evaluation and change in an effort to respond to this fast track requirement. The manufacturing precepts that allow for "doing it over" or "fixing it later" can no longer be accepted if we are truly targeting a quality product for delivery to our customers in the minimum possible time.

This is not to say that we should not exercise every effort to salvage or recover a part that can be repaired. However, it must be understood that the cost of imperfection in this market segment involves an ingredient that is not recoverable. And that ingredient is TIME! Once this basic concept involving the need for time compression is understood and accepted, it becomes clear that extreme procedures must sometimes be implemented (relative to normal production practice) to ensure a 100% yield from every operation in the process.

CAD's Impact

The incorporation of electronic engineering data in the form of CAD solid models is presenting challenges and opportunities to the foundry industry. The challenges primarily involve learning a new language and acquiring additional hardware in order to use this data efficiently. The opportunities lie in an increased ability to participate with the customer earlier in the design, to enhance the effectiveness and castability of their designs, and to expand utilization of the data once it is received at the foundry.

QuickCast has given the foundry industry a new capability to serve their customers' need for rapid delivery of functional metal prototypes. Furthermore, it has also opened the doors to the world of virtual manufacturing. Solid CAD modeling capabilities, coupled with RP&M machines, have enabled more manufacturing operations to be performed and perfected in a virtual setting than ever before. Once the image of the item to be produced is described in digital form, the opportunities for the foundry engineer to complete supplemental operations, such as thermal analysis for gating placement, can now be realized. Eventual incorporation of entire gating systems for individual parts, and the placement of these parts on a tree, is not an unlikely prospect in the virtual world.

QuickCast has eliminated the physical necessity of one critical operation in the normal casting process: *toolmaking*. Consequently, it has allowed the pattern making function to be initiated digitally and then duplicated physically by SL. By extending the methodology, a second pattern assembly operation can also be accomplished in this virtual setting as well. QuickCast has the potential to deliver to the foundry an entire assembly of parts ready for the ceramic shell operation. The opportunities for time compression, increased accuracy, and increased quality are tremendous. The "foundry of the future" is being constructed today. The building blocks are already available and the marketplace is demanding that construction of this foundry proceed to reality as soon as possible.

7.2 Investment Casting Processes

Investment castings are produced by one of two techniques, depending on the method employed to form the disposable ceramic mold built around the pattern as well as its associated gating system. The techniques are referred to as "solid mold" (or "flask casting") and "shell casting."

Solid Mold

The solid mold investment casting process uses expendable patterns most commonly made of wax, produced from female cavity dies. The patterns are assembled on a common gating system creating what is generally referred to as a cluster or a tree. The tree is then dipped into a precoat material to reduce surface tension and increase wetability. Next, the tree is placed in a metal flask, open at the top, and a plaster-of-paris-like slurry material is poured around it. Vacuum and vibration are often used to facilitate the slurry settling around the tree, thereby avoiding entrapped air bubbles on the surface of the patterns.

Heating the Molds

The molds are then air dried for a period of time, depending on the total mold mass and the water content of the particular slurry used. This air drying cycle usually takes from 6 to 8 hours. The wax from the patterns and gating system is then melted out of the mold, either by heating the assembly in a steam autoclave, or by flash firing the entire assembly in a gas furnace. The molds are brought to a temperature of roughly 200 to 300° F (95 to 150° C) for the melt-out cycle, and then to a burnout temperature of 1,200 to 1,400° F (650 to 760° C). Upon completion of the burnout cycle the molds are stabilized at the appropriate metal pouring temperature of 550 to 650° F (290 to 340° C) for aluminum, ≈1,600° F (870° C) for brass, or ≈1,900° F (1,040° C) for ferrous alloys.

The temperature rise of the mold is controlled at 100 to 160° F (55 to 90° C) per hour to keep entrapped water in the mold material from flashing into steam and damaging the mold. The combination of the meltout cycle and the burnout cycle, followed by heating and stabilizing the mold at the pouring temperature, will completely eliminate all wax and any other combustible materials from the mold. Molten metal is poured into the mold with sufficient superheat to completely fill the mold before solidification begins. The metal can simply be poured, or the mold filled with the assistance of vacuum, pressure, vibration, or centrifugal force. Sometimes combinations of these methods are used.

Removing the Castings

Once solidified and allowed to cool, the castings are removed from the mold. Various methods are used, including vibrating the tree to break away the ceramic mold material. High-pressure water, in the range of 2,000 to 6,000 psi (13 to 40 MPa), can also be used to hydro-blast the mold material from the tree. After eliminating the majority of the exterior mold material, the castings can be placed in a caustic salt bath to remove any additional mold material from difficult areas such as blind holes and cored passageways. A combination of the methods described is commonly used.

Having successfully removed the tree from the solid mold, the castings are separated from the gate system with a radial cut-off wheel or a band saw. The gate tabs are ground close to the part, and any required heat treatment cycles are performed before the part is straightened. The straightening operation is necessary to remove any distortion that the casting may have experienced during cooling. Next, specified post-straightening heat treatment cycles are completed. The casting is then surface finished prior to any nondestructive testing (NDT) or x-ray inspection that may be required before final shipment to the customer.

Shell Investment Casting

The shell investment casting process involves essentially the same procedures as the solid mold process, except for the techniques used in creating the disposable ceramic mold. Shell gating systems differ from those in the solid mold method, due to the considerable difference in the thermal mass involved in the two mold forms. The ceramic shell approach allows a much faster solidification rate due to the considerably reduced thermal mass associated with the mold. Size and shape of the gating system must be altered accordingly.

Additionally, in the shell process the tree is required to provide structural support not only for the patterns attached to it, but for the weight of the encapsulating ceramic shell during its construction phase. A mandrel is attached to the tree to facilitate manipulation during processing. The tree is dipped in a precoat material prior to applying the primary or face coat, as in the solid mold process. The primary coat is generally applied by dipping the tree in a tank filled with the ceramic slurry or "slip." The tree is then withdrawn and rotated to allow an even coating of the slurry to be formed on the assembly.

Once the tree has received a thin encapsulating layer of slurry, a fine refractory grain is applied to the surface of the liquid slurry, forming a strong ceramic envelope. The tree is then placed in an environment where the ceramic layer will quickly dry, allowing subsequent coats to be applied in a similar manner. After the primary coats have been applied, the "back-up" coats are formed with slurry and grain of a larger size to build the shell thickness more rapidly. When the desired shell thickness is achieved, the mold is dewaxed and processed in a manner similar to that of the solid mold or flask method described previously. Schematic representations of both the "solid mold" and "shell" investment casting processes are shown in Figure 7-1.

Investment Casting Advantages

Investment casting has several important advantages relative to other casting methods. Many of these advantages result from the fact that ceramic investment molds, which hold the metal during casting, are one-piece molds. Furthermore, the molds do not need to be recovered after casting. As a result, no special considerations such as parting lines and draft angles have to be taken into account. This greatly simplifies part design and minimizes subsequent machining requirements.

Also, one-piece investment casting molds allow for greater part complexity. Many of the geometric limitations encountered with multipiece mold processes are avoided altogether. Investment casting allows production of thin wall sections, and typically yields parts with smooth surface finishes. For most applications, the surface finish of a cast part made from a wax pattern is acceptable after sandblasting. Investment accuracies are excellent relative to other casting processes. Design tolerances vary depending on part geometry. In general, linear tolerances of 0.003 in/in (0.003 mm/mm) are routinely achievable, with tighter tolerances definitely possible. More detailed tolerance information and design guidelines are available from investment casting foundries[2] or other general references.[3]

7.3 Tooling for Investment Casting

Various Molds

Female cavity molds or dies, suitable for reproducing wax patterns, are made from a number of materials and manufacturing methods. Renewable soft metal tools are produced from a master male pattern. The master is parted and the mold halves are created by pouring a soft metal, usually kirksite, about the master pattern. This type of mold is generally used for low-volume production applications.

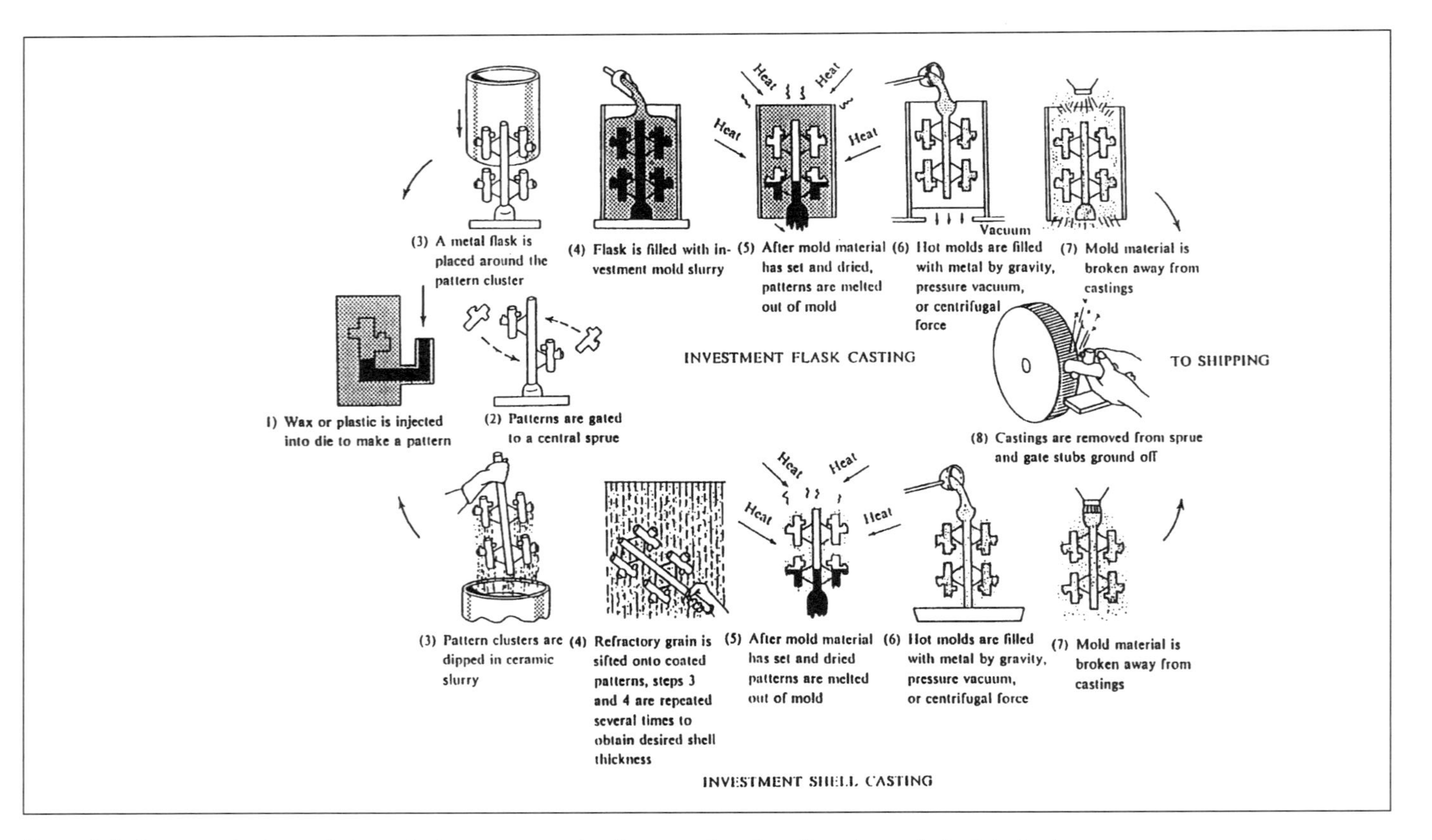

Figure 7-1. Basic production techniques for investment casting (courtesy American Foundrymen's Society, Inc.).

Rubber Molds. Rubber molds are typically made in a similar manner. A male master is initially fabricated, and then a reverse is taken to produce the cavity. Rubber has the substantial advantage of flexing and releasing from minor undercuts. However, a significant drawback to a rubber mold is the extended cycle time due to the low thermal conductivity of rubber, and the resulting inability to rapidly extract heat from the molten wax. Rubber molds are viable for only the most limited production requirements involving 25 to 50 pieces.

Epoxy Molds. Epoxy molds are produced in the same manner as soft metal and rubber molds, utilizing a male master and then taking a reverse. The epoxy often has material added to it to improve surface hardness and abrasion resistance. Secondary metal "fillers" can also improve the thermal transfer capabilities of the epoxy material. Epoxy molds are often enhanced by incorporating cooling water jackets to further reduce cycle time and to control dimensional repeatability. Aluminum-filled epoxy tools fabricated from RP&M masters are already seeing increased use in those applications that require low to medium production quantities, typically involving up to 1,000 wax patterns.

Spray Metal Tool. Another female cavity mold produced from a male master is a spray metal tool. Microscopic droplets of molten metal are sprayed against a male master where they rapidly solidify, forming a thin facing skin. The metal skin is backed with another material, usually metal-filled epoxy. Spray metal tools can incorporate the same cooling methods as epoxy molds. Due to their abrasion-resistant metal skin, spray metal molds are suitable for medium to high production quantities of wax patterns in the 2,000 to 4,000 piece range.

By far, the most prevalent wax injection tool is the machined aluminum die. Aluminum tooling stock produces core and cavity sets, with excellent surface finish, close dimensional tolerances, long tool life, and short cycle times. Mold life in excess of 10,000 wax patterns is not uncommon.

7.4 RP&M Investment Casting Patterns

The requirement for wax injection tooling in investment casting is one disadvantage of this process. Generating tooling is both expensive and time consuming. More importantly, this expense must be incurred up-front, before first article production. For these reasons, investment casting foundries were quick to turn to RP&M as an alternate means of rapidly providing accurate patterns for casting.

The use of RP&M patterns to fabricate preproduction samples has given casting designers a powerful tool for design verification. Several design iterations can be cast and tested prior to tooling commitment. This translates into significant financial savings since the costs associated with tooling rework, resulting from design changes, are minimized. The design cycle time is also significantly reduced since no machining and tool reworking operations are required between design iterations. Furthermore, the use of RP&M patterns to produce preproduction samples ultimately reduces the production cycle time itself, as these samples allow the investment casting foundry to optimize gating strategies and shrinkage factors prior to full-scale production using conventional wax patterns.

While providing investment casting designers with a practical method to reduce cost and cycle time, cast prototypes have also attracted considerable interest from designers of parts that are ultimately destined for production by other manufacturing processes. Since RP&M models are not yet available in metal, designers have turned to investment casting as a conversion technology to enable them to quickly transition from RP&M models to accurate functional metal prototypes. Thus, investment casting has become an important tool for any designer or manufacturing engineer seeking the shorter design cycle times of these new technologies, but limited by the available selection of RP&M materials.

7.5 QuickCast Patterns

An engineer's satisfaction with a particular solution is almost never complete. The criteria for acceptance is quite often influenced more by production schedule and budget constraints than the ultimate functional design being achieved. The quick turnaround of metal parts from RP&M patterns (2 to 3 weeks) has given the design engineer the luxury of spending more time in the CAD medium, achieving a greater design maturity than previously possible with conventional casting practices. This rapid casting process also allows more iterations of the physical part to be produced, tested, modified, and reproduced when necessary, than previously possible for a given time frame.

QuickCast Advantages

Expeditious. Short lead times are now possible for small quantities. Making additional pieces to compensate for possible machining scrap or alternate machine feature incorporation becomes unnecessary due to the fast turnaround time. RP&M now has the ability to: (1) quickly generate patterns without tooling for small lot sizes, (2) reduce the time necessary to yield functional parts, and (3) discover any manufacturing problems in the process. All these attributes are valuable.

Knowledge gained from RP&M efforts can also enhance the quality of conventional production tooling. This can occur by incorporating refinements in shrinkage factors based on results from actual configurations, rather than relying entirely on calculated predictions. Prototype casting results are also useful in tool design to accommodate production wax gating systems for the part configuration, while maintaining die functionality.

Easily Transported. The robust nature of QuickCast makes it practical to transport the pattern to customer or foundry sites for evaluation and acceptance prior to committing to the casting process. When properly packaged, these patterns can be transported great distances with little or no degradation. The robustness of QuickCast patterns is vastly superior to conventional investment casting wax pattern materials, where cracking and deformation during shipment are often encountered.

More Robust. Furthermore, QuickCast patterns are generally more robust than comparable wax configurations at both ends of the spectrum, from the very small and thin, to the very large and thick. This condition exists as a result of the mechanical properties of the epoxy resin itself, and also because the part volume is defined by a thin, internally supported skin. Void ratios have recently exceeded 80% for QuickCast 1.1 patterns. Hence, the part volume is more than 80% air. This results in an extremely lightweight pattern. The weight-to-volume ratio, being significantly less than that for a comparable wax pattern, also reduces the need for secondary pattern support structures in the gating systems of large parts, a common practice with wax patterns.

SL Most Widely Used

SL models continue to be the most widely used RP&M patterns for investment casting. QuickCast SL patterns are used as direct replacements for conventional shell investment wax patterns with a minimum of model preparation. The general processing scheme is the same as that for the conventional investment casting process. However, some modifications are required to account for the differences in behavior between wax and QuickCast patterns.

Cross-linked Polymers Don't Melt. These differences are the result of the physical properties of the photopolymer resins used in stereolithography. One such difference is the fact that cross-linked photopolymers do NOT melt. When SL patterns are heated, they tend to soften somewhat, but they never soften enough to flow. Consequently, QuickCast patterns cannot be removed from the ceramic investment shell using the conventional steam autoclave meltout cycle.

In processing QuickCast patterns, the steam autoclave cycle is used only to remove the wax runner and gating system attached to the SL model. The actual SL pattern is eliminated in the next step, which calls for firing the shell to yield a stronger cohesive ceramic body. By modifying the process conditions of this operation, the firing cycle can be utilized to simultaneously achieve the dual purposes of ceramic fusing and QuickCast pattern burnout.

To facilitate pattern burnout, conditions are adjusted to produce a rich oxidizing environment within the furnace. Typically, a minimum oxygen concentration of 10% is required. (An afterburner may also be necessary to minimize furnace emissions.) Burnout can be further enhanced by modifying the gating system during mounting to maximize air circulation within the molds. Ideally, molds should also be cooled down to room temperature after firing to allow for mold inspection and removal of any residual ash, if necessary. This is also a modification of the traditional casting process, which typically involves cool down of the molds only to the casting temperature, followed by direct metal casting.

Greater Stress. The coefficient of thermal expansion (CTE) is another important physical property of SL photopolymer resin materials having an impact on the casting process. Cured acrylate and epoxy SL resins both have a positive CTE about an order of magnitude greater than typical ceramic shell materials. Consequently, when heat is applied to the investment molds during autoclaving, or flash firing, the SL patterns will expand more than the ceramic. This thermal expansion will then exert considerable stress on the surrounding shell. In conventional investment casting, wax patterns also exert a similar expansion stress. However, since the wax has a very low melting point (about 105 to 125° F [40 to 50° C]), the stress buildup is quickly relieved as the wax reaches its melting point and exits the mold. With QuickCast 1.0, the expansion stress is not relieved until the combustion temperature of the resin is reached (roughly 750° F [400° C]). Thus, the ceramic molds must withstand greater stresses, over larger temperature ranges, when using QuickCast 1.0 patterns.

Initial experiments with solid acrylate SL patterns showed that traditional ceramic shells lacked the strength necessary to counter this expansion force. Catastrophic shell failure was the result. Some success was achieved using heavier shells, and shells made from higher strength materials. It was later shown that solid SL patterns were really only successful using the solid mold process. However, QuickCast eliminated the need for either of these approaches. The quasi-hollow QuickCast pattern was designed to collapse inward on itself during thermal expansion, rather than exerting an outward stress on the shell.[4] Enormous shell strength was, thus, no longer required. Nevertheless, a brief description of solid mold casting is included for historical perspective and completeness.

7.6 Solid Mold Investment Casting

As noted earlier, the solid mold technique involves the same general processing steps as shell investment casting, with the exception of the molding materials and the manner in which the mold is built up around the patterns. In solid mold casting, the pattern and gating assembly are placed inside a steel flask. The investment material, rather than being built up layer by layer, is poured into the flask forming a complete solid block around the pattern in one step. Solid molds, as the name implies, are considerably heavier and stronger than shell molds. Materials used in solid mold casting are typically similar to refractive materials used in plaster formulations.

Thicker and Heavier

The thicker and heavier nature of solid molds explains their success with solid acrylate SL patterns, based on their ability to better withstand the expansion forces of these patterns. An additional advantage of this process is a somewhat shorter processing time, since the investment mold is formed in one step, as opposed to layer-by-layer fabrication in shell casting. On the other hand, solid molds require longer firing times, and gradual heating to avoid thermal stresses.

Shortfalls. The heavier nature of solid molds, combined with the properties of solid molding materials, account for some of the shortfalls of this process. For example, the mechanical properties of castings are largely determined by the cooling rate of the metal after pouring. Solid molds are considerably thicker than shell molds. Thus, solid mold cast parts cool much more slowly and are therefore metallurgically and mechanically inferior to shell castings. Solid mold materials also have higher CTEs than shell materials. Hence, solid molds undergo greater thermal expansion and contraction deflections during the firing and cooling cycles. Dimensional control is thus more difficult, and solid mold castings are less accurate than shell castings, particularly for larger parts.

The superior accuracy and mechanical properties of shell investment castings provided an impetus for patterns that were more compatible with shell casting, despite the early success of the solid mold process using solid acrylate SL patterns. The initial testing of QuickCast in 1992, by 3D Systems, Cercast Group, Solidiform, Inc., and Precision Castparts Corporation, vastly improved the compatibility of SL patterns with the shell investment casting process.[5,6]

7.7 QuickCast and Shell Investment Casting

QuickCast is a build style that results in a quasi-hollow honeycomb-like SL pattern. This build style, in combination with the development of epoxy resins and the optimization of the internal hatch structure of QuickCast parts, has resulted in a more compatible SL investment casting pattern. Furthermore, all of this has been accomplished without sacrificing accuracy.

Advantages and Problems

QuickCast patterns can be shell investment cast more easily than earlier solid SL patterns for two primary reasons. First, these patterns have been found to exert less expansion pressure on the ceramic shell than solid patterns.[5] It has been suggested that this is probably a result of the quasi-hollow interior. As the QuickCast pattern is heated, and expansion pressure is exerted on the shell, the pattern begins to collapse inward rather than pushing outward, since at elevated temperatures the interior hatch structure is weaker than the surrounding ceramic shell. Second, the quasi-hollow pattern contains considerably less material to burn out than solid patterns, for any part geometry. This facilitates a more efficient burnout, and results in less residual ash in the fired mold, yielding castings of higher quality.

Surface Openings. While QuickCast 1.0 greatly improved the quality of prototype castings, some problems remained. One disadvantage of a quasi-hollow pattern common to all QuickCast patterns is the ease with which ceramic investment slurry can penetrate the interior if there are any openings on the surface. The outside surface of a QuickCast 1.0 pattern is fairly thin, and is susceptible to such openings. Holes on the pattern skin can be the result of improper support removal, resin dewetting during model building, or intentional "vent holes" and "drain holes" that have not been sealed. Another problem encountered with QuickCast 1.0 was difficulty with pattern drainage, particularly in thin wall sections. Incompletely drained features become completely solid on postcuring. The advantages of the quasi-hollow build style are lost, and these features will then behave in the same way as the older solid stereolithography patterns.

However, even with these initial problems, QuickCast has undoubtedly taken SL casting into a new era. Since the introduction of QuickCast 1.0 in June 1993, about 9,000 patterns, covering a broad range of sizes and geometries, have been successfully cast. Some of the sample castings presented in the literature,[6,7] as well as Chapters 6 to 9, are a testament to the success of QuickCast. It is important to note, however, that the successful production of a high-quality certifiable investment casting is not guaranteed. The overall foundry yield of successful shell investment cast QuickCast 1.0 patterns has been around 80%. The probability is higher for small simple parts, but may be lower for large complex parts with difficult to cast features. A foundry procedure for the preparation of patterns, as well as their processing, is listed in the following section. This information is presented to help users maximize the likelihood of fabricating a high quality casting on the first attempt.

7.8 QuickCast Foundry Procedures

After building a QuickCast pattern, it is important to ensure that pattern drainage is complete, to avoid unnecessary solid pattern sections. As many openings as necessary can be made in the pattern skin for this purpose, but these must be completely sealed after drainage. Gravity drainage is the most widely used method, although assisted drainage has also been employed using air pressure or vacuum. Low speed centrifuges, previously developed to extract honey from beehives, are producing excellent results.

Skin Integrity

When using a QuickCast pattern, the first foundry operation involves verifying the skin integrity. The holes, put in the pattern to allow venting and draining of the resin entrapped during the build cycle, must be closed. Sealing waxes of different compositions are readily available in the investment foundry and have been used with success. Experimentation with other materials, including epoxy photopolymers thickened with finely ground, solidified epoxy resin, and subsequently cured with handheld ultraviolet sources, are also showing promise.

Closing the Leaks. All holes must be closed but one. A specially machined brass hose barb is attached to this hole and then sealed tight with wax. Positive air pressure at 5 to 10 psi (0.035 to 0.070 MPa) is initially applied to the internal cavity. The external surfaces are then scanned by an operator using an automotive stethoscope. The stethoscope magnifies the sound of the escaping air and makes identification of any leaks rather easy. Once all obvious leaks are repaired and verified, the process is repeated, except this time a vacuum (roughly 10 inches of mercury) is pulled, and the external surface is again scanned with the stethoscope. The pattern should experience a 1- to 3-minute "no-leak" condition, under substantial vacuum, before being accepted.

Sound Frequency. Electronic stethoscopes are also being evaluated to increase sensitivity for the detection of small leaks. As an alternative to air and vacuum, some thought has been given to experimentation in high-frequency sound transducers to flood the interior of the pattern with a specific sound frequency. The electronic stethoscope would then be set to detect the frequency being used. The procedures outlined are certainly more elaborate than simply putting the pattern in a bucket of water, pressurizing with air, and watching for bubbles. Unfortunately, since the smallest amount of water is detrimental to the Ciba-Giegy epoxy resinsSL 5170 and SL 5180, pressure and vacuum testing are required.

Assembling the Wax Gating System

Once the pattern skin integrity has been verified, the wax gating system is assembled to the pattern and, in turn, the pattern is assembled to a tree. The general positioning of feeding gates is determined by standard foundry procedures. Some additional gating may be required for larger QuickCast patterns. These gates are used to provide air circulation channels within the ceramic mold rather than for metal feeding. The wax material used to construct these gates melts out quickly and leaves clean, open channels through which air can pass to the SL pattern during the rest of the burnout cycle. The position of these gates is determined so that air circulation is maximized to ensure a good oxygen supply and a clean pattern burnout.

QuickCast patterns are shelled using standard wax pattern procedures. Prior to burnout, each pattern has one or more holes opened into its interior at a wax gating location(s). This prevents a pressure buildup, due to expansion of trapped air, that would damage the mold. Next, the mold is either autoclaved or flash-fired to remove the wax gate material. The subsequent high temperature burnout operation at 1,830 to 2,100° F (1,000 to 1,150° C) simultaneously consumes the pattern material and fuses the ceramic shell. After the firing cycle, the molds are returned to room temperature. Any residual ash is removed with compressed air or an alcohol rinse, and the molds are then processed in accord with standard casting procedure.

7.9 Minimum Requirements for Quality Castings

The casting process dictates that certain part features are more readily cast than others. The foundry and design engineers should collaborate, whenever possible, to incorporate as many of the established casting parameters as possible into the part design. Maximizing the use of constant wall thickness, reducing the occurrence of isolated mass sections, and especially incorporating the use of generous fillet radii and corner radii, are strongly recommended.

The investment casting process is capable of reproducing fine detail. Surface detail approximations, as a consequence of low resolution solid CAD files, should be minimized. Furthermore, STL files should be generated at the highest possible resolution to reduce faceted surface artifacts on the casting. The use of SLC files may further improve pattern detail definition, especially on contoured surfaces.

Many cast configurations require a subsequent straightening operation to ensure that the casting is within tolerance. To aid the foundry in determining the true surface relationships, reverses of part geometry and datum plane surfaces can be obtained directly from the solid CAD model and subsequently reproduced using SL. The reverse forms then serve as excellent guides to check the straightening operation, and in some cases they have been cast in metal to function directly as straightening fixtures.

To compensate for expansion of the ceramic mold as it is brought to pouring temperature, as well as the volumetric shrinkage of the particular metal alloy being cast, a combined shrinkage factor is applied to the part configuration. The most common practice is to incorporate this factor in the build parameters of the SL process. An overall global shrinkage factor is established by the foundry process engineer, after taking into consideration the part configuration, as well as the SL build orientation. The shrinkage factor used in this method is inevitably a compromise because the build parameters only allow scaling in the primary coordinate directions (X, Y, Z).

Absolute factor analysis may result in as many as 20 to 50 discrete shrinkage factors applied for a given configuration. When the goal is a casting with the highest possible dimensional accuracy, all shrinkage values for the part description should be used as input data to define a multiple shrinkage factor compensated solid CAD model. The QuickCast pattern should then be generated from this CAD representation.

Foundry process engineers should utilize gating techniques that take into consideration the unique combustion requirements of the epoxy resin. This requires the creation of sufficient entry and exit channels to reduce the possibility of oxygen starvation during the burnout cycle. The size and placement of these "chimneys" is dependent on the amount of free oxygen available during the burnout cycle. If the stoichiometric ratio for the resin is not substantially exceeded, the polymer can be converted to a form that is not readily combustible. Noncombustible ash deposits in the mold will produce surface as well as internal defects in the casting.

Modifications to Ensure Proper Production

Regardless of the end use of a QuickCast pattern, its design should respect investment casting guidelines to ensure the successful production of a quality casting. If the pattern is intended as a prototype of a part destined for investment casting, this should be automatic and is not usually an issue. However, for parts designed for eventual production by other manufacturing methods, some modifications may have to be made to ensure the proper production of investment cast prototypes. Some general recommendations follow. Additional details can be obtained from investment casting foundry design guides.

Shrinkage Factors. Most metals, on cooling from the liquid to the solid state, will shrink volumetrically. To account for this shrinkage, investment casting patterns are always oversized by a specific shrinkage factor, as mentioned previously. The shrinkage factor will depend on the type of metal to be used, the part geometry, and, to a lesser extent, the properties of the shell mold. In conventional investment casting, it is fairly common to use as many as five or six different shrinkage factors for various part features. In SL casting, the application of separate shrinkage factors for individual pattern features is difficult. Thus, most foundries will recommend an "overall" global shrinkage factor to simplify the model building operation. It is very important that a foundry be contacted prior to building the SL pattern to obtain the appropriate shrinkage factor for a specific part.

Sharp Corners. Sharp corners on investment castings are high stress points that are susceptible to cracking during straightening and heat treatment operations. An additional problem arises for sharp internal corners between heavy and thin sections. After casting, the molten metal cools more slowly in heavy sections and more rapidly in thin sections.

Since metal shrinks during cooling, the thick and thin sections will pull away from a connecting joint at different rates. Cracks or distortions may form at such locations. To get around this problem, all sharp internal corners connecting thick and thin sections should have generous fillets. A minimum fillet radius of R = 0.04 inch (1 mm) is recommended. Larger radii may be required for heavier sections. Ideally, fillet radii should be built into the SL pattern. However, wax radii can be added manually at the foundry if needed. It is also important to apply radii to all sharp external corners. Sharp corners on a pattern can lead to shell thinning. This phenomenon can then cause stress concentrations to occur at external corner locations during burnout. Excessive local stress can subsequently result in catastrophic failure of the shell.

Blind Holes and Through Holes. It is often difficult to get sufficient shell material within small holes. This may result in a shell, weak in these areas, collapsing when fired. In general, holes that can be machined after casting (e.g., holes for screws) should be eliminated from the pattern. If they must be included, length/diameter ratios should be limited to L/D = 1 for blind holes, and L/D = 2 for through holes.

Cored Passages

In conventional investment casting of wax patterns containing long passages and channels, it is difficult to ensure that a sound shell coat is built up on interior sections. This is especially true if the passages are narrow. Preformed ceramic cores are often used to generate such features. A preformed core having the shape of the desired passage is placed in the mold and wax is injected around it. The shell is then built up over the complete structure, with the ceramic core essentially becoming an integral part of the shell. However, cores are not easily combined with QuickCast patterns. Cored passage features are perhaps the most difficult to produce, and a foundry should always be consulted for prototypes involving such features. Depending on part geometry, a foundry may recommend openings along the core to allow more shell material to enter the feature, or prestuffing the core with ceramic core mixes or, as a last resort, may suggest eliminating the passage from the prototype casting altogether, with the passage to be machined later.

Wall Thicknesses

QuickCast 1.0 patterns having wall thicknesses of less than 0.060 inches (1.5 mm) are difficult to drain and may not always remain quasi-hollow. On the other hand, solid walls are difficult to cast and may cause shell cracking as discussed in Chapter 6. Recent results for QuickCast 1.1 patterns show good drainage down to 0.050-inch (1.3-mm) wall thickness. The QuickCast 1.1 build style, which is strongly recommended in general, is therefore especially useful for parts with thin sections. Nonetheless, as a possible result of casting mold flow limitations, the foundry should be consulted for any parts involving wall thicknesses less than 0.060 inch (1.5 mm). The foundry will make appropriate recommendations based on the wall size and geometry.

Nearby vertical walls should be spaced far enough apart to allow adequate shell material to build up between them. The spacing required will depend on wall size and geometry. When the walls are thin and susceptible to incomplete drainage as described above, these features are high-risk areas and require close attention.

7.10 Case Studies of Casting Applications

Case Study—Cryptographic Housing

Figure 7-2 shows a large casting developed by Kurt Anliker and Todd Stahlhut at the U.S. Naval Air Warfare Center (NAWC). The project was part of an effort to develop a limited number of on-ship cryptographic housings. Such mission-specific hardware is often required in very limited quantities, where both the cost and time required to fabricate hard tooling for traditional production processes cannot easily be justified. Fabrication of these parts from QuickCast patterns, followed by shell investment casting, was selected as an alternative to conventional processes.

Figure 7-2. Large casting developed at U.S. Naval Air Warfare Center. The project was part of an effort to develop a limited number of on-ship cryptographic housings.

Because of the rather large size of this component (overall dimensions: 30 × 18 × 10 inches [762 × 457 × 254 mm]), the QuickCast pattern was built in multiple sections. The CAD file was sliced into seven sections and each part was subsequently built at NAWC on their SLA-250 using SL 5170 resin. The various QuickCast patterns were assembled using an epoxy glue. Two fully assembled patterns were built within a period of 10 days. Both patterns were subsequently cast at Cercast in A356 aluminum alloy within three weeks. The total project savings just for these two parts were estimated to be in excess of $15,000, when compared to other production methods. Additionally, the time saved through the use of QuickCast on this project has been estimated to be 4-6 months.

Case Study—Electronics Box

The electronics box shown in Figure 7-3 houses components for a high-performance fighter jet prototype. The housing was initially designed to be manufactured as an investment casting. Prior to fabricating hard tooling for conventional wax pattern casting, two prototypes were generated via QuickCast to enable design verification. The patterns were built in one piece on an SLA-500 and were subsequently cast in aluminum by Cercast. The resulting castings had a dimensional accuracy within ±0.001 inch/inch (±0.001 mm/mm) for the longest di-

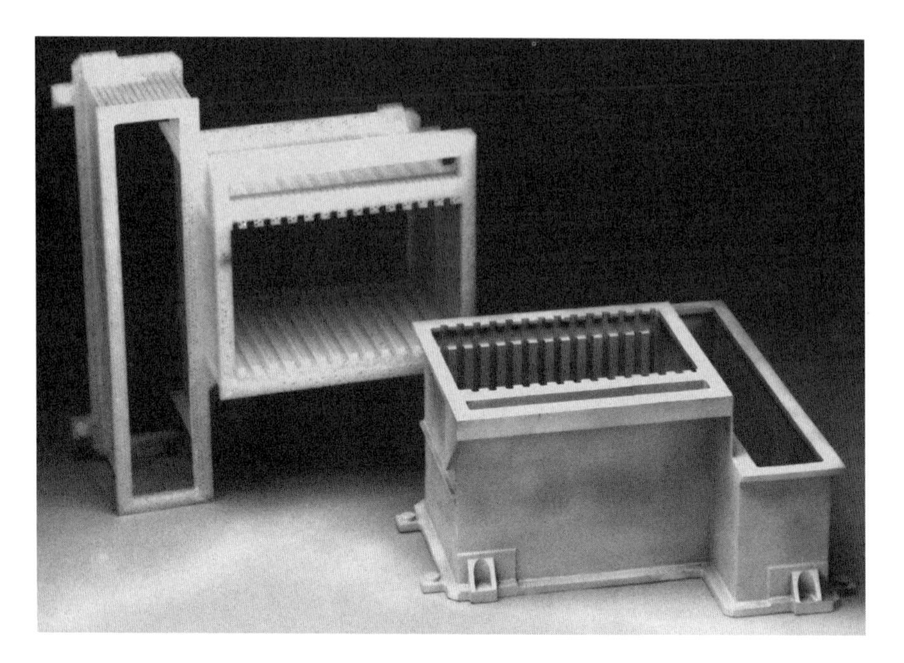

Figure 7-3. This electronics box houses components for a high-performance fighter jet prototype. The patterns were built in one piece in an SLA-500.

mensions, and met Grade B aerospace x-ray requirements. Both castings were successfully tested under actual operating conditions. Several fine-tuning iterations were made prior to fabricating hard tooling for production.

Case Study—Stingray Optical Chassis

The chassis shown in Figure 7-4 provides support structure for components of a targeting device built by Lockheed-Martin Aerospace Corporation, Orlando, Florida, for their Stingray program. The casting shown is among the largest and most complex parts produced via QuickCast (16 × 17 × 18 inches [406 × 432 × 457 mm]). Seven prototype castings of this part were fabricated. The patterns were built on an SLA-500 using SL 5180, and the castings were produced in aluminum by Cercast Group, Montreal, Quebec, Canada.

With a lead time of 36 weeks and a cost of $250,000, the conventional process was too slow and too expensive. These cast prototypes were fabricated to complete the part design before the production of hard tooling. Tony Corpuz, director of Lockheed-Martin's rapid prototyping laboratory, estimates that this single project saved $78,000 and 24 weeks. The design for this optical chassis ultimately underwent seven iterations. Had these changes been made after hard tooling was produced, the additional costs and time delays associated with tooling rework would have made this project unprofitable.

Figure 7-4. The chassis shown here provides support structure for components of a targeting device built by Lockheed-Martin Aerospace Corporation for its Stingray program. This casting is among the largest and most complex parts produced via QuickCast.

Case Study—Subframe Support

The part shown in Figure 7-5 is a subframe support structure intended for ultimate use on the International Space Station. As with most parts destined for space applications, size and weight are critical. Consequently, the production of this part via QuickCast became an attractive option. It was uncertain, however, that the strict specifications of the international space program could be met via QuickCast. The requirements for this specific part included dimensional tolerances of ±0.002 inch/inch (±0.002 mm/mm), excellent metallurgical/mechanical properties, as well as Grade B radiographic requirements.

To test the feasibility of this project, Bob Davis and Steve Irwin of the Hamilton Standard Space & Sea Systems Division of United Technologies Corporation selected two SL patterns to be produced on an SLA-500 using epoxy resin SL 5180 and the QuickCast 1.1 process. The patterns were later invested using Cercast's high mechanical properties process. Both of these castings were completed successfully in A357 aluminum alloy and, more importantly, the castings were found to meet all the desired characteristics, including accuracy, as well as superior mechanical properties. The project convinced Hamilton Standard that QuickCast small-lot production is indeed viable. Future production runs of the subframe support will be produced via QuickCast exclusively—no other production process will be utilized.

Case Study—Four-cylinder Engine Block

The ability to fabricate and test a completely new automotive engine design, with all its auxiliary components, is an expensive and time-consuming procedure.

Figure 7-5. Two illustrations of subframe support for international space station. Top photo shows QuickCast pattern; bottom is metal casting.

To keep abreast of consumer expectations, as well as respond to and create competitive pressure, new prototyping methods have to be developed and tested. The Mercedes-Benz Division of Daimler-Benz AG has initiated a program of physical design verification on prototype engines using SL components for initial form and fit testing, where applicable. After completing the preliminary design reviews, functional metal components are now produced using QuickCast.

The first major structural component to be cast by this process was a complete four-cylinder engine block for the new Mercedes-Benz "A class" car. Arnold Zieger, Berthold Mueller, and Juergen Weber were responsible for the project. The design was developed in CATIA by Mercedes-Benz at its engine plant in Stuttgart, Germany. The CAD solid model was forwarded to the 3D Systems Tech Center in Darmstadt, Germany. There, a one-piece pattern of the entire engine block was subsequently built on an SLA-500 using epoxy resin SL 5180 in the recently released QuickCast 1.1 build style.

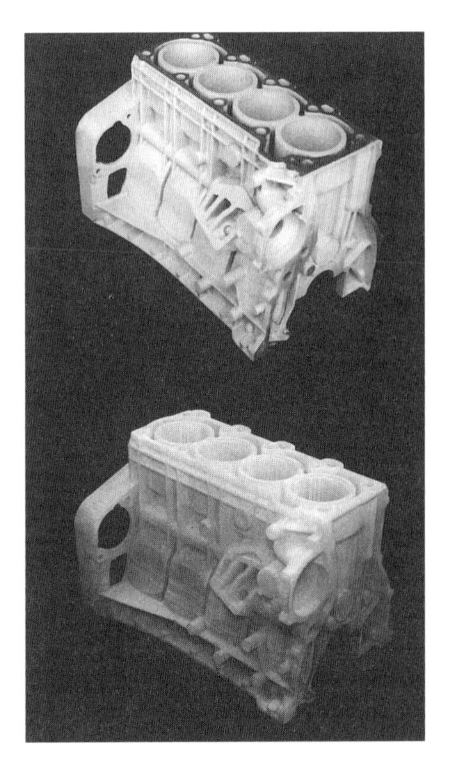

Figure 7-6. This four-cylinder engine block was the first major structural component cast from a QuickCast 1.1 pattern for Mercedes-Benz.

The pattern was then forwarded by air freight to Solidiform, Inc., Ft. Worth, Texas, for shell investment casting. LaMar Daniels, Director of Research and Development at Solidiform, oversaw the special procedures required to produce the 12 × 13 × 18 inch (305 × 330 × 457 mm), 40 pound (18.1 kg) aluminum casting in just five weeks. The completed engine block, shown in Figure 7-6, incorporated cast-in water jackets, 5/8 inch × 18 inch (16 × 457 mm) cored passageways, and exhibited grade B radiographic quality in all areas evaluated. The A356-T6 aluminum cylinder block was entered by Mercedes-Benz in the 1994 European Stereolithography Users Group Excellence Award Competition, taking first place.

Case Study—Torque Converter

Increased fuel economy is a never-ending quest at Ford Motor Company. One of the primary areas for improvement, other than engine efficiency, is the reduction of drive train losses. For some time, the torque converter in automatic transmissions has been a target for further improvement. Considerable effort has been expended to create sophisticated computer simulation models to test new design advances.

The computer models have generated some potentially significant improvements, but the numbers are just numbers until verified by physical performance tests on a dynamometer. Quickly producing metal duplicates of the computer-generated solid CAD models for functional testing was a challenge. Product Design Engineer Eli Avny of Ford Motor Company Automatic Transmission and Engineering Operations decided to try QuickCast.

One major component of the torque converter is the turbine, which consists of a 10.375 inch (264 mm) diameter × 2 inch (51 mm) high × 0.070 inch (1.78 mm) thick outer shell, with 23 contoured integral blades sandwiched between the inner shroud and the outer shell. Consolidated Technologies Inc. (CTI), of Xenia, Ohio, was commissioned by Ford to produce the prototype turbines. CTI assembled a team of suppliers to perform the CAD modeling, SL pattern building, shell investment casting, final machining, and inspection. Bastech Engineering, of Dayton, Ohio, incorporated the casting producibility enhancements into Ford's computer model, and generated the STL files. The QuickCast pattern and the resulting aluminum casting are shown in Figure 7-7.

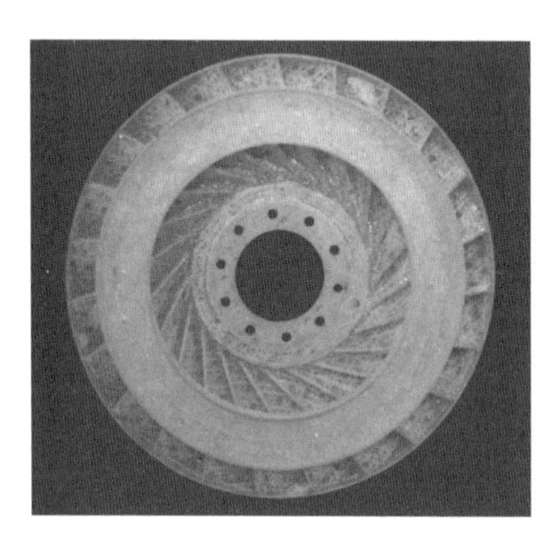

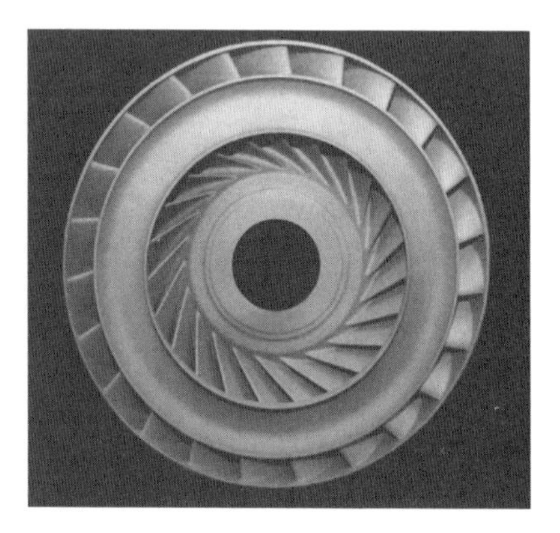

Figure 7-7. Ford torque converter. Left photo shows QuickCast casting; right depicts resulting aluminum casting.

Total elapsed time from the instant the STL files were released to produce the QuickCast pattern, until receipt of the A356-T6 aluminum turbine for final machining and assembly at Ford, was just three weeks. Of this, one week involved Solid Concepts, Valencia, California, building the QuickCast pattern on an SLA-250 using epoxy resin SL 5170. Solidiform, Inc., Ft. Worth, Texas, then produced the investment cast turbine over the next two weeks. Machining and inspection were accomplished in one day by Royer Technologies, Springboro, Ohio. Ford performed the final assembly and performance testing the following week. Release of the STL file to actual performance data on an operating prototype torque converter took just four weeks. The program continues to produce follow-on iterations for evaluation.

Case Study—Receiver Structure

QuickCast has allowed Owen Baumgardner of Texas Instruments, Inc., Defense Systems & Electronics Group, Lewisville, Texas, the luxury of further maturing a model in CAD. This is now possible because of the extremely fast turnaround capability foundries have recently achieved when generating metal configurations from QuickCast patterns, relative to conventional methods. Such was the case for this large, complex receiver structure for an aerospace electronic application. The designer was able to incorporate more functional enhancements into the configuration as the subcomponent designs matured.

The A356-T6 casting is shown in Figure 7-8. The casting was produced by Solidiform, Inc., just two weeks after receipt of the customer-supplied SL 5180 pattern. This two-week interval involved only one-tenth the time normally required for production of hard tooling needed to injection mold conventional wax patterns. Not only was the program able to forestall \$86,000 in tooling costs until after six functional design iterations were completed using QuickCast produced castings, but the costs associated with the tooling rework that would have been necessary for each of these iterations was avoided altogether.

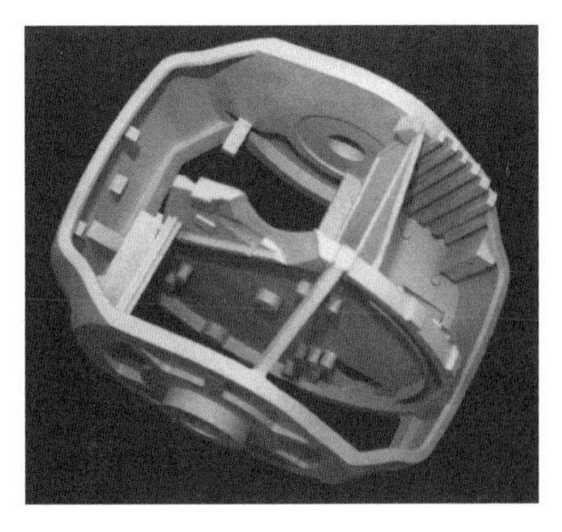

Figure 7-8. This casting for a receiver structure was produced by Solidiform in just two weeks after receiving the SL 5180 pattern from the customer.

Case Study—Plastic Injection Mold Core and Cavity

Obtaining plastic configurations molded in production material under production molding conditions in less time than conventional mold construction practice is a "rapid tooling" goal at the Delphi Division of General Motors Corporation. A solid CAD model of a glove compartment door for a new production vehicle was generated using GM-CGS. The original CAD model was sent by Tom Greaves, of Delphi, to a plastic mold shop where reverses of the CAD model were taken to create the core and cavity inserts.

QuickCast 1.1 patterns for the inserts were built by Hop Nguyen and Bryan Bedal of the R&D Department at 3D Systems, and were shipped to the Delphi mold shop where initial CMM analysis was performed on the patterns prior to shipment to Solidiform, Inc. The 3 × 12 × 20 inch (76 × 305 × 508 mm) inserts were cast in A356-T6 aluminum within three weeks and shipped back to the mold shop for final CMM evaluation. The QuickCast core and cavity patterns, as well as the cast tooling inserts, are shown in Figure 7-9. The inserts have been machined, polished, and subsequently mounted in a master mold base to successfully obtain test shots with production injection molded material.

Case Study—Rubber Duct Joint Core and Cavity Tooling

Metal molds for rubber injection molding are yet another application of "rapid tooling" now being successfully implemented. Pro/ENGINEER™, and its associated module Pro/Mold™, was used to create the split core and cavity segments

Figure 7-9. QuickCast core, cavity patterns, and cast tooling inserts for a glove compartment door.

for this flexible rubber duct joint. The solid CAD models and QuickCast patterns were generated at Texas Instruments, Inc., Defense Systems & Electronics Group (DSEG), Lewisville, Texas, and shipped to Solidiform, Inc. where they were cast in A356-T6 aluminum in two weeks. The QuickCast patterns and the resulting investment cast core and cavity tooling inserts are shown in Figure 7-10.

According to Owen Baumgardner, Manager of Fabrication Producibility at TI, DSEG, the cast inserts received minimal machining prior to being assembled in a master mold base for rubber injection. The initial shots, as well as the subsequent production run, satisfied all the engineering requirements without modification. This "rapid tooling" effort resulted in savings of 10 weeks and $4,800 relative to the conventional approach. The time, cost, and quality objectives for this project were all met, and provided a basis for follow-on projects using the same methodology.

Case Study—Straightening and Check Fixtures

To rapidly and accurately fabricate many cast configurations, it is also necessary to rapidly and accurately generate associated production straightening fixtures and check fixtures. This case study involves a contoured missile door having none of its surfaces parallel to any of the primary planes. A Boolean reverse of the contoured surface was performed, with the surfaces extended to allow for placement of tooling buttons. The CAD model was integrated onto a solid base block about 1.5 inches (38 mm) thick, to provide enhanced stiffness.

One QuickCast 1.1 pattern and one solid ACES model of the check gage were built on an SLA-250 using epoxy resin SL 5170. The QuickCast pattern was shell investment cast at Solidiform, Inc., in A356-T6 aluminum. This casting then functioned as a straightening fixture, as shown in Figure 7-11. The ACES model was verified on a CMM and served as a check fixture. When a missile door casting is placed against the buttons of the check fixture, the surface contour can be tested with a feeler gage.

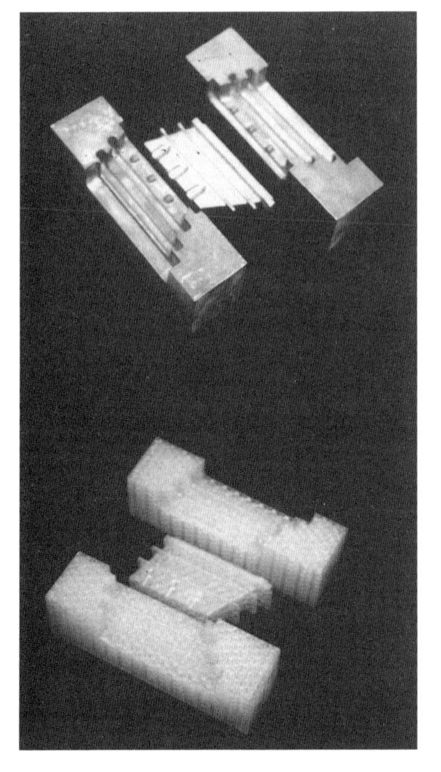

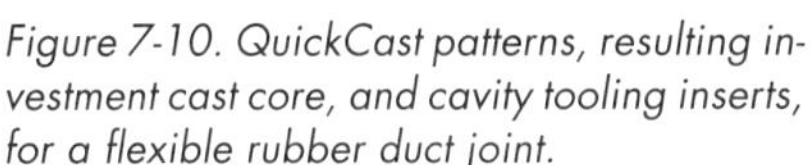

Figure 7-10. QuickCast patterns, resulting investment cast core, and cavity tooling inserts, for a flexible rubber duct joint.

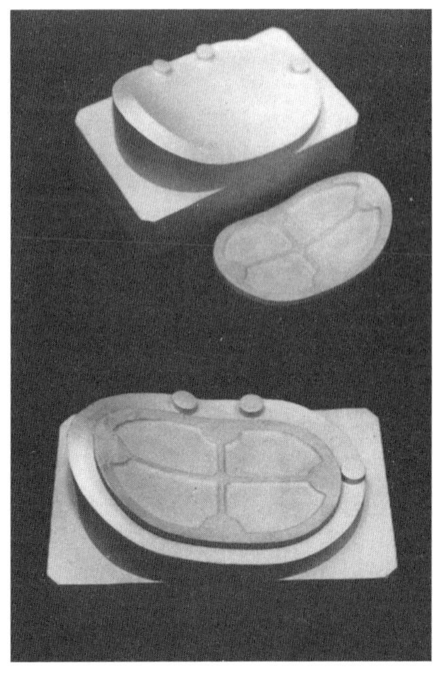

Figure 7-11. This shell investment casting, made from a QuickCast pattern, functioned as a straightening fixture for a contoured missile door.

Case Study—Transmission Oil Transfer Tube Gage

On occasion, a steel oil transfer tube became unseated at one end when press fit into an aluminum transmission case. Both the "drop-length" of that endform and the contour of the tube were suspect as potential sources of the assembly problem. As a result, manufacturing engineering at Ford Motor Company procured a go/no-go gage for tube configuration that could be utilized to quickly diagnose if the tube was out-of-print with respect to these characteristics.

A CAD solid model of the tube was created from part print specifications using CADDS5 software. Because the end result needed to be a usable gage, the tube solid was subtracted from a solid block in CAD, thus leaving a void where the tube was previously located. The next step was to section the CAD model, along the centerline of the tube, to produce a "female half" tube gage. Finally, some of the excess portion of the block was removed to reduce mass. The finished solid model was completed in approximately two days.

The database was then forwarded to Solidiform, Inc. to review various aluminum casting considerations. The QuickCast pattern was built by 3D Systems Tech Center and subsequently cast by Solidiform, Inc. in A356-T6 aluminum. The gage is shown in Figure 7-12, along with a mating production oil transfer tube, and is now available to check production tubes.

7.11 Future R&D for the Foundries

Cross Platform, Solid CAD Translators

The ability to translate a solid model from one CAD system to another is extremely important. Foundries, due to financial constraints, will primarily support only one CAD system. Thus, translation capability is critical to quickly quote, shrinkage compensate, and gate a customer-supplied part file. The STEP initiative may be one answer to this problem.

Windows™-based STL Viewer

The STL file provides a common denominator for the viewing of part files from many CAD systems. The STL file quite often provides the stereolithographer enough detail for quotation purposes and it may, in many cases, serve the needs of the foundry as well. This capability would best be available on a PC, and would ideally have the ability to perform planer cuts in order to find widespread use in the foundry environment.

"Multiple Shrinkage Factor Compensated" Solid CAD Model Generator

End users and foundries both need the ability to generate a solid CAD model of a part that already includes the appropriate, geometry dependent, multiple shrinkage factors for different cast alloys and different foundry practices. Only when the foundry engineer is able to incorporate all the various shrinkage factors that are currently used in the generation of investment cast tooling, can the expectation of equaling or improving conventional casting tolerances be realized. This will be especially important for QuickCast tooling applications.

Intelligent Internal Hatch Generator

An intelligent QuickCast internal hatch generation system is needed. The basis for QuickCast involves a thin photopolymer skin defining the part geometry, supported by a collapsible internal resin hatch structure. This concept was well founded, and its viability has been established in practice through production of thousands of successful investment castings.

Although the procedure has worked well, there is still variability in the process from one configuration to another, as evidenced by shell cracking that periodi-

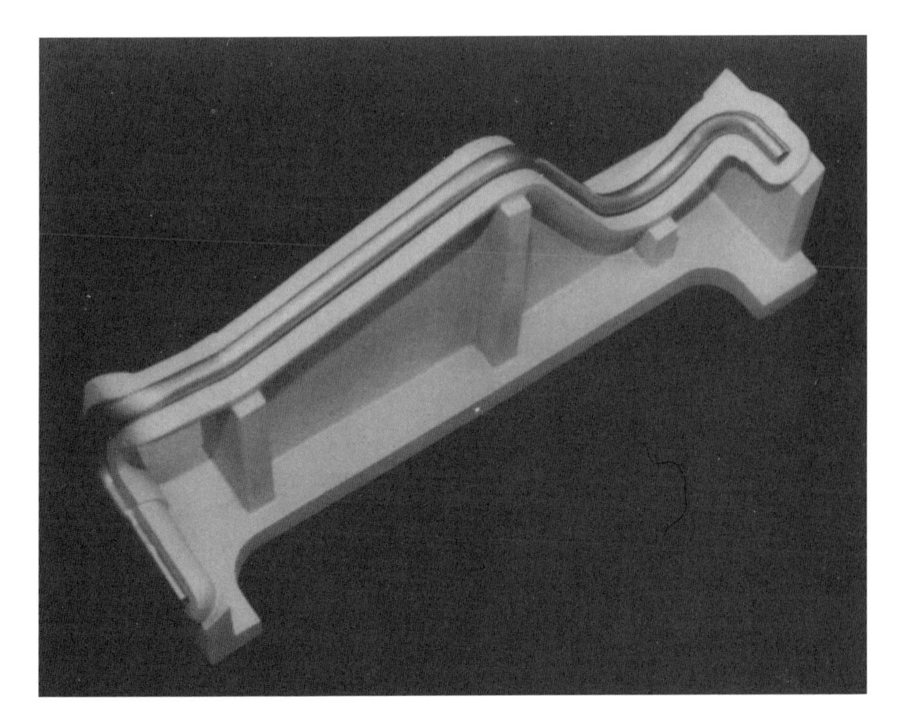

Figure 7-12. Gage used to check steel transmission oil transfer tubes.

cally occurs when the user employs all established practices. The cracking is consistently found in areas where solid resin has a thickness greater than about 0.050 inch (1.27 mm). It is in these locations that part geometry, in conjunction with the hatch spacing selected, as well as its orientation, can generate thermal expansion stresses in the ceramic shell exceeding the flexural strength of various commercial shell formulations.

Based on the fundamental "collapsing internal hatch structure" concept of QuickCast, it would be desirable to analyze the part geometry and control part border widths, to ensure that no undrained solid walls having a thickness greater than 0.050 inch (1.27 mm) can occur. Additionally, using a finite element analysis (FEA) program that takes into consideration part geometry and the pattern's structural strength, it should be possible to create a quasi-hollow internal structure with variable hatch spacing.

This variable hatch structure would utilize narrow QuickCast hatch spacings in some regions and wide hatch spacings in others. The intent here is to ensure that the ceramic shell stress during burnout always remains below the critical breaking stress for a particular shell system. To maintain adequate pattern rigidity and dimensional stability prior to burnout, minimum QuickCast flexural strength requirements would also need to be established.

It can be seen that this methodology could result in some configurations yielding a structure without any internal hatch at all. The part geometry, in conjunction with a particular skin thickness, might satisfy the strength requirements previously outlined. Patterns approaching 100% void ratio have been built and successfully cast. The potential exists for QuickCast patterns not only to equal their wax counterparts with respect to ceramic mold stress generated during pattern elimination, but, in fact, to substantially reduce the stresses encountered.

Pattern Structural Integrity Validation

FEA is often performed on proposed casting designs. The mechanical property values used in the analysis are, appropriately, those of the cast alloy. To minimize the possibility of dimensional integrity being compromised in the casting due to pattern distortion, the casting design should also undergo FEA evaluation using resin material values. Internal hatch structure could be added or deleted in patterns as required. Interim external support structure could also be added where necessary, and later removed after the cast configuration is complete.

Part Resolution/Build Time Tradeoff

It would be very helpful for users to have a software program that could calculate the dual requirements of "build time" and "part resolution." Inputting a maximum acceptable build time value, together with a minimum acceptable part resolution value, would result in a build procedure where surfaces perpendicular to the resin plane would call for the maximum allowable layer thickness. This would result in the shortest build time. However, when the part geometry is at an angle to the resin plane, and high resolution is desired, then the minimum allowable layer thickness would be employed. Combinations between the extremes for both variables could be established, enabling the individual user to select the most appropriate build time/part resolution tradeoff for their particular application.

Meniscus Smoothing

When excellent surface resolution is needed (e.g., for rapid tooling), another technique, in conjunction with minimum layer thickness, would be of value. Meniscus smoothing utilizes the natural capillary attraction of the liquid polymer for cured versions of itself. This action can reduce the "stair-step" effect due to finite layer thickness. After a layer is completed, the part is raised above the liquid level, allowing a meniscus to form. Appropriate laser exposure of the residual meniscus would cure this resin, thereby greatly reducing stair-stepping errors.

Stereolithography with Statistical Process Control

SL process variables are well defined, lending themselves to real time statistical process control methods. Provided all process variables are controlled, and yield an acceptable Cpk value, then the process as a whole is under control. Minimal dimensional verification would then be necessary to conclude that an accurate reproduction of the CAD solid model had been achieved in the SL process. Predictive process control is likely to become the hallmark of the RP&M world, where the goal should be nothing less than 100% first-time yield.

References

1. McCaughin, J., *The History of Investment Casting*, J.F. McCaughin Co., Rosemead, CA.
2. Engineering Staff, *Casting Design Guide*, Cercast Group, Montreal, Quebec, Canada, Internal Report, 1990.
3. Greenwood et. al., "Ceramic-Mold Casting," *Tool and Manufacturing Engineers Handbook*, 3rd Edition, Dallas, D., editor, Society of Manufacturing Engineers, Dearborn, MI, 1976, pp. 21.6-21.10.
4. Jacobs, P. and Kennerknecht, S., "Stereolithography 1993: Epoxy Resins, Improved Accuracy & Investment Casting," Proceedings of the SME Rapid Prototyping and Manufacturing '93 Conference, Dearborn, MI, May 11-13, 1993.
5. Blake, P., Baumgardner, O., Haburay, L., and Jacobs, P., "Creating Complex Precision Metal Parts using QuickCast™," Proceedings of the SME Rapid Prototyping and Manufacturing '94 Conference, Dearborn, MI, April 26-28, 1994.
6. Sarkis, B., "Rapid Prototyping for Non-ferrous Investment Casting," Proceedings of the SME Rapid Prototyping and Manufacturing '94 Conference, Dearborn, MI, April 26-28, 1994.
7. Kennerknecht, S. and Sarkis, B., "Rapid Prototype Casting (RPC): The Fundamentals of Producing Functional Metal Parts from Rapid Prototype Models using Investment Casting," Proceedings of the Fifth International Conference on Rapid Prototyping, University of Dayton, Dayton, OH, June 12-15, 1994, pp. 291-300.

CHAPTER 8

QuickCast Applications

"Talk does not cook rice."

Chinese Proverb

In all industrial engineering disciplines, prototype parts are required to test design and manufacturing approaches before committing a design to production. These prototypes are used to test the form, fit, and function of the intended components. Typical RP&M parts have been used successfully to provide form and fit prototypes in a shorter cycle time and at a lower cost, resulting in improved designs. Unfortunately, additional prototype requirements exist that could not be accomplished through the early rapid prototyping technologies. These requirements involve functional testing of parts, designs, and manufacturing processes that require components made of end-state materials. To accomplish this, additional costs are incurred to either produce tooling or utilize CNC machining. In addition to functional prototype requirements, industry continues to seek new processes to generate production quality in production quantities.

The QuickCast process has enabled Texas Instruments to accomplish both requirements, resulting in significant benefits. The benefits realized to date include, but are certainly not limited to: (1) the ability to produce functional prototype parts in the desired end-state aluminum material, (2) generating quality investment cast components for initial production requirements, as well as (3) the ability to integrate investment castings into the initial stages of design and tooling to produce large-scale production quantities. These benefits have resulted in lower design and production costs, decreased cycle time, and improved design and part quality.

8.1 History of SL and Investment Casting at Texas Instruments, Incorporated

Texas Instruments' Defense Systems and Electronics Group has utilized RP&M technology since 1989 in the mechanical product design and development pro-

---

By Paul Blake, Mechanical CAE Systems Engineer, and Owen Baumgardner, Manager, Fabrication Producibility, Texas Instruments Incorporated, Lewisville, Texas.

cess. The immediate benefits realized concentrated on: (1) design verification (form, fit, and function), (2) concept communication models, and (3) investment casting pattern generation. These three applications have been enormously successful and have demonstrated substantial payback.[1]

Along with these benefits, TI developed the techniques to utilize SL patterns in two different investment casting processes. These include development of the solid mold process in 1990-1993, and the testing and use of QuickCast in 1993-1995. Also, tooling applications from RP&M parts were explored as early as 1990. Today, SL pattern development has been accepted division-wide as a way of doing business in the manufacturing of investment castings for the initial stages of production programs. Castings are routinely produced in four weeks with no initial monetary investment in tooling. This compares to a 12- to 26-week cycle time and an average $25,000 tooling expense using conventional methods. Nearly 11,000 parts, involving all three applications, have been produced at Texas Instruments on seven SLA machines (four SLA-250s and three SLA-500s) since 1989. Of these parts, over 4,000 have already been used for casting pattern generation.

Two Requirements

Texas Instruments has two unique requirements that facilitate the application of RP&M to the investment casting process, both of which receive immediate payback. Furthermore, this payback is independent of the technique used. The first requirement involves the relatively small initial lot sizes characteristic of many Defense Systems & Electronics Group projects. Engineering development programs are not often required to produce high quantities of deliverable systems in the initial stages of production. Typical requirements dictate delivery of 1 to 12 systems for customer test and evaluation before high rate production is started. In the past, parts were fabricated using traditional machining methods in order to meet these low quantity requirements, because the use of castings did not provide sufficient payback.

The second requirement is that TI's material applications are limited in scope. Over 95% of the investment castings fabricated for TI are made of aluminum. This fact, coupled with the low rate production run, places TI at an advantage over typical automobile, medical, and aerospace companies with significantly different material and lot size requirements. These advantages show RP&M as a way to integrate investment castings into the system design earlier in the production cycle, without additional cost or cycle time penalties.

Initial Problems. Immediately upon the arrival of the first SLA-250 system in 1989, test patterns were built in XB 5081-1 resin (now SL 5081-1) and sent to several well known investment casting foundries. Initial foundry results were not favorable, as the solid SL pattern would undergo greater thermal expansion than the ceramic investment shell during the pattern "burnout" cycle. This would then result in a cracked and useless shell.

In the early 1990s it was decided to attempt using the older "solid mold" technique for investment castings. Because the investment tree and ceramic mold were backed by a steel flask, the expansion differential did not produce undesirable results. Production castings were fabricated at the time.

From 1990-1993, a total of 720 cast parts were produced for 140 different designs using solid SL patterns in conjunction with the solid mold process. Part size ranged from 1 × 0.5 × 1 inch (25.4 ×12.7 × 25.4 mm) from the SLA-250, to as large as 18 × 20 × 20 inches (457 × 508 × 508 mm) from the SLA-500. Factors were applied to each pattern allowing for shrinkage of both the SL resin and the investment cast metal. The resulting solid mold investment castings were typically Grade "D." Figure 8-1 shows a representative casting produced using this method, as described in Reference 2.

Figure 8-1. Initial investment casting from the solid mold technique; circa 1990.

At the time the solid mold technique was utilized, TI worked around several limiting factors, including: (1) the small number of foundries capable of production using the solid mold process, (2) problems consistently achieving castings better than Grade D, and (3) the inability to cast other materials (i.e., steels and high temperature alloys).

While the solid mold technique had been successful, during a 24-month period TI transitioned from solid patterns, using the solid mold process, to the QuickCast build style using the shell investment casting method. The QuickCast process eliminated the three constraints and broadened the base of manufacturing applications.

8.2 QuickCast Casting Examples

In the spring of 1993, Texas Instruments began using the QuickCast build style and epoxy resin as part of the SLA-250 QuickCast beta test program. Upon successful completion of the testing, all additional SLA-250s at TI were converted to the epoxy resin XB 5170 (now SL 5170), and the use of the QuickCast build style. In the fall of 1993, TI began testing the QuickCast process on some SLA-500 machines that had been converted to the epoxy resin SL 5180, and are currently also used to build QuickCast patterns.

While the QuickCast process was refined as problems were uncovered and solved during the beta test period, the fundamental process was immediately successful. Several factors contributed to this success. These included a good working relationship with the foundries, using parameters provided by 3D Systems with only minor deviations, applying practical manufacturing knowledge, and finally a corporate culture and team willingness and desire to try something new.

The following are specific examples of successes using QuickCast to produce deliverable castings, and also for developing a number of tooling applications.

Example 1: Electronics Tray (See Figure 8-2)

Application:	Deliverable Cast Part
Time Period:	3rd Quarter 1993
Machine:	SLA-500
Resin:	LBM 5353 (now SL 5180)
Quantity:	40 of 2 different configurations
Size:	6 × 14 × 2 inches (152 × 355 × 50 mm)
Cast Material:	A356-T6
Tool Savings:	≈$3,500
Cycle Time:	≈4 weeks
Cycle Time Savings:	≈12 weeks

This example is typical of components produced using QuickCast. The Grade C aluminum castings are now used in system test and qualification. This part highlights the benefits of the QuickCast process. First is the rapid fabrication of parts without tooling. Second is postponing the tooling cost until the part has passed all qualification tests and is ready for final production. Third, the expense normally needed to update the tool is not incurred as the design is changed. And fourth, time savings of about 12 weeks are realized relative to the traditional tooling cycle required for investment casting.

Example 2: Platform (See Figure 8-3)

Application:	Deliverable Cast Part
Time Period:	2nd Quarter 1994
Machine:	SLA-500
Resin:	SL 5180
Quantity:	12 of 2 different configurations
Size:	14 inch (355 mm) diameter; 6 inch (152 mm) depth
Cast Material:	A356-T6

Tool Savings:	≈$27,000
Cycle Time:	≈4 weeks
Cycle Time Savings:	≈12 weeks

This is another part in a long line of investment castings produced by TI. Items to note include the improved surface finish, undercut, and compound curve part geometry.

Example 3: Printed Wiring Board (PWB) Chassis (See Figure 8-4)

Application:	Deliverable Cast Part
Time Period:	1st Quarter 1994
Machine:	SLA-500
Resin:	SL 5180
Quantity:	6
Size:	10 × 6 × 8 inches (254 × 152 × 203 mm)
Cast Material:	A356-T6
Tool Savings:	≈$8,500
Cycle Time:	≈5 weeks
Cycle Time Savings:	≈10 weeks

The ability to cast flight hardware with PWB card guide features directly from QuickCast patterns has significant merit to the design and manufacturing engineers at TI. While card guide part features had been produced with the solid mold technique, the QuickCast process has improved the consistency and reliability of the investment castings. This has increased confidence in our ability to maintain the feature tolerances required.

Example 4: Housing (See Figure 8-5)

Application:	Design Verification Model/Investment Casting Pattern
Time period:	3rd Quarter 1994
Machine:	SLA-500
Resin:	SL 5180
Quantity:	12
Size:	18 × 19 × 16 inches (457 × 483 × 406 mm)
Tool Savings:	≈$85,000
Time Savings:	≈22 weeks

Note the complex geometric configurations on this part, including thin walls (≈0.070 inch [1.8 mm]), numerous bosses, and PWB card guides. Also note certain dark areas on the model that required manual wax patching to seal any holes or torn surface skin areas. This "pinhole" problem happened while removing supports from the QuickCast 1.0 model.

Example 5: Electronics Housing (See Figure 8-6)

Application:	Design Verification Model and Investment Casting Pattern
Time Period:	3rd Quarter 1994
Machine:	SLA-500
Resin:	SL 5180
Quantity:	10
Size:	17 × 8 × 18 inches (432 × 203 × 457 mm)
Tool Savings:	≈$65,000
Time Savings:	≈22 weeks

Once again, note the unique geometric features on this model. The feature of greatest interest is the "pin fin" configuration on the left side wall. Pin fins are used in airborne electronic chassis for cooling purposes. The pin fins are typically 0.125 inch (3.175 mm) in diameter and 0.75 inch (19.05 mm) in length and can number in the hundreds on any part. Also note the thin walls and the inability of the resin to drain from the back portion of the left side wall. This consideration will be discussed later.

8.3 QuickCast Tooling Examples

Example 1: Straight-Pull Investment Wax Tool (See Figure 8-7)

Application:	Utilize QuickCast to generate an Investment Wax Tool
Time Period:	2nd Quarter 1993
Machine:	SLA-250
Resin:	XB 5170 (now SL 5170)
Quantity:	1 each
Size:	7 × 7 × 1.5 inches (178 × 178 × 38 mm) (the QuickCast tooling patterns)
Size of Wax Pattern:	1.5 × 3 × 1 inches (38 × 76 × 25 mm) (the final wax pattern)
Cast Material:	A356-T6
Cost Savings:	≈$1,000
Cycle Time:	≈4 weeks
Cycle Time Savings:	≈4 weeks

This project was executed in May 1993. It used QuickCast parts, not in the deliverable geometric configuration, but rather in a tooling geometry configuration (i.e., as a "negative" rather than a "positive").

The resulting cast parts were then used as a tool to produce wax patterns for conventional investment casting. This tool was built as a proof-of-concept example in order to test shrinkage factors, cleanup requirements, and accuracy. Therefore, a simple geometry was selected for a two-part tool.

The SLA part incorporated the appropriate shrinkage factors to accommodate the entire fabrication process, including the resin, the shell, the metal, and the wax. The resulting castings underwent minimal cleanup procedures in the corners of the tool to provide a smooth surface finish. Also, additional machine stock was allowed on each face for a final machine cut to ensure a clean interface between the two tool halves.

Wax parts produced from this very early version of "QuickCast tooling" have been used in the production of subsequent investment castings. Overall, this test was successful in proving the concept that QuickCast negatives can be used as tool patterns for larger scale manufacturing applications.

Example 2: Electrical Connector Straightening Fixture (See Figure 8-8)

Application:	Straightening and Inspection Fixture
Time Period:	2nd Quarter 1993
Machine:	SLA-250
Resin:	XB 5170 (now SL 5170)
Quantity:	3 QuickCast parts, 1 Casting
Size:	3 × 6 × 4 inches (76 × 152 × 102 mm)
Cast Material:	A356-T6
Tool Savings:	≈$4,500
Cycle Time Savings:	≈3 weeks

This example demonstrates the versatility and diverse applications of RP&M. The first problem to be solved was how to inspect the accuracy of a part's compound curve features. An additional problem was how to "straighten" the casting to ensure that it possessed the correct curvature subsequent to a heat treating cycle. Both CMM inspection and the development of straightening tools would have been expensive and time prohibitive.

The solution to both of these problems was to build a series of three QuickCast parts to be used as fixtures in both roles, for straightening and for inspection. Three fixtures were produced using QuickCast. One of the three patterns was shell investment cast in A356-T6 aluminum, so that it could be used as a straightening fixture. The other two QuickCast parts were used as inspection go/no go fixtures.

Both fixture designs were generated using the original part design geometry to produce a negative of the compound angled surface. Additional modeling was completed to produce backing and pin features. The pins were placed at three edge locations to be used as datums for the part to butt against as a check of the surface curvature. The parts were then used in the following manner:

Part 1: Investment cast aluminum straightening fixture used at the foundry. The fixture was used to "straighten" the final part in order to assure that the correct curvature had indeed been achieved subsequent to the heat treatment process.

Part 2: QuickCast inspection fixture built with SL 5170 resin, employed at the foundry to inspect the finished product. This was used as a go/no go gage.

Part 3: QuickCast inspection fixture, also built with SL 5170, but utilized in Texas Instruments' incoming Quality Control Shop. Each casting that TI receives undergoes a specified level of inspection. In this case, the TI stereolithography facility provided a duplicate copy of the inspection fixture to the TI QC inspection shop. The QuickCast parts were sufficiently accurate to assure compliance with the inspection criteria.

Example 3: Precision-set Sand Casting Multi-piece Tool (See Figure 8-9)

Application:	Proof of Principle Sand Casting Tool
Time Period:	3rd Quarter 1993
Machine:	SLA-250 and SLA-500
Resin:	XB 5170 (now SL 5170) and LMB 5353 (now SL 5180)
Quantity:	11 parts
Size of the QC Parts:	Range from 1 × 1 × 1.5 inches to 18 × 15 × 8 inches (25.4 × 25.4 × 38 mm to 457 × 381 × 203 mm)
Cast Material:	A356-T6
Cost Savings:	≈$27,000
Cycle Time:	≈6 weeks
Cycle Time Savings:	≈18 weeks

These parts were developed as a proof-of-principle test to demonstrate the capability of QuickCast patterns to be used as tooling for precision-set sand castings. Precision-set sand casting differs from typical sand casting in that a chemical reaction is used to "set" the sand before pouring the molten metal into the cavity. Also, the grain size of the sand particles used in this process is much finer than that used in traditional sand casting. This technique provides the ability to cast more complex features, with significantly smoother surface finish than conventional sand casting.

The test part selected was a complex optics housing. Wall thickness is typically 0.080 inch (2 mm) with an overall diameter of 8 inches (203 mm) and a height of 16 inches (406 mm). The tooling required was multi-piece due to the complex geometric features. The QuickCast parts produced included:

- The cope and drag matchplate;
- Two core boxes; and
- Five removable detail slides.

These parts were used as replacements for typical sand casting tooling. The modeling of parts was completed using Parametric Technology Corporation's "Pro/Mold" CAD software. This CAD package facilitates reversing the pattern in addition to providing both core print definition, as well as slide definition and placements. SL resin and casting shrinkage were incorporated in the QuickCast patterns.

The QuickCast build style and epoxy resin were necessary for this test. The improved material properties of the epoxy resins increased part durability, so the patterns could withstand the additional stresses encountered during sand casting.

8.4 Additional Benefits and Considerations

In addition to the case studies cited herein, other QuickCast process benefits and considerations have been realized and need to be taken into account. These include the following:

1. The initial benefit is the improved accuracy of the parts. Before using the epoxy resins, part accuracy ranged from ±0.010 to 0.030 inch (±0.25 to 0.76 mm). Furthermore, part accuracy was often geometry dependent. Current accuracy with the epoxy resins has improved to a range from ±0.003 to ±0.008 inch (±0.075 to ±0.2 mm).
2. Durability of the patterns has improved as noted above. This has led to using the QuickCast build style for engineering evaluation models as well as for castings.
3. A decrease in resin usage has been realized. Because each part is built in a quasi-hollow manner, less resin is required. Unfortunately, resin prices have risen, somewhat offsetting the potential overall savings.
4. The overall postcure cleanup labor requirements have been shortened due to the increased strength of the resin. The strength increase reduces the number of supports required to produce each part, resulting in a 50% reduction of labor costs to postprocess the SL parts. Greater savings have been realized since QuickCast 1.1 improvements were made that greatly reduced the need for hand labor to "touch-up" and "patch" the models.
5. The actual cycle time required to postprocess a part has increased due to the time required to drain the resin from the parts. While this is not labor intensive, it does add as much as 12 additional hours to the overall post-processing requirements. It is critical to drain all the resin from the parts to ensure

Figure 8-2. This casting of an electronics tray is an example of successful uses of QuickCast to produce deliverable castings.

Figure 8-3. Another example of a QuickCast pattern and resulting investment casting.

Figure 8-4. QuickCast pattern and resulting aluminum casting.

proper pattern burnout at the foundry. Proper pattern drainage is only accomplished through gravity, time, and patience. Unfortunately, even with the lower viscosity epoxy resin and the quasi-hollow characteristic of QuickCast 1.0, certain part geometries, including thin walls and large radius features, do not always drain properly and special foundry considerations are required to produce acceptable parts.

6. As previously noted, small holes may be introduced into the part when the supports are removed. These "pinholes" must be patched using the SL resin or foundry compatible wax to ensure that the ceramic slurry or moisture

does not penetrate the model. Either of these scenarios will cause inclusions, resulting in an unacceptable surface finish or, in the worst case, the pattern may fail completely in the casting process. Fortunately, with the release of QuickCast 1.1, the pinhole problem has been essentially eliminated.

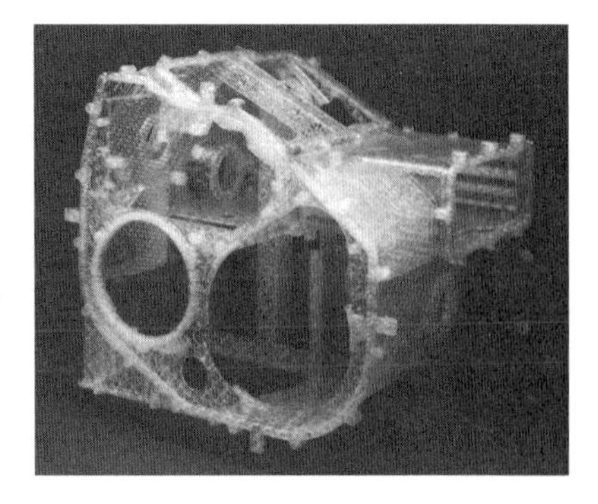

Figure 8-5. Note complex geometric configurations on this part, including thin walls, numerous bosses, and PWB guides.

7. The QuickCast epoxy resin parts are also susceptible to a large amount of creep over a short period of time depending upon the amount of humidity in the environment. TI has experienced as much as 0.25 inch (6.35 mm) of creep distortion in thin wall parts over a period of a few weeks. While the quasi-hollow epoxy resin patterns are initially more durable than the previous acrylic parts, over time, the parts will not hold acceptable tolerances in the presence of high relative humidity. Consequently, maintaining a low (<50% relative humidity) level in the SLA and also in the part

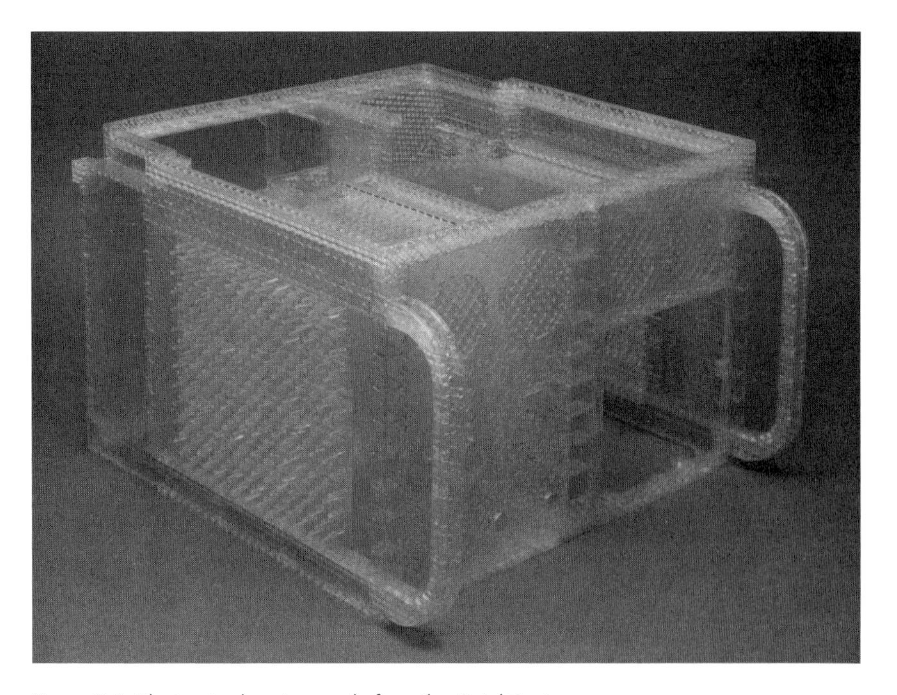

Figure 8-6. Electronics housing made from the QuickCast process.

Figure 8-7. Wax investment casting tool generated in aluminum from QuickCast patterns.

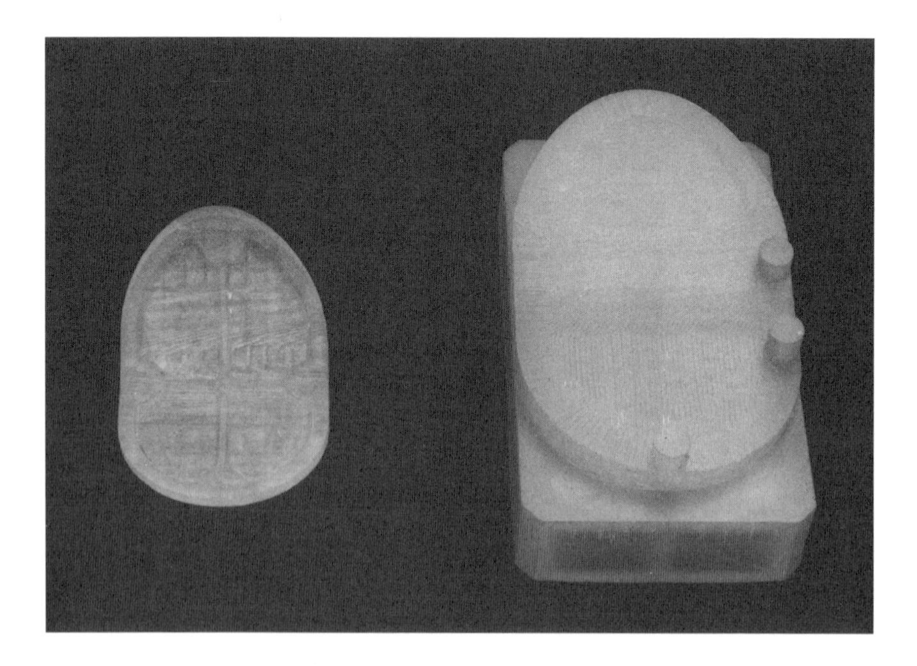

Figure 8-8. Epoxy resin straightening and inspection fixture.

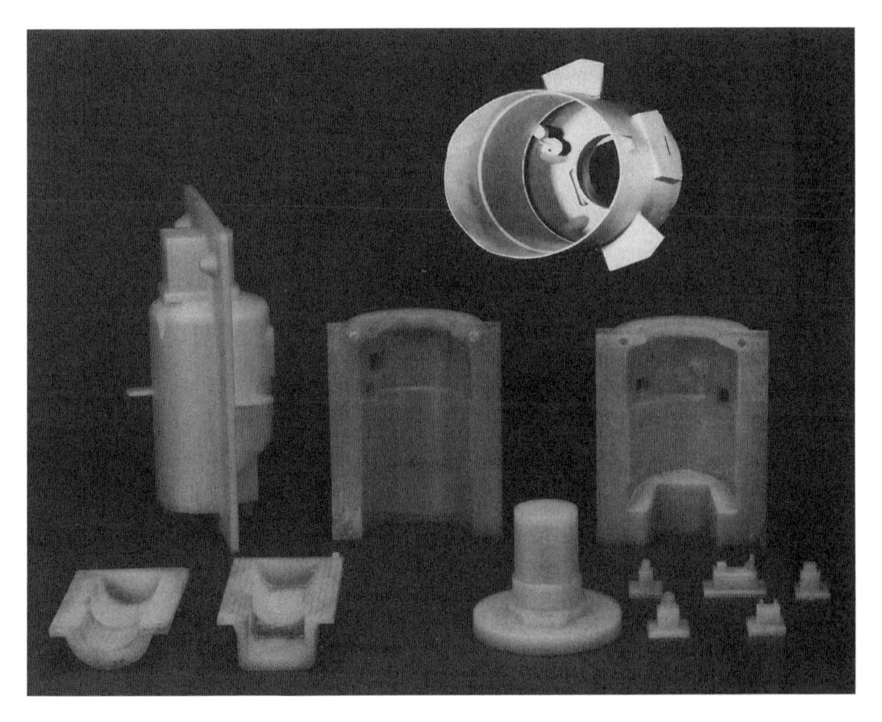

Figure 8-9. QuickCast patterns for a precision-set sand casting multi-piece tool.

finishing area is absolutely critical to QuickCast pattern accuracy. It should be noted that this is not true for solid ACES epoxy resin parts.

8.5 Future Outlook and Conclusions

Ten years ago, the RP&M technology that we use today was only a vision. Today, we envision where these initial technological breakthroughs will lead us 10 years from now. As the overall CAE and RP&M industry develops, turnkey design and manufacturing solutions will be utilized to design and produce products in shorter cycle times at reduced costs. These technologies include the coupling of embedded engineering and manufacturing (tool design, investment and die casting requirements), design rules in the CAE systems, and RP&M process improvements, including the ability to produce ceramic, stainless, and titanium parts, and also to fabricate tool steel tooling as easily as aluminum.

Today, Texas Instruments has successfully incorporated the QuickCast build and casting technique into a production engineering and manufacturing environment. TI transitioned in 1993 from using solid mold investment castings to the QuickCast build style, utilizing shell investment casting processes. This transition resulted in significant cost and cycle time savings to TI. In the future, as the technology matures, the considerations noted will be overcome, resulting in even greater savings. The QuickCast technique has enabled TI to utilize numerous foundries across North America, improve customer satisfaction by reducing cycle time and cost for both engineering evaluation models and investment casting patterns, and also to achieve the additional benefits of generating tools for production requirements.

References

1. Blake, P. and Baumgardner, O., "Solid Modeling and StereoLithography as a Solid Freeform Fabrication Technique at Texas Instruments Incorporated," Proceedings of the Solid Freeform Fabrication Symposium, University of Texas, Austin, Texas, August 1990, p. 170.
2. Blake, P., "The Applications and Benefits of Rapid Prototyping at Texas Instruments Incorporated," Proceedings of the First European Conference on Rapid Prototyping, University of Nottingham, England, July 6-7, 1992, pp. 267-288.

CHAPTER 9

RP&M Applications at Sandia National Laboratories

"There is no treasure without toil."

Aesop's Fables

Sandia National Laboratories (SNL) is a multi-program laboratory operated for the U.S. Department of Energy. As an engineering laboratory responsible for design and manufacture of a variety of prototype devices, mostly electrical and electromechanical, a need continually exists for production in a more timely and efficient manner. Over the years, our manufacturing processes have evolved from labor-intensive, manually operated machine tools to computer-aided machining centers and wire-feed electrical discharge machines. Despite these advances in computer numerically controlled (CNC) machining, many components at SNL still require extensive and time-consuming fabrication and assembly.

9.1 Sandia Turns to RP&M

For these reasons, the lab began investing in RP&M technologies in 1990. Initially used to fabricate polymer-based parts for design verification, these systems became more valuable when we realized that their products could readily be used as patterns for investment casting. The lead times and tooling expense needed to produce wax investment casting patterns is often prohibitive. Being able to use RP&M patterns directly for investment casting would dramatically reduce the time for first article delivery. Also, the time saving is a key factor in many final applications because cast components can often be produced, tested, and rede-

By Clint L. Atwood, Rapid Prototyping Team Leader, Michael C. Maguire, Ph.D., FASTCAST Project Leader, and Michael D. Baldwin, Member of the Technical Staff, Sandia National Laboratories, Albuquerque, New Mexico.

signed faster than one could normally acquire a wax injection tool. When married with investment casting, RP&M has produced some dramatic benefits in the fabrication of cast metal parts.

Sandia began generating investment cast prototypes in 1985. Cast primarily from aluminum or stainless steel, the first items were often produced many months later because of the time required to procure the wax injection tool. Once the tool was available, further time was needed to determine proper methods of gating the part to produce a defect-free casting. Often this required several gating and rigging design iterations. In an effort to speed the design and fabrication process while awaiting the wax injection tool, several processes were used. The first method involved hand-built patterns from sheet wax, formed to closely resemble the component. While crude in appearance, and far from achieving the required dimensional accuracy, these items usually served to evaluate mechanical behavior (vibration, shock, and crush tests), but little else.

The next method used CNC machining to produce patterns from blocks of wax. While the machining of wax was more accurate, the time associated with it was not significantly less than if a metal part had been machined. However, the production of a cast component was generally considered a valuable aid in validating the performance of the final cast design. Not until RP&M technologies began to quickly produce accurate patterns, processed for investment casting, were they considered competitors to CNC machining for prototype fabrication.

RP&M Replaces Wax

RP&M patterns lived up to this expectation so well at Sandia, use of more traditional wax injection tools just about died out during a two-year period. This was accomplished by using patterns from RP&M machines to produce shell investment castings in lot sizes of less than 20 units. Not only have the fabrication technologies for investment casting been dramatically impacted, the ability to simulate the casting process has also seen pressure from RP&M technologies.

As mentioned, the time to produce castings from RP&M patterns has been cut dramatically. However, with RP&M, the more traditional methods used in investment casting to conduct gating trials with several wax patterns are not practical. Hence, a strong desire for accurate, reliable, user-friendly, computer simulations of the casting process, in order to help solve difficult gating problems, has become more important than ever. For this reason, SNL began a program called FASTCAST to help integrate both computational and RP&M technologies into investment casting.

9.2 The Investment Casting Process

Investment casting, often referred to as the lost wax process, is a precision casting procedure used to fabricate metal parts from almost any alloy. Although its history to a large extent lies in the production of art, the most common recent use of investment casting has been the production of complex, tightly toleranced, often thin-section castings of high quality.

New Pattern Needed Each Time

Unlike sand casting, where a single reusable pattern can serve to produce a large number of molds, a new pattern is required for every investment casting. These patterns, typically produced in injection molding machines, are made from wax specifically formulated for this use. Once a wax pattern is produced, it is assembled with other wax components to form a metal delivery network, called the gate and runner system. The entire wax assembly is subsequently dipped in a ceramic slurry, covered with a sand stucco coat, and allowed to dry. Multiple dipping and stuccoing is repeated until a shell of approximately 1/4 to 3/8 inch (6.35 to 9.5 mm) thickness has been applied.

Once the ceramic has dried, the entire assembly is placed in a steam autoclave to remove most of the wax. After autoclaving, the remaining wax that previously soaked into the ceramic shell is burned out in a furnace. At this point the shell is empty. It is then usually preheated to a specific temperature and filled with molten metal. The hot mold assists in the filling of intricate shapes and thin sections. Once the casting has cooled sufficiently, the shell is chipped from the mold and the desired casting cut from the gates and runners. As can be seen, the process requires that a pattern be destroyed for each metal casting produced. A more detailed step-by-step process description is presented in Table 9-1.

Table 9-1. Steps to Produce an Investment Casting

1. Design casting	9. Fire shell
2. Design gating system	10. Pour casting
3. Design and procure die	11. Remove shell
4. Produce pattern	12. Remove gates
5. Produce gate/runner assembly	13. Hot/cold straighten
6. Fabricate shell	14. Heat treat
7. Dry shell	15. Inspect
8. Dewax	16. Ship

In a direct sense, the onset of RP&M has had an impact on this process by eliminating only one of the steps listed earlier, namely, design and procurement of the wax injection die. However, it has been the cost and time associated with manufacture of this tool that has contributed substantially to the perception of investment casting as an expensive and time consuming process, not at all suited for the rapid production of functional metal prototypes. In the past, a complex injection tool for a structural casting in the 5-pound range could cost upwards of $50,000 and require 30 to 50 weeks of lead time.

Compounding this problem was the observation that, more often than not, the final design was, in fact, not so final. Changes to the die before it ever left the toolmaker were not uncommon for SNL. With RP&M processes now shortening the time to make the first pattern from several months to as little as one day, the investment casting industry is slowly modifying commonly held opinions regarding its agility.

9.3 Implementing RP&M in Investment Casting

Although the primary pattern material for traditional investment casting is wax, other materials have been used for specialized applications. Hence, substituting something other than wax for the pattern is not unique, but the changes to the process that this entails present some distinct challenges to the foundry.

Stereolithography

A 3D Systems model SLA-250 has been in operation at SNL since 1990. An SLA-500 was also installed in late 1993. These systems were originally employed as design tools to facilitate the prototyping process. The SLA-250 was used initially to quickly fabricate models from solid CAD designs. These models were typically involved in proof-of-concept demonstrations, in mechanical design reviews, for interference fit-checks, and as visual aids for other methods of manufacturing. While the process was not a replacement for traditional machining, it represented another tool used to speed the product development cycle. SL also offered advantages such as enhancing the agility with which designs were improved or modified. These advantages included unattended operation, ease of fabricating complex parts, increased design flexibility, reduced lead times and fabrication costs, as well as more efficient design iteration.

The evolution of SL has brought about significant changes in the application of RP&M technology. Parts produced with SL have matured from "show-and-tell" models to high quality expendable patterns for investment casting, and also as soft tooling for sand casting. Nonetheless, in most cases, early efforts at SNL to use solid SL patterns for investment casting failed.

SNL Decides on Shell Molding. Whereas wax patterns would melt and soak into the ceramic material prior to expanding and cracking the shell in thermal dewaxing, cross-linked photopolymer SL patterns do NOT melt. Thermal expansion of solid acrylic patterns is sufficient to fracture ceramic shells during dewax and burnout, rendering the shells useless. Although some foundries were capable of producing crack-free molds employing the older "solid mold investment casting" method, SNL decided to pursue shell molding as its sole method for investment casting.

QuickCast Patterns in the Investment Casting Process

Recent advances in the implementation of: (1) the QuickCast process, (2) epoxy resins, and (3) new software have had a significant impact on applications for stereolithography. Using the QuickCast build style with epoxy resin, it is now possible to build quasi-hollow patterns for direct shell investment casting. This build style ameliorates the thermal expansion problem during burnout by reducing the amount of resin in the pattern. With proper treatment, the quasi-hollow patterns generate much less stress on the ceramic shell. As with any new manufacturing process, there are many variables associated with using QuickCast patterns as a replacement for traditional wax patterns in investment casting.

Sandia is currently in the iterative process of defining the steps necessary to consistently produce good castings with QuickCast patterns. Our experience indicates that modifications from the traditional lost-wax processing techniques, such as careful pattern preparation prior to dipping, special burnout cycles, and mold venting are all required to produce a quality casting.

Our experience with QuickCast patterns can be characterized by first examining the steps in their preparation, then how the mold is dewaxed and, finally, how the pattern is eliminated. Most QuickCast parts require some hand work to prepare them for use as patterns. Typically, hand sanding "stair-stepped" surfaces and removing support structures are part of the process, whether or not the SL parts are used as investment casting patterns. However, it is essential that any sharp exterior corners or thin edges from the support structure be removed, to minimize stress on the shell during burnout.

Draining the Pattern. QuickCast patterns can be drained of uncured liquid resin since they are constructed of interconnecting hollow cells. However, all holes in the pattern's exterior skin must later be filled in the wax room prior to dipping. The reason for this is twofold. First, any holes in the skin will allow ceramic slurry to flow into the pattern. This will obviously result in extensive inclusions in the casting. Secondly, both the dewax and burnout steps can create significant gas overpressure inside the pattern. While the polymer in the interior pattern cells has been removed, the presence of air in the sealed pattern can be sufficient to cause a "ballooning" effect sufficient to crack the shell. This expansion is eliminated by careful use of vents in the pattern. However, any individual cells not properly vented can also cause localized shell failure.

Before the gating system is attached to the pattern, all holes intentionally added to assist in resin draining must be filled. In addition, any holes that resulted from support removal or sanding through the surface skin must also be patched. Identification of holes is most easily accomplished by pressure or vacuum testing the pattern.

Venting the Pattern. Once the pattern has been fully prepared, it can be attached to the gating system. Whereas most investment casters design the wax runners and gates to facilitate draining the melted wax out of the mold in a steam autoclave, the use of QuickCast patterns places an additional requirement on the wax distribution system. As mentioned, the patterns must be vented to the atmosphere. This is critical during the early stages of burnout, prior to thermal decomposition of the polymer. Venting of the pattern can be accomplished several ways.

Hollow gates and runners can be constructed and used to provide an uninterrupted gas path back to the pouring cup. The hollow runner can also be vented to the atmosphere at an appropriate location, prior to the runner being attached to the pouring cup. Individual vents, placed directly on the pattern, can also be used. An example of such a vent is shown in Figure 9-1. The pattern is initially perforated as shown in Figure 9-1(a); the subsequent wax vent is placed over the hole, Figure 9-1(b); the shell is then constructed, and finally removed in the region of the vent, Figure 9-1(c).

However the pattern is vented, it is critical to perforate it at the location of the gate or vent, and then seal the wax gate or stub directly over the hole. While it may seem contrary to previous advice requiring that all holes in the pattern skin be sealed, if the pattern is not vented directly at specific locations, then shell cracking via the ballooning effect is likely. The holes used to drain excess resin from the pattern also must be sealed because generally they are not where the gates or vents will be located. Unless all drain holes are sealed, smaller pinholes cannot be located by vacuum or pressure testing. By perforating the outer skin of the pattern, and then sealing a wax gate or vent over the hole, the penetration of the pattern by ceramic slurry is precluded.

Gates and Runners. If the pattern is vented using hollow gates and runners, it is important to make sure that a continuous gas path exists from every isolated hollow section of the QuickCast pattern, through the gating system, to the atmosphere. For example, using preformed hollow runner systems is of no value if they are sealed at the end. In fact, the heated air trapped in thin hollow wax runners can often crack ceramic shells if not vented. When solid gates and runners are used, placement of small wax stubs directly over perforations in the patterns can produce the desired venting. However, these vents must be exposed to the atmosphere prior to dewaxing, and then patched with ceramic mortar following the final burnout. Most foundries prefer not to have large open areas to patch since this increases the likelihood of unintentional small inclusions.

Facecoat Adhesion. Most facecoat slurry systems contain surfactants to promote wetting the slurry on wax patterns. However, it has been observed that the slurry may not adhere to SL resin patterns as well as it does to wax. As a consequence, the resulting surface on the casting may have an orange peel effect, but this is generally not as severe a problem as defects produced from gross facecoat delamination. The cast surface finish resulting from a pattern experiencing facecoat delamination is unacceptable. While there have been no published studies to determine what, if any, modifications to the slurry would be necessary to preclude this effect, it would seem prohibitive to have a separate slurry composition solely for QuickCast patterns. The solution to this dewetting problem at SNL has been to promote facecoat adhesion by lightly spraying the SL pattern with an aerosol glue. The spray is applied just prior to dipping and is then allowed to dry. Several types of spray adhesives are available at most artist supply stores.

Shell Cracking. It has been our observation that whether shell cracking is a significant or a minor problem is strongly dependent on the type of shell system

(a)

(b)

Figure 9-1. Sequence of venting a QuickCast pattern in a small localized area: (a) Predrilling holes to assure venting of the pattern during burnout, (b) placing a wax stub to form the vent and covering the holes prior to dipping the pattern in ceramic, and (c) appearance of vent after shell construction.

(c)

used. Most foundries experience some degree of shell cracking in autoclaves, even for wax patterns. This is usually the result of not being able to properly drain and vent wax from certain regions.

Shell cracking in these cases is often readily patched using ceramic mortar, with no deleterious effects. That RP&M patterns produce higher stresses on the shell is generally accepted, but whether or not this is sufficient to cause catastrophic shell cracking varies with each ceramic system. SNL's traditional shell system consists of two primary coats of alumino-silicate/colloidal silica slurry with silica stucco, followed by a fused silica/colloidal silica backing slurry and fused silica stucco. A total of six backing coats are used for traditional wax patterns.

When initial shell cracking was encountered with RP&M patterns, the number of backing coats was increased to eight, and chopped ceramic fibers (typically 3M Nextel™) were incorporated into the fourth backing coat. This improved the shell strength, as verified by mechanical testing, to the extent that shell cracking was largely eliminated when coupled with effective pattern venting. Some methods observed in other foundries consist of incorporating chopped stainless steel wire or chopped stainless steel wool. Our testing of chopped steel fibers indicated that they were almost as effective at improving shell strength as the ceramic fibers.

While traditional dewaxing with a steam autoclave is considered the most economical and environmentally friendly method for investment casting, it has become increasingly clear that its use can be problematic for RP&M patterns. Autoclaving is designed to remove the majority of the wax from the mold, and subsequent burnout is then employed to eliminate the residual wax. Even with the venting methodologies described earlier, SNL's experience with autoclaving under several different temperatures and pressures has yielded no consistent method to eliminate shell cracking. At pressures as low as 15 psig and temperatures of only 220° F (104° C), shell cracking was still observed. These lower temperature cycles were used with completely hollow gating systems. They generally removed most of the wax, but increased cycle times.

Flash-fire Dewaxing. For this reason, the consensus at SNL, and indeed in industry, is that flash-fire dewaxing is the method of choice for RP&M pattern burnout. Unfortunately, the reason that flash-fire dewaxing was abandoned in favor of autoclaving many years ago was that the considerable, even choking, amount of smoke resulting from direct wax combustion was eliminated by removing the wax with steam prior to final burnout. While some might consider a return to flash-fire dewaxing a giant leap backward, these furnaces have improved dramatically in recent years through the addition of afterburners owing to strict air quality standards adopted by most states.

While flash-fire dewaxing may be the ideal method to dewax and burnout RP&M patterns, most foundries cannot readily invest in such a furnace solely to process RP&M molds. The practice at SNL, prior to adopting newer generation flash-fire dewaxing, was to minimize the amount of wax by using hollow gates and runners, then employing heat guns or oxy-acetylene torches to remove as much wax as possible prior to final burnout. Even so, the pattern material can generate a significant amount of combustible gas, so fully aspirated gas-fired furnaces are highly recommended to minimize the amount of smoke during final burnout. Venting the mold to improve pattern combustion during the burnout cycle is critical once the chance of shell cracking has passed. In order to ensure better air flow through the shell, there should be a reasonable number of vents.

Burnout. After the pattern has been burned out, the mold may need to be patched. Any vents introduced earlier will require sealing with ceramic mortar. If the hollow gates and runners were vented up through the pour cup, then it is possible to go directly to the casting operation from burnout. However, our experience with direct pouring has been limited since it is necessary to cool the molds after burnout to allow inspection and patching.

Whether or not separate burnout and casting preheat cycles can be used will typically depend on the type of shell system employed by the foundry, and its tolerance for repeated thermal cycling. Burnout cycles for the patterns tend to be higher than traditional processing. For example, burnout of traditional autoclaved molds at SNL was conducted at 1,300° F (700° C) for four hours, but for SL patterns the burnout temperature is usually 1,700° F (930° C) or higher for periods extending to eight hours or more. Recent experience with temperatures as high as 2,100° F (1,150° C) indicates that QuickCast patterns can be fully eliminated with little or no residual ash, in less than two hours.

Once the mold has been filled with metal, there is virtually no difference from traditional processing, with perhaps one exception. If the shell has been reinforced with fiber or is excessively thick, its removal may be slightly more difficult than usual.

Selective Laser Sintering

When the concept of Selective Laser Sintering (SLS) was introduced by the University of Texas, Sandia became involved in the SLS Technical Associates Group to follow and help guide the development of the SLS technology. This process was inherently attractive to investment casters because it could produce patterns directly from investment casting wax. As a result, the various process modifications necessary for SL patterns would not be required. In 1991, Sandia used a segment of an investment casting called a "firing set housing" to benchmark different RP&M patterns for investment casting. Subsequently, in March 1992, Sandia became a beta test site for a preproduction Sinterstation 2000 Selective Laser Sintering machine produced by DTM Corporation, Austin, Texas. Efforts focused on building the best wax patterns possible, using the SLS process.

At that time, the SLS process produced the most accurate RP&M wax patterns for direct shell investment casting. With the SLS process, we were able to demonstrate some of the benefits of using RP&M processes to fabricate prototype patterns for investment casting. At the end of the beta program, Sandia purchased a Sinterstation 2000 SLS machine. A decision was later made to transition from wax to polycarbonate powder when generating investment casting patterns. The advantages of using polycarbonate material are: faster build time, more robust parts, improved dimensional stability, and absence of a support structure during the build process. Sandia is currently developing techniques to make investment casting molds from polycarbonate patterns.

SLS Patterns in the Investment Casting Process

While in principle the wax patterns built in a Sinterstation could be processed in a manner identical to traditional injection-molded wax patterns, the reality of handling and finishing these extremely fragile parts made their use difficult. This problem was compounded in thin sections. Eventually, the introduction of polycarbonate to SLS made the handling and finishing problems much easier.

Rough Surface. SLS polycarbonate patterns are rather porous. Exact levels of porosity vary with the build parameters. This is generally not a critical factor in using SLS patterns for shell investment casting. Since the patterns are porous, they must be sealed prior to being dipped in ceramic slurry. Although the slurry will not penetrate through the polycarbonate as it would with an unsealed hole in a QuickCast pattern, the colloidal silica binder in the facecoat slurry typically seeps into the pattern, thereby weakening the primary coats of the shell. The resulting SLS surface finish is unacceptably rough.

Hence, the standard procedure for SLS polycarbonate patterns has been to soak the pattern in seal wax (an investment wax used for dipping gating systems to create a smooth surface). This is done for a sufficient interval to allow the wax to thoroughly penetrate the polycarbonate structure. Although the wax is molten and the pattern heats to the temperature of the molten wax (typically 190° F [88° C]), negligible distortion has been observed in the patterns.

Heat Gun. Upon removing the SLS patterns from the seal wax, excess material is removed by shaking or spinning. The patterns are then cooled back to room temperature. Any excess wax is then scraped from the surface. A heat gun is used to lightly melt the wax on the surface of the pattern. This melting "sweats" the wax up to the pattern surface, and tends to smooth low areas between the polycarbonate particles. Extensive use of the heat gun, however, will bring too much wax to the surface, reducing pattern dimensional accuracy. Finally, the patterns are hand sanded or finished with an abrasive pad. As with QuickCast 1.0 patterns, autoclaving SLS patterns tends to produce shell cracking, so its use is not recommended. Direct flash-fire burnout is again the method of choice to eliminate both the wax gating and the pattern. Since the SLS patterns are not hollow, placement of gates and care in sealing holes is not an issue, but placement of vents is again important. Vents are routinely positioned on SLS patterns similar to the one shown in Figure 9-1(b). The vents serve to provide air access to the pattern as it undergoes combustion, minimizing the amount of residual ash or carbonaceous residue in the mold.

Unlike resin-based SL patterns, polycarbonate is a thermoplastic and tends to flow out of the mold in a manner similar to wax. However, molten polycarbonate is substantially more viscous than wax, and is not removed as readily. With most of the SLS patterns tested at SNL, only about half of the polycarbonate flows out of the mold, with the remainder subject to combustion during burnout. The material that remains in the mold, especially in areas where it may collect or puddle, has been shown to attack the shells. While the exact mechanism is still uncertain, it is suspected that, like wax, the polycarbonate may, to some extent, soak into the shell, causing facecoat delamination. The surface of the resulting casting has the appearance of a cracked desert lake bed.

Depending on the orientation of the pattern, the addition of vents in these reentrant areas is encouraged to assist drainage. Burnout times with SLS polycarbonate patterns tend to be longer than with SL QuickCast patterns. A typical SLS burnout cycle involves soaking for a few hours at 1,600 to 1,800° F (870 to 980° C), until the primary combustion of the residual polycarbonate material is complete, followed by about 2 to 4 hours at about 2,100° F (1,150° C). This generally removes nearly all ash residue from the mold. If possible, excess air during the burnout process will further assist in removing any residual ash.

9.4 Case Studies of RP&M Patterns in Investment Casting

The following sections will describe the production of specific investment cast parts using both the SL and SLS processes. Frequently, questions are asked regarding the actual benefits accrued from use of RP&M relative to traditional methods. The following examples offer answers to questions about traditional machining or investment casting processes used in parallel with RP&M, and their impact on lead time, and actual hands-on time (hours billed).

Case Study of Firing Set Housing

The firing set housing is a component that has been designed and redesigned many times at SNL for different applications over a period of several years. Before the integration and implementation of RP&M processes, traditional design iterations required either machining from a block of PH13-8Mo precipitation hardened stainless steel or fabricating an aluminum injection molding tool to produce wax patterns for investment casting. For most of the castings produced at SNL, certain areas were stocked for machining weld preps and other tightly toleranced features. CNC machining of wax injection mold tooling was very costly and required long lead times. Since the designs were modified on a fairly regular basis, changes to the tooling were common, further adding to lead time and cost.

For the firing set housing, the SL design model, shown in Figure 9-2, was used for dimensional verification, assembly verification, and design review. Several design iterations were completed before a prototype stainless steel housing was needed for testing and other evaluation. These SL models provided designers the ability to tune manufacturability issues with greater speed than ever before possible. After the design was verified for form and fit, an SLS wax pattern, shown in Figure 9-3, was fabricated on the Sinterstation 2000. The pattern was invested in ceramic to create a mold and the prototype was cast in stainless steel. The design was changed even further in subsequent iterations. By this time, polycarbonate was the pattern material of choice for SLS. Figure 9-4 shows a CAD solid model, a polycarbonate SLS pattern, and the rough casting produced from that pattern. Figure 9-5 illustrates the detail involved in constructing the gating system for this part. Several vents described previously are shown toward the top of this figure.

Figure 9-2. SLA design verification model for a firing set housing.

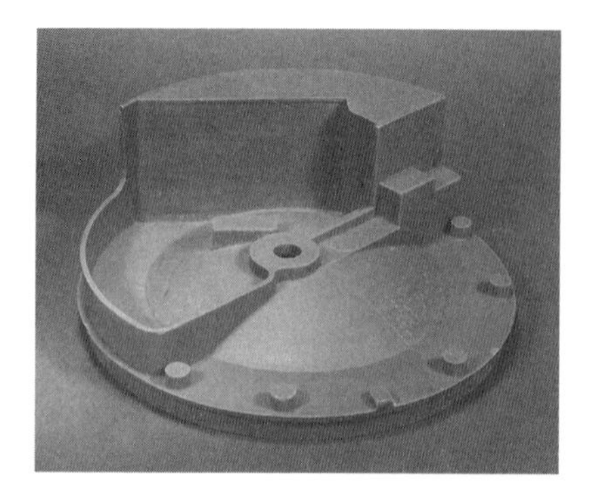

Figure 9-3. SLS wax pattern for firing set investment casting.

Figure 9-6 represents fabrication time (hands-on time) and lead time comparisons using three different processing methods to fabricate prototype firing set housings. Lead time, represented in weeks, is associated with the design and acquisition of tooling and fixturing. The fabrication time, in hours, represents the actual processing time (either hands-on or unattended operation of equipment) required to fabricate the part.

Figure 9-4. CAD solid model, polycarbonate pattern, and rough stainless steel casting produced from SLS pattern.

Case Study of a Synthetic Aperture Radar Reflector

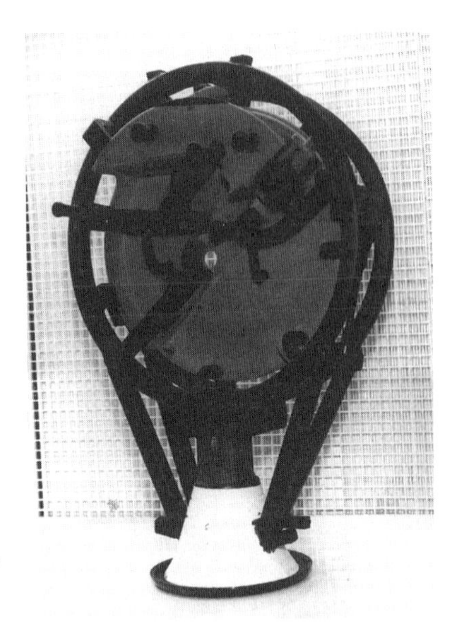

Figure 9-5. Polycarbonate pattern with gates and runners attached.

Sandia offers state-of-the-art capability in the design and development of synthetic aperture radar (SAR), from system design through system integration, to data collection and analysis. The prototype SAR, similar to the one shown in Figure 9-7, was installed on a Twin-Otter aircraft providing a low-cost testbed. Applications for the SAR are: reconnaissance; surveillance and targeting; environmental monitoring; treaty verification and nonproliferation; interferometry; navigation and guidance; and others.

The initial application of SL for this project involved validating the design of the new parabolic reflector. Since it was unclear if the unit would function as expected, the designers employed SL to produce a model that could be used as a substrate for copper plating the reflective surface prior to committing to an actual aluminum reflector. The copper plated reflector is shown in Figure 9-7, in place on the SAR assembly. Once the performance of the unit was validated, CNC machining, as well as shell investment casting from a QuickCast pattern, was used to produce prototypes, as shown in Figure 9-8. The following represents actual data collected by the engineer who designed the prototype SAR unit.

Figure 9-9 represents data similar to that shown in Figure 9-6. In this case, only RP&M and CNC machining were used for comparison. Note that the reflector is less complex than the firing set housing. Consequently, the fabrication time required for the SAR reflector is significantly less than that for the firing set. Hence, in this comparison the differences between RP&M and CNC machining for the reflector are not as dramatic as for the firing set. This smaller time differential highlights an important point. Namely, RP&M is not always expected to be the most efficient or economical route for producing every component. It is, however, vitally important in reducing production time in numerous applications. For limited quantities of geometrically complex, thin-walled investment casting prototypes, RP&M essentially becomes the only economically viable method. For simpler shapes, CNC machining will remain the process of choice.

9.5 Dimensional Accuracy of RP&M Patterns

Investment castings can only be as accurate as the patterns from which they are produced. Patterns made from SL and SLS machines continue to improve in

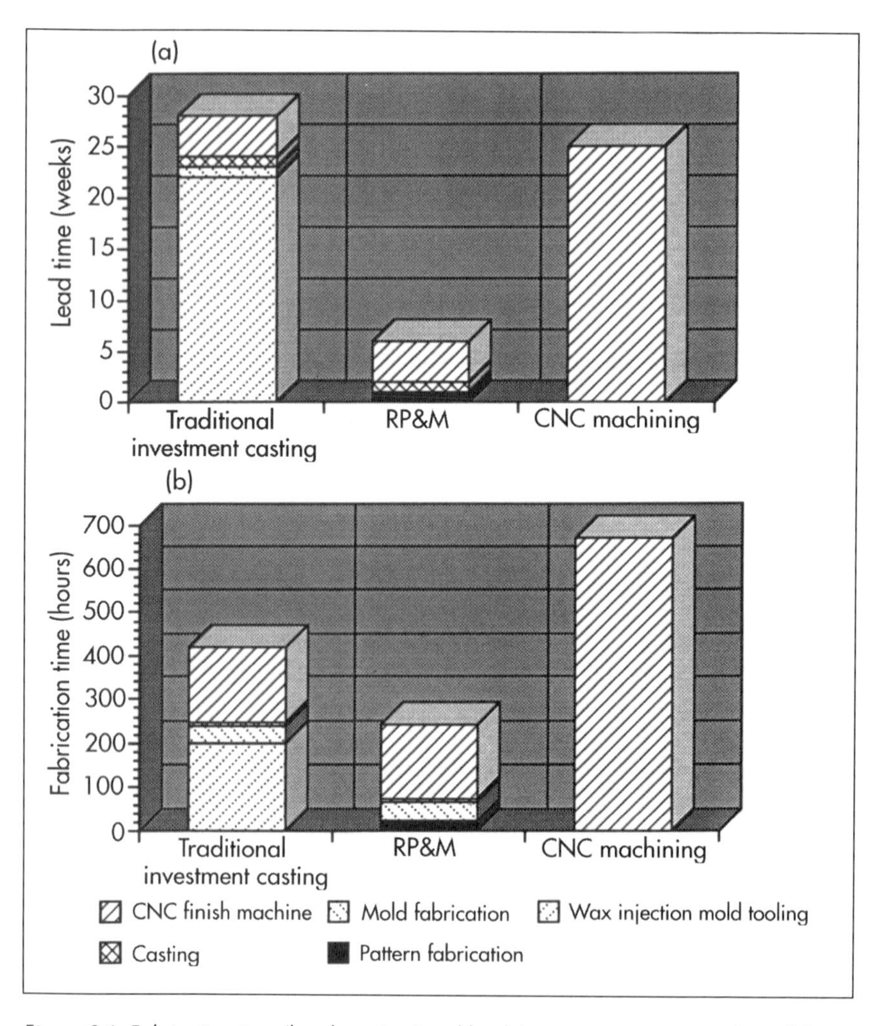

Figure 9-6. Fabrication time (hands-on time) and lead time comparison, using three different processing methods to fabricate prototype firing set housings.

both dimensional accuracy and surface finish. These improvements are directly related to process advances in software and hardware, as well as postprocessing techniques. For nearly two years, Sandia has only used parts fabricated from these processes as patterns for investment castings.

In an effort to establish the expected accuracy of patterns built using both SL and SLS, measurements are documented from a benchmark part. This part is a bicycle crank arm, shown in Figure 9-10. The crank arm was initially fabricated as part of a technology transfer program to support a small manufacturer of bicycle components. This part was first fabricated on the Sinterstation 2000 in Sep-

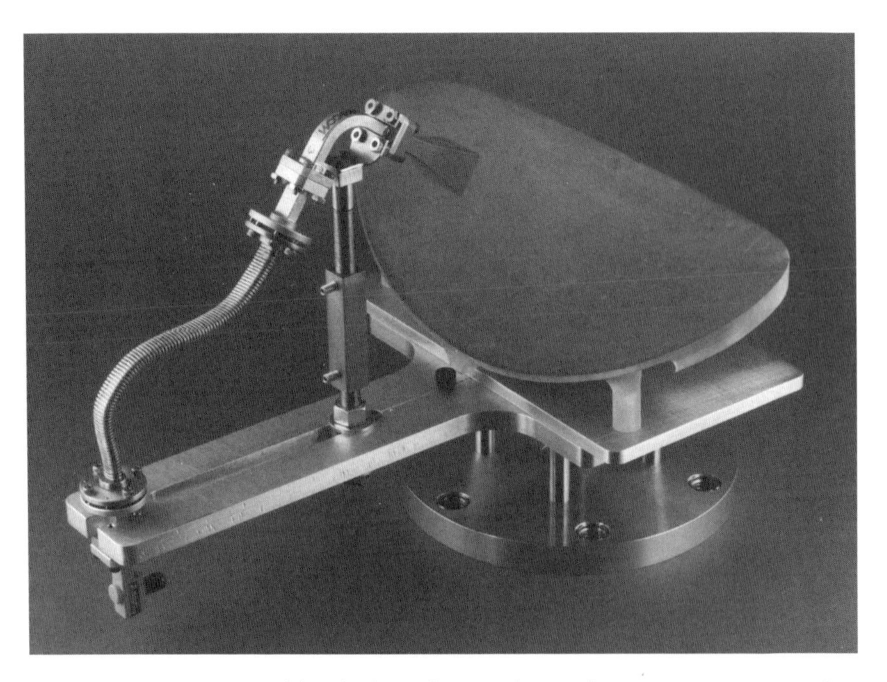

Figure 9-7. SLA design model with physical vapor deposited copper coating mounted on prototype SAR apparatus. The model was used in an actual radar test to validate the reflector's design.

tember 1992. Since then it has become our *de facto* accuracy test part. The geometric shape of the crank arm is representative of a part of average complexity.

The measurement results for SL acrylate and epoxy resins, generated in the QuickCast 1.0, QuickCast 1.1, and ACES build styles, are displayed in Figure 9-11, while data for the SLS wax and polycarbonate patterns are shown in Figure 9-12. All measurements were taken after the parts were removed from the machine, and normal postprocessing techniques were used to remove support structures and other excess material. These parts were measured in the Sandia metrology lab using a Zeiss coordinate measuring machine. A total of 21 diameters, 21 X-

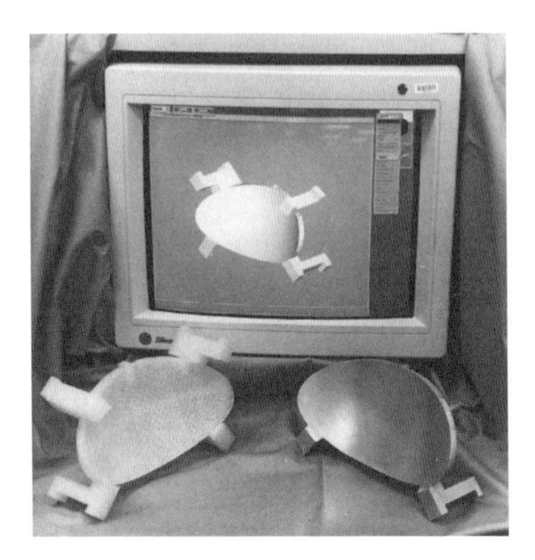

Figure 9-8. CAD solid model, SLA QuickCast pattern, and a semifinished aluminum casting of the parabolic SAR reflector.

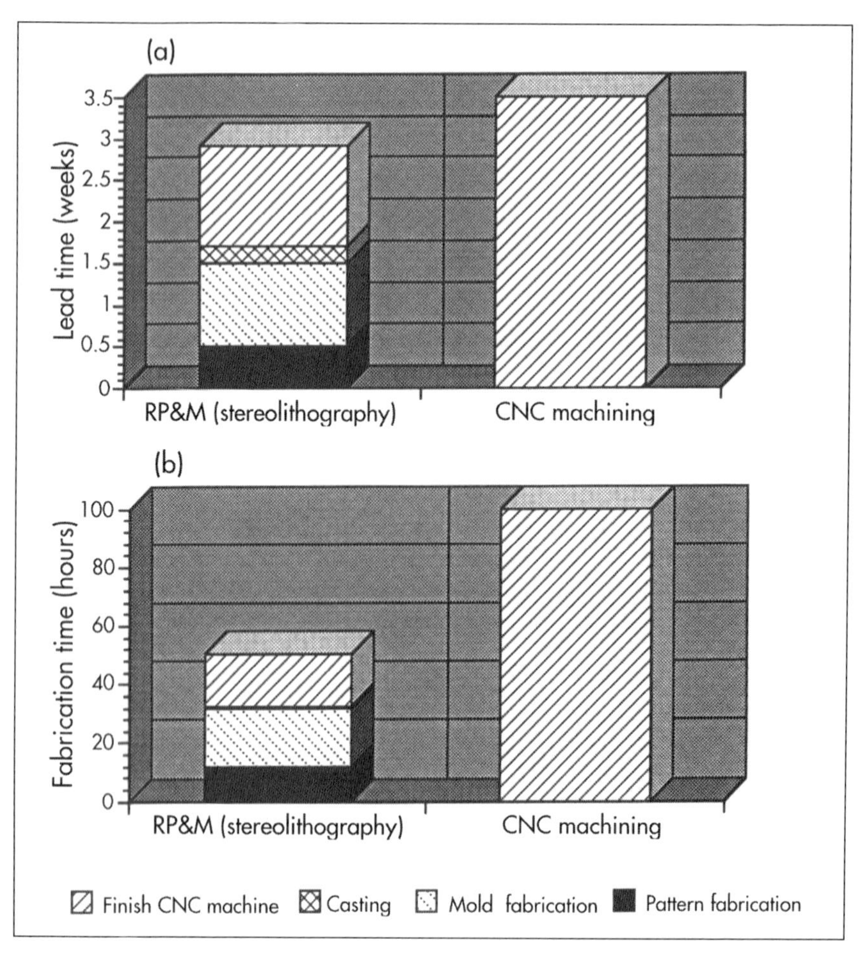

Figure 9-9. Fabrication and lead time comparison, using only RP&M and CNC machining for comparison.

axis and Y-axis locations, 30 radius points, and 38 Z-axis measurements were taken from each part.

This measurement data has been compiled over a two-year period and is an update of previously reported data.

9.6 Summary

It is clear from our experience at Sandia that RP&M is valuable for reducing lead time in the fabrication of prototype components. In some circles, the term "prototype" is considered to be somewhat derogatory. This sentiment is often heard

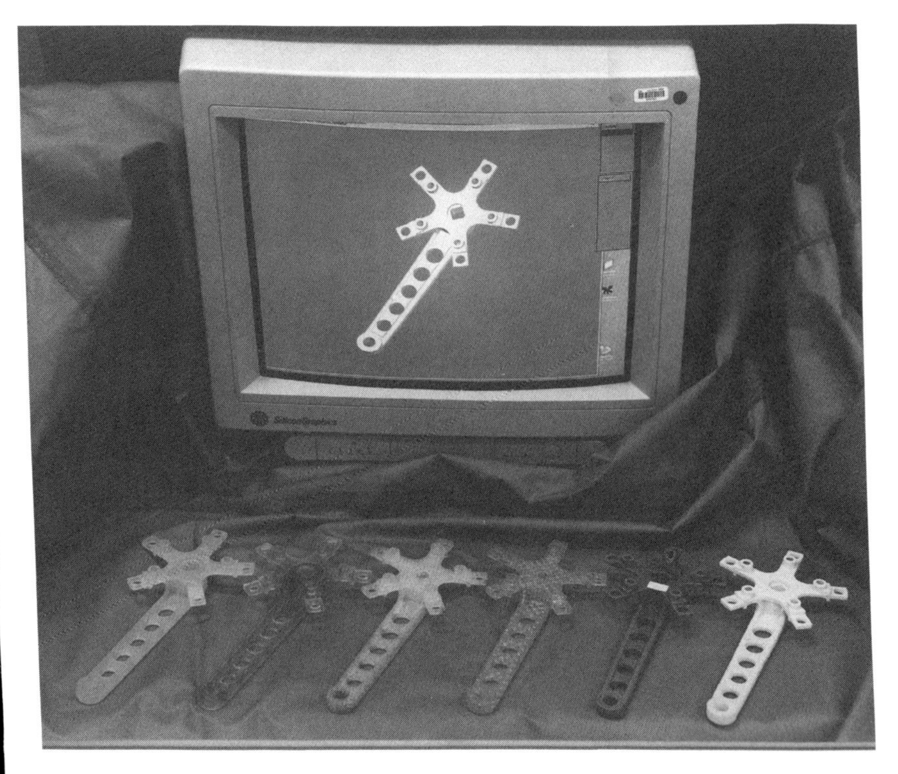

Figure 9-10. CAD solid model and, from left, SL acrylic model, ACES model, SL QuickCast 1.0 model, SL QuickCast 1.1 model, SLS wax model, and SLS polycarbonate model.

from the manufacturing sector, where some prototypes may be fabricated from processes very different than those used in the final production environment. In these situations, it is thought that relatively little is learned in prototype fabrication that would be of real assistance in discovering actual problems likely to arise in high-volume production.

However, for investment casting this criticism is invalid. While the pattern materials may be different, in most cases the resulting fired ceramic molds are identical to those for higher volume production investment castings made from wax patterns. Hence, issues such as soundness and quality of the casting can, in fact, be properly addressed from the very start. The mechanical properties, as well as the performance of these prototypes, are expected to be identical to production components. For these reasons, Sandia National Laboratories has made RP&M processes an integral part of its manufacturing capability, and we anticipate both the capabilities and the importance of RP&M to continually expand.

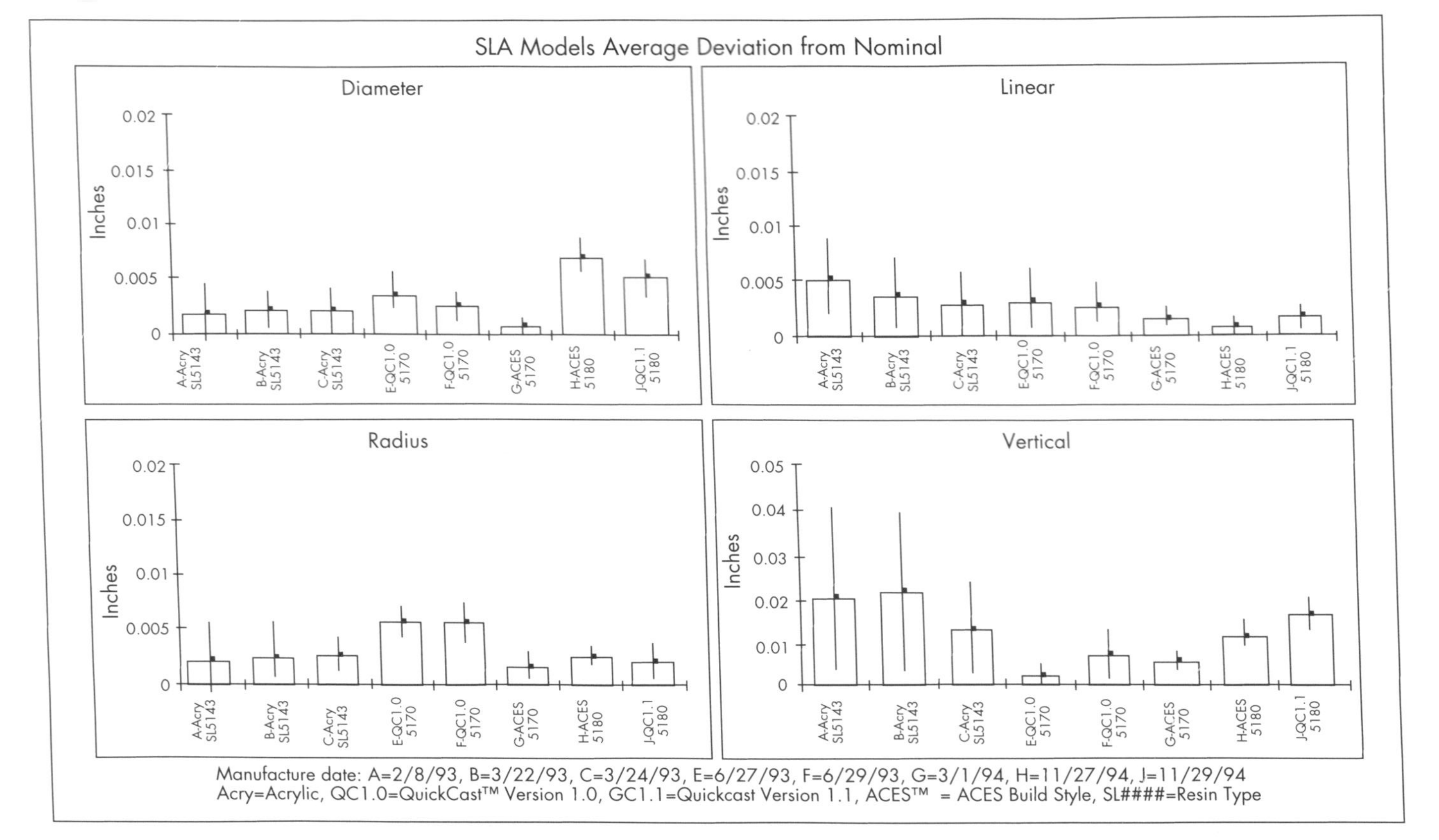

Figure 9-11. Measurement accuracy reported as deviation from nominal dimension for stereolithography pattern. Error bars represent ±σ deviation.

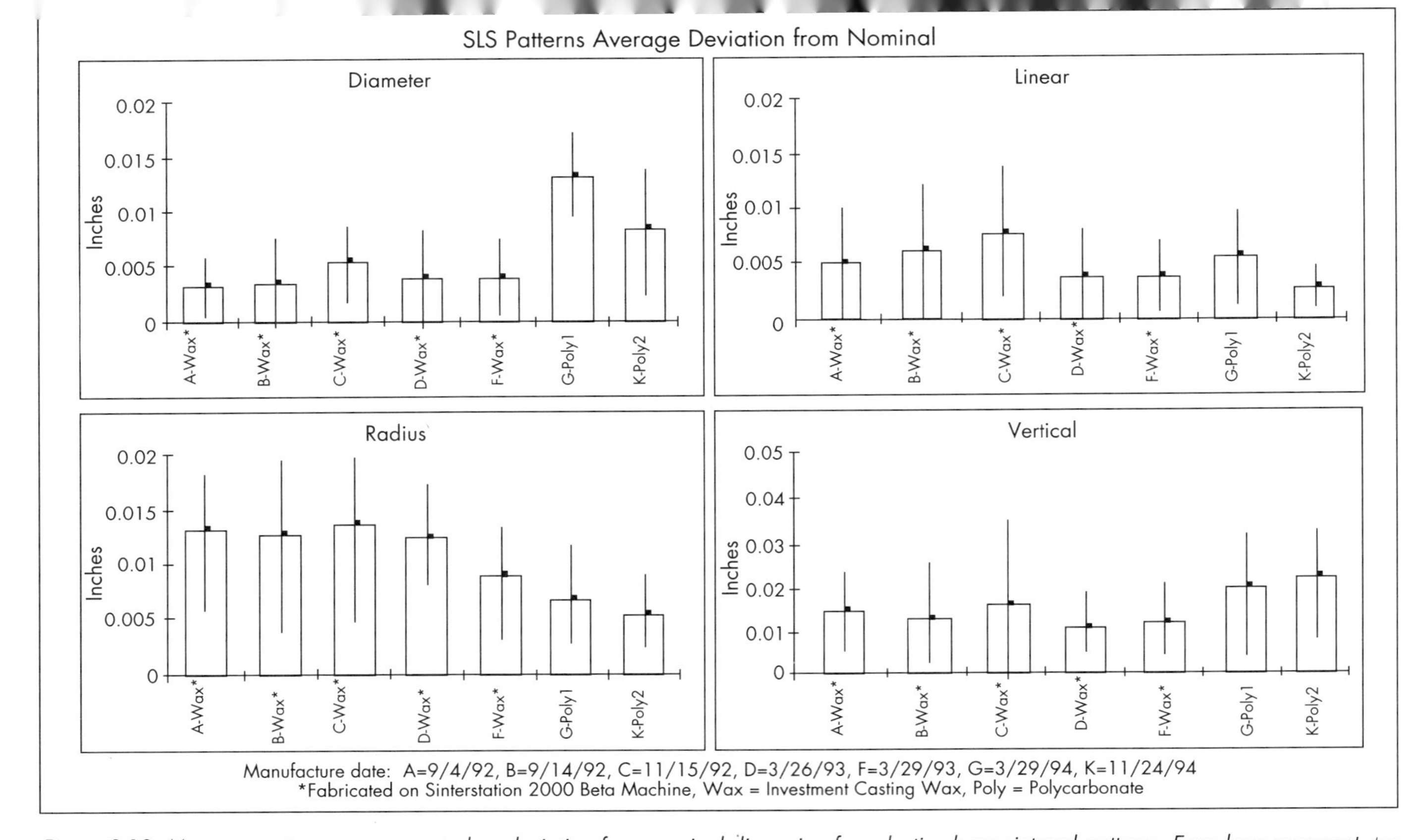

Figure 9-12. Measurement accuracy reported as deviation from nominal dimension for selective laser sintered patterns. Error bars represent ±σ deviation.

Acknowledgments

The authors gratefully acknowledge the extraordinary efforts of certain SNL staff in the preparation of the information presented in this chapter. Special thanks go to B. Pardo, E. Bryce, G. McCarty, L. Gonnsen, and S. Giron. This work, performed at Sandia National Laboratories, is supported by the U.S. Department of Energy under contract number DE-ACO4-94AL85000.

CHAPTER 10

Soft Tooling Applications of RP&M

"Give us the tools and we will finish the job."

Sir Winston Churchill, 1941

10.1 What is Soft Tooling?

Before we can discuss RP&M techniques and the applications possible for soft tooling, it is necessary to define the term *soft tooling*. The definition varies as there are different ways of considering the techniques and the applications. The main definitions follow.

1. Tools required for small production quantities. This obviously depends on the product and its application. In some industries, such as defense and aerospace, a large production run can be a few thousand. Whereas in the consumer products industry, a large run will be many millions and a few thousand is just a prototype quantity.
2. Soft tooling is normally associated with low cost. Hence, the terms soft and low-cost are often used interchangeably.
3. The most obvious definition for soft tooling corresponds to the type of material used for manufacturing the tool. With this definition, hard tooling is often referred to as that made from hardened tool steels. Materials with lower hardness levels are considered soft (e.g., silicone, rubber, epoxies, low melting point alloys, zinc alloys, aluminum, etc.).
4. It is also possible to define soft tooling by the method of manufacture. Most hardened steel tools are manufactured by either conventional machine cutting processes or variations of electrical discharge machining (EDM).

By Philip M. Dickens, Ph.D., Director of the Centre for Rapid Prototyping in Manufacturing, University of Nottingham, Nottingham, England.

The Need for Soft Tooling

Although it is possible to make models very quickly by the various RP&M techniques, many engineers do not consider them true prototypes. Designers often require a prototype to be made in the end use material, by the same process that will be used in final production. This requirement is often laid down as company policy, so that production personnel can be confident of the design as well as the material. Models from existing RP&M techniques are made from a limited variety of materials having poor to marginal mechanical and thermal properties compared to the end use materials.[1] It is often not possible to use an RP&M model for functional testing. However, soft tooling can provide limited quantities of true prototypes in the desired end use materials. Furthermore, using RP&M as a starting point, soft tooling can also be realized quite rapidly.

Processes Suitable for Soft Tooling

There are many manufacturing processes that can make use of soft tooling, such as:

1. Plastics: injection and compression molding, vacuum casting, vacuum forming, glass reinforced plastic lay-up, blow molding, extrusion, etc.
2. Metals: sand casting (patterns, matchplates, coreboxes), wax patterns for investment casting, sheet metal forming, die casting, and hot and cold forging.
3. Ceramics: slip casting, powder compacting in presses, and isostatic forming.

10.2 Soft Tooling Techniques

Many soft tooling techniques exist. This section covers the predominant methods.

Castable Resins

This is, without doubt, the simplest and least expensive technique for producing a soft tool. It consists of mounting a pattern within a mold box, setting up a parting line with plasticine, and then painting/pouring resin over the pattern until there is sufficient material to form half a tool. There are many tooling resins available, each with different mechanical and thermal characteristics. The resins are often loaded with aluminum powder or pellets to improve thermal conductivity of the tool, and to reduce the cost of resin.

Castable Ceramics

There have been a few examples published where ceramic materials have proved very successful as tooling materials. The simplest and cheapest are based on cement/sand mixtures poured over a pattern. It is usually necessary to reduce the water content in the mixture to avoid excessive shrinkage as the material sets. Small quantities of plasticizer render the mixture fluid and help assure complete filling of cavities. It is necessary to have a mixing machine to evenly distribute the material constituents, and a vibration table to help pack the material after it has been poured. Some materials are also poured under vacuum to avoid air bubbles. Two companies involved in producing low-shrinkage, high-compression strength versions of these materials are: CEMCOM Corporation in the United States, and Densit in Denmark (see Page 288 for a case study involving Pitney Bowes using CEMCOM materials).

Spray Metal Tooling

Very popular for producing soft tooling, metal spraying is one of the most common processes. It involves spraying a thin metal shell (about 0.08 inch [2 mm] thick) over a pattern and then backing the shell with epoxy.[2] Choosing one of the several different metal spraying techniques depends on the pattern material and the material to be used for the shell. With most RP&M techniques, the models produced have a low glass transition temperature (the temperature at which the material starts to change to a soft amorphous structure). Therefore, it is important to keep the pattern temperature as low as possible when spraying. If the temperature of the model increases sufficiently, it will soften and distort. This will obviously produce an inaccurate tool.

High or Low Melting Points. The two most popular spray metal techniques for use with RP&M models involve low melting point alloys (lead/tin-based) distributed with a gun similar to a paint sprayer, or metal deposition with an arc system (the so-called "Tafa process"). The arc system feeds two wires into a gun and an electric arc is struck between them. This causes the wire material to melt. The exhaust jet from a compressed gas atomizes the molten material and sprays it onto the pattern. The higher the melting point of the wire material, the more difficult it is to keep the pattern cool. Therefore, it is customary to spray zinc- or aluminum-based alloys directly onto RP&M models. It is possible to spray higher melting point materials onto RP&M models, but it is necessary to be a little creative.

One technique is to metallically coat the model using electroless plating or metal vapor deposition. Once there is an adequate metallic coating on the model, the heat will be transmitted more readily across its surface.[3] Another possibility is to spray a low melting point material such as a zinc-based alloy onto the model, remove the model, and then spray a higher melting point material into the cavity shell.[4] Obviously, it would be possible to continue this procedure, moving up the temperature scale. Unfortunately, the multi-step process becomes expensive, and reproducing fine detail becomes progressively more difficult.

High Stresses. One problem associated with metal spraying is that it produces shells with high internal tensile stresses. It is possible to counteract these stresses by simultaneously shot peening the sprayed shell. Steel shot is fired at the shell during the spraying process. This induces compressive stresses that tend to counteract the tensile stresses intrinsic in the shell.

Metal spraying is often used on models that have large, gently curved surfaces. Unfortunately, spraying into a narrow slot or a small diameter hole is very difficult. Thus, it is common to make brass inserts for these features, locate them in the model, and then spray around them. When the model is removed, the inserts are permanently fixed to the shell. The brass inserts are also stronger than the shell material, which tends to break easily if formed as a tall thin feature. Spray metal tools can produce from 1,000 to 5,000 parts, depending on the process, the material being formed, and the amount of tender loving care given to the tool.[5]

Clamping and injection pressures for spray metal tools are usually less than those employed with steel or aluminum tools. This may have an effect on the mechanical properties of the injection molded part. The thermal conductivity of a metal-sprayed tool is typically less than a machined aluminum or steel tool. This will also affect the mechanical properties of injection molded materials, and will increase the cycle time. Additionally, some filled plastics are much more corrosive and abrasive on the tool face. This can be partially overcome by techniques such as plating the tool surface with nickel or chrome, or using aluminum or steel inserts.

Electroforming

Electroforming, a relatively little known process, is used to make tools for many applications. The technique involves electroless/electroplating a thick shell (several millimeters) onto a master pattern. The master pattern is then removed and the shell backed with a suitable material to generate half the tool. This process is very popular for producing tools having complicated patterns that are intended for molding shoe soles. Here, the original shoe model is made from wax. It is usual to take rubber negatives off the wax master and then electroform onto the negatives. As the rubber is flexible, it is easy to remove from the shell.

Problems and Solutions. A common material for electroforming is nickel, which has good thermal conductivity, good strength, and excellent abrasion resistance. The process gives very faithful reproduction of the master, but can be limited when plating into deep, narrow slots. Electroplating builds up more material on exterior corners and less on internal corners. Thus, a narrow slot could be closed at the top before sufficient material has been plated at the bottom. This problem can be partially overcome by reducing the electroplating current, but the time to produce the shell will increase. To start the plating process, the surface must be made electrically conductive. A simple technique to achieve this requirement is to spray a conducting lacquer onto the model.

Silicone Rubber Molds

Silicone RTV (room temperature vulcanizing) rubber is an extremely versatile material. Although somewhat expensive, it can be readily molded around a master to produce a cavity. There are two basic types of silicone rubber used to make molds, and the method of mold fabrication differs slightly for each. Silicone RTV material is available in either opaque or transparent form, with the transparent type being more expensive.

With the transparent material, the RP&M master model is suspended within a box. The silicone rubber is then poured into the box so that it fully surrounds the model. After the silicone rubber has solidified, the parting line is cut with a scalpel and the master removed, leaving the required soft tooling cavity. It is then possible to mold a variety of materials within this cavity. One of the most popular is polyurethane, which is commercially available with a variety of mechanical properties and can replicate elastomers, nylon, acrylic, etc. Polyurethane is usually poured into the silicone rubber cavity under vacuum to avoid the formation of air bubbles in the molded component.

A silicone RTV rubber tool will generally produce about 20 polyurethane parts before it starts to break up. The exact number depends on the amount of detail in the tool and the type of polyurethane being molded. Flexible polyurethanes require longer postcure times within the mold, which is placed in an oven at about 150° F (65° C). This prolonged heating unfortunately dries out the surface of the silicone rubber and renders it more brittle. Once embrittlement occurs, fine detail on the inner surface of the mold starts to break off and subsequent moldings reflect this loss.

One or Two Steps. Opaque silicones are either cast in one step as above, or in two separate stages. When cast in one step, it is more difficult to cut to the parting line around the master, as it is not possible to see through the silicone. To complete the cut, the operator starts at the sprue and carefully works along the parting line until the mold is separated. A Polaroid photograph of the model within the molding box is a useful aid in choosing the correct direction for cutting.

The alternative to casting in one step requires the model to be set up with the parting line provided in plasticine. One half of the mold can then be produced. After the silicone has solidified, it can be inverted, the plasticine removed, and the second half poured. This process generally takes longer to produce a mold, but the opaque silicone rubber is much less expensive.

Spin Casting

A variation on the opaque silicone RTV rubber process is available from Tekcast Industries, Inc.[6] The tools generated in this process are made from vulcanized rubber, with several models located radially in a disc-shaped tool. When the tool is produced, it is possible to cast polyurethane or zinc-based alloys. To aid filling, the tool is rotated so that centrifugal force pressurizes the cavity. This is an excellent process for forming small zinc castings that will be ultimately manufactured in large quantities by die casting.

The Keltool Process

It is arguable whether the Keltool process should be considered "soft tooling," as the resulting tools are generated in bronze, stellite, or even A6 tool steel. There is only limited information about how the tools are actually made, but the procedure appears similar to a process described by Arie Ruder.[7] A metal powder/binder mixture is poured around a positive silicone RTV submaster, taken from an original RP&M negative master pattern. At an appropriate temperature and pressure, the binder sets and holds the part as a green compact similar to a conventional powder-compacted part. The part is then sintered at a high temperature to burn off the binder and fuse the powder particles. Infiltration of a lower melting point metal, typically copper, can help fill in the spaces between the particles, thereby reducing the overall shrinkage during sintering.

The tools from this process provide excellent lifetime (greater than 1 million injection molded ABS components have been manufactured from a single Keltool), good detail definition, and a high quality surface finish. The two disadvantages are: (1) about 0.8% linear shrinkage must be allowed for in the sintering process, and (2) only relatively small tools (about 6 × 6 × 4 inches, or 150 × 150 × 100 mm) are currently possible with high accuracy.

Orbital Abrasion

Equipment cost for this process is very high. The technique involves molding a material loaded with abrasive against a master. This abrasive molding is then used to wear away a carbon block until an electrode is produced. It is difficult to preserve very fine detail with this process due to the orbital motion of the abrading tool. The technique appears better suited to larger, gently contoured tools, or where multiple electrodes are required.

Metal-coated RP&M Models as EDM Electrodes

Although this process is still at the research stage, initial work in the 1990s showed the possibility of electroless plating SL models, and then using these as electrical discharge machining electrodes.[4,8] This has the advantage of being a very direct route to produce tooling in a variety of materials. Initial work suggests that for every inch the electrode penetrates the workpiece, there will be approximately 0.005 inch (0.125 mm) of material removed from the electrode.

Therefore, only relatively thin metal coatings are required on the model. The model could be used several times after subsequent plating if deep cavities are required. There are other techniques that could also be utilized to provide a conductive metallic coating on the model, such as chemical vapor deposition and electroforming.

10.3 Tooling Surface Finish Requirements

There are many different types of surface finishes required on tooling, depending on the process used, material being processed, and aesthetic requirements of the final product. Some of these include:

Finishes Polished for Optical Clarity. Many injection molded plastic parts are produced in transparent material for use as eye shields, inspection windows, drill guards, etc. The tools to produce these items require a mirror-like surface finish; otherwise, irregular features in the tool surface will impair the material's optical clarity. Some metal parts also require a mirror finish, such as the parabolic housings for automobile headlights. Achieving this type of surface finish requires considerable work and patience.

Matte and Spark Finishes. These finishes can be achieved on RP&M models by spraying the model with primer. This finish will then be transferred to the tool surface on replication.

Leather-Grain Finishes. The easiest method of obtaining this finish is to cover the RP&M model with a leather-grained material before making the tool. Obviously, an allowance is needed for the thickness of the material when making the model. Levels of surface roughness required on tooling will vary greatly, but are generally between 1 micro inch (0.025 micron) center line average (CLA) for super fine mirror finishes sufficient for optical clarity and 250 micro inches (6.3 microns) CLA for rougher spark finishes.

10.4 Surface Finish Studies of RP&M Patterns

All existing commercial RP&M techniques produce parts by adding layers on top of each other. This produces surface features peculiar to these processes. The effect on a larger scale is similar, but there are some differences when the surfaces are observed with magnification. The general surface features of RP&M models show a definite layered effect in the vertical direction, which can be seen by eye. Obviously, if thinner layers are used, they become more difficult to see. A consequence of building in layers is known as "stair-stepping" and is especially pronounced on surfaces at a small angle to the individual layers.[9] This effect is seen in Figure 10-1, and is similar to the problem of machined cusps found on milled surfaces.

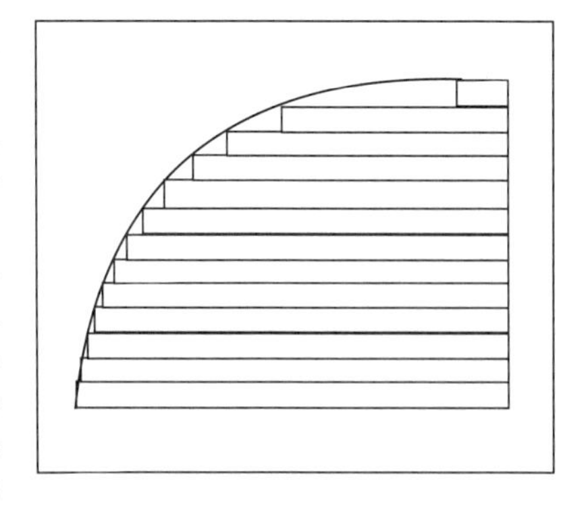

Figure 10-1. Stair-stepping effect from layers on curved surfaces.

Part Orientation

RP&M parts can be oriented in many different ways. This may lead to very different surface finishes on various faces due to the stair-stepping effect. Generally, it is much better to arrange the part geometry so that curved surfaces are as vertical as possible. For example, turbine blades should be built to stand on the root of the blade.

Building parts in the correct orientation reduces the effort required to finish them. Optimum orientation of parts generally requires a level of user experience, as well as common sense. There are mathematical approaches being developed that will show a graphical simulation of the surface finish. These are similar to graphics on finite element analysis packages, where different stress levels are shown in various colors on the surface of a model. There is a direct relationship between the surface roughness of a model and the layer thickness used to build it.

Many finishing techniques are possible with RP&M models, such as hand sanding, abrasive blasting, tumbling, vibratory finishing, abrasive flow finishing, painting, electroless plating, and electroplating. The vast majority of RP&M parts are finished with either hand sanding using abrasive paper or abrasive bead blasting. These are simple techniques, but they can be very time consuming. Major disadvantages are: (1) quality of the finish is extremely operator dependent, and (2) it is not always possible to maintain model accuracy with hand finishing.

Each RP&M process has its own unique surface features.

Stereolithography. When a Gaussian laser beam scans the surface of a photopolymer resin, it produces a cured region having a parabolic cross-section. Therefore, vertical walls are actually a series of parabolic forms stacked on top of each other.[10] Consequently, each layer has a small undercut. Horizontal surfaces are first scanned on the borders, then the interior areas are hatched with sequential X and Y scan vectors. Border and hatch scans appear as raised lines. These features can be seen in Figure 10-2. Furthermore, stereolithography requires supports, which leave witness marks after removal.

Figure 10-2. Surfaces on a stereolithography model (100%, 300% magnification).

There can be major surface finish differences between models made from acrylate resins and epoxy resins. Epoxy models, which have much smoother flat upfacing surfaces than acrylate models, are not normally cleaned in a resin stripper. Excess resin is simply wiped away with absorbent towels and cotton buds. After postcure, this produces a passive meniscus smoothing effect between layer features, further reducing the surface roughness of epoxy parts.

Solid Ground Curing. These models have edges shaped differently from those produced by stereolithography, as can be seen in Figure 10-3. These features are left following excess resin removal after each layer. The meniscus effect remains as a series of feather edges.

There are no supports used in this process, so no witness marks are left. Each layer is milled with a fly cutter, usually making surfaces reasonably smooth.

LOM. These models show a witness of laser burns on all surfaces, a faceted finish on side walls, and a tile pattern on curved and horizontal surfaces (Figure 10-4).

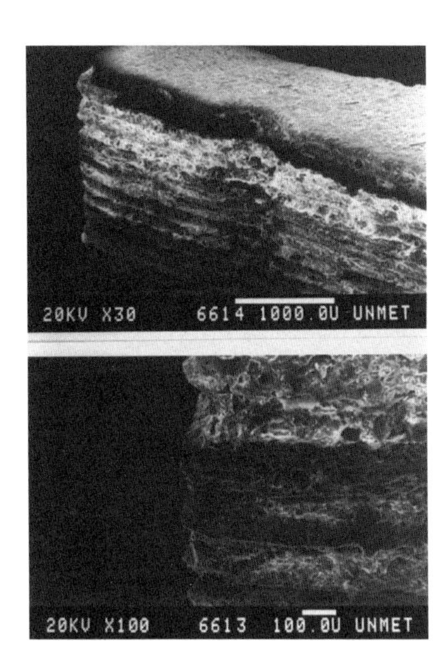

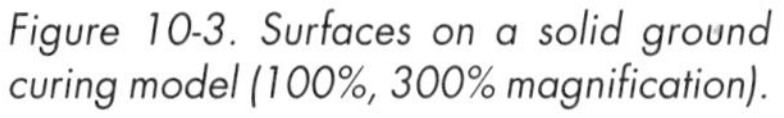

Figure 10-3. Surfaces on a solid ground curing model (100%, 300% magnification).

Figure 10-4. Surfaces on an LOM model (100%, 300% magnification).

There are differences in surface finish between vertical surfaces and horizontal surfaces. A better finish is more easily obtained by abrading LOM models across the grain. Abrading across the top of sheets can cause the layers to fur.

SLS. As the powder particles tend to be partially fused, gaps or interstices are left between them (Figure 10-5), yielding a porous surface that needs to be sealed.

The common methods of sealing SLS surfaces are either using a wax coating, or painting the surfaces with Superglue™. Variations can also exist in surface porosity due to nonuniform heat transfer within the powder bed.

FDM. Vertical surfaces show the shape of the extruded beads, in the form of cusped grooves (Figure 10-6). Feather edges can sometimes be seen on horizontal surface fills where bead runs have come together and not fully filled a layer. Some models also exhibit vertical seams produced by stepper motor delays in the Z direction.

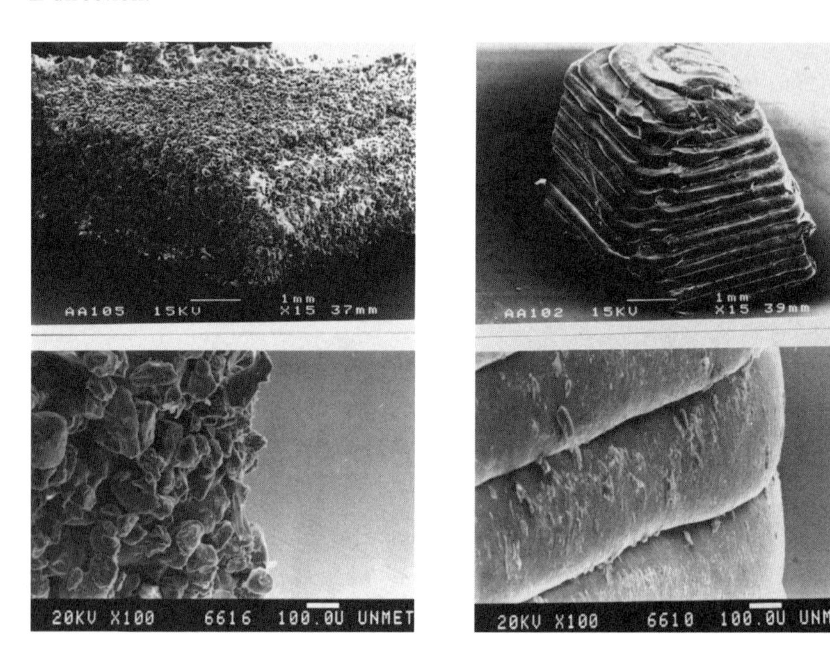

Figure 10-5. Surfaces on an SLS model (100%, 300% magnification).

Figure 10-6. Surfaces on an FDM model (100%, 300% magnification).

10.5 Case Studies

Silicone RTV Experience at Rover Group Ltd.

This case study involves the use of vacuum-cast polyurethane resins formed in silicone RTV rubber molds to fabricate nine heater housings. The rapid prototyping facility at Rover Group Ltd.'s Design Center was opened in June 1991. Installation of an SLA-500 was seen as a major breakthrough in generating design prototypes. This covered a wide range of applications, including design verification and form, fit, and function.

One group to realize significant savings was Trim and Hardware, based at Canley near Coventry, England. In November 1991, just five months after the commissioning of the SL facility, a project was undertaken consisting of 12 heater parts. One of these parts was the actual heater casing in two halves. The CAD data was supplied in a single surface format. This meant the engineer had only to design the surface from which he needed the data. When the stereolithography department received the CAD information, they created the other surface using the STL file translator software. This saved the engineer considerable time in surface generation.

Once the SL model was completed in two halves, which took approximately 30 hours, it was used to check air flow by passing colored smoke through the assembly. A silicone RTV mold was then made directly from the SL model. Using the silicone mold, nine complete vacuum-cast polyurethane heater assemblies were generated. These components were subsequently built into complete heaters and used as prototypes for functional testing on running engines. A time and cost comparison for the heater casings can be seen below.

	Traditional Route	SLA and Silicone Tooling	Savings
Time	15 weeks	4 weeks	11 weeks
Cost	$73,764	$6,695	$67,068

A cost comparison for the entire heater project, which took only four weeks, or about one third of the time required by conventional methods, is summarized below.

	Traditional Route	SLA and Silicone Tooling	Savings
Cost	$314,488	$58,426	$256,061

Although financial savings are certainly very important in developing an item, time saved was considered by Rover to be the most important factor.

Silicone RTV Experience at The University of Nottingham

This case study describes the use of vacuum-cast polyurethane resins in silicone rubber molds to manufacture a car radio handset. One member of the University of Nottingham's Centre for Rapid Prototyping in Manufacturing is Maxon Europe Ltd. They had a particularly tight deadline to produce prototypes for demonstration to a tooling company in Korea so that they could explain their design. The original design was produced as a solid CAD model in AUTOCAD, using version 12 with the Advanced Modelling Extension (AME) 2.1. As the design was rather involved, with complex surfaces and blend radii, AUTOCAD had great difficulty in completing the solid model. This meant that an STL file could not be produced and therefore an RP&M model was not possible. The subsequent events occurred in the following order:

Friday. The 2D paper drawings were received at the University of Nottingham in the afternoon.

Sunday. CAD modeling was started using Unigraphics version 10.2.

Monday. The CAD model of the handset, with all the button holes as shown in Figure 10-7, was finished in the afternoon and the STL file produced. The SL

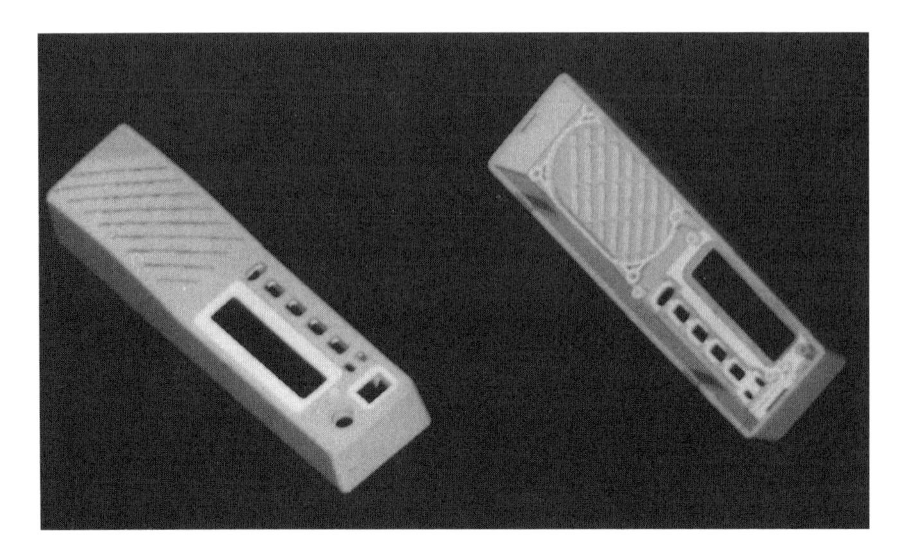

Figure 10-7. CAD solid model of the handset.

model was then built overnight in epoxy resin SL 5170, using the ACES build style.

Tuesday. The SL model was cleaned of supports and postcured. Outside surfaces of the model were finished with abrasive paper to smooth the surface, and then primed with spray paint. The spray paint gives a matte finish as required on the mold surfaces, and helps prevent the resin from inhibiting silicone rubber curing. This is not unusual. There are many materials that will inhibit curing of silicone rubber, such as Superglue activators, glues on tapes, and some car body fillers. At this point, a second iteration of the CAD model was produced, and the corresponding SL model was generated overnight.

Wednesday. Cellophane tape was used to mask the windows in the body of the handset. This tape was also added all around the parting line so that the mating faces of the mold, next to the part, would be smooth. Edges of the tape were marked with a black permanent marker pen. This helped the technician see the tape through the transparent silicone RTV rubber. The silicone rubber tool was made in one piece using the clear material. The silicone rubber was then cut and the tool split. Expanding pliers were used to help split the rubber, making it relatively easy to cut very close to the black edge of the tape. The initial polyurethane moldings were completed on Wednesday.

Thursday. Final polyurethane moldings were produced in the morning, yielding four complete car radio handsets. In the afternoon, the engineer from Maxon collected the SL models, as well as cast parts. The models helped the tooling company see exactly what was required of the tool.

The vacuum cast polyurethane prototypes were used in a trade show after being assembled with the other parts. Customer response at the trade show indicated that the buttons were the wrong size, and a third iteration was incorporated into the final design.The vacuum cast materials, while different from those used in production, provided a close simulation for acoustic tests.

10.6 Plaster Molds: Advantages and Disadvantages

Plaster mold casting (Figure 10-8) is used as a prototyping process to generate simulated die castings. There are several variations, but they usually proceed along the following stages:

1. An RP&M master is produced as close to the final shape of the die casting as possible. It is not absolutely necessary to incorporate draft on the walls at this stage, but it can be helpful.
2. A silicone RTV rubber negative is molded over the master.
3. A second silicone rubber is molded into the first. This provides a silicone rubber positive of the original model.
4. Plaster is molded around the silicone rubber positive to provide a plaster cavity.
5. Molten metal is poured into the plaster cavity.
6. After solidification of the metal, the plaster mold is broken away and the casting fettled and cleaned.

The rubber version of the master is required so that it can be easily withdrawn from the plaster mold. It is also possible to mold epoxy off the RP&M master, then pour plaster over this pattern.[11] The epoxy molds will have a greater life than those made from rubber.

The lead time, from generating the RP&M master model to obtaining 10 castings, is claimed at eight days, with two weeks needed to produce 30 to 50 castings. However, 3 to 4 weeks appears a more typical delivery time.[11] The cost of prototyping with this process is about 2 to 5% of the die cost, and is considered very good insurance.[12]

Advantages of plaster mold casting are:

- low mold costs;
- good surface detail;
- large size range; and
- relatively easy technique.

Disadvantages are:

- low cooling rates and therefore reduced mechanical properties, leading to parts with a yield strength 20% lower than a conventional die casting;
- the need to make a new mold for each casting;
- as the surface of the casting is not chilled (unlike die cast parts), it is relatively soft; and
- slightly different alloys are used for plaster casting compared to die casting.

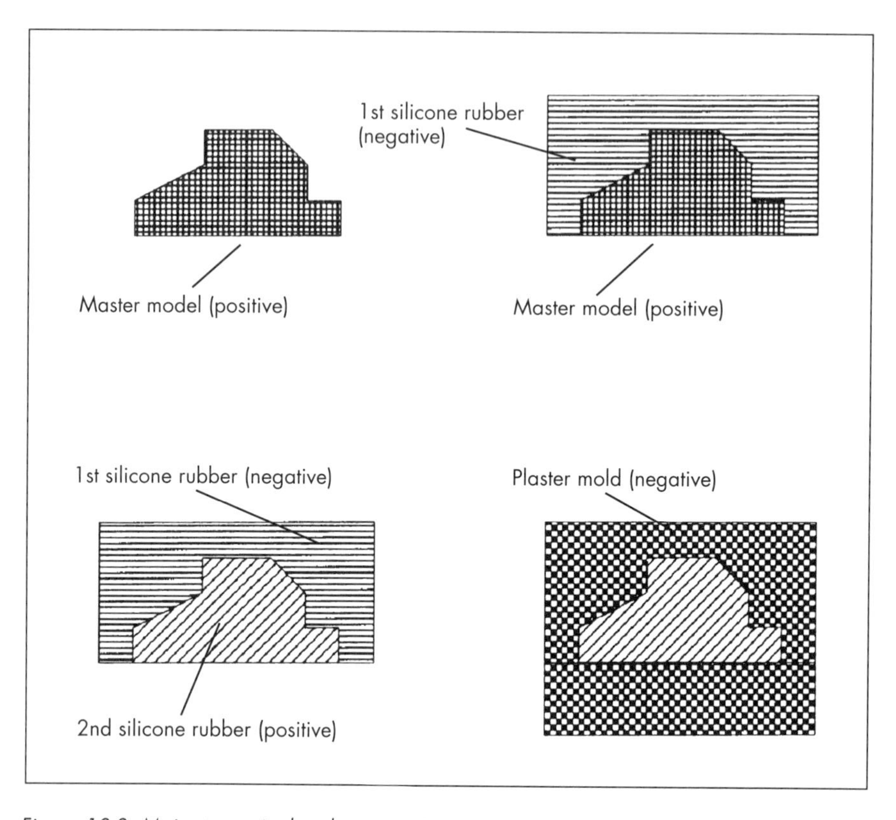

Figure 10-8. Main stages in the plaster casting process.

10.7 Spray Metal Tooling

This section includes information presented by Scott A. Martyniak at the SME Rapid Prototyping and Manufacturing '93 Conference.[13]

Spray metal tools have been used in many applications including sheet metal forming, injection molding, compression molding, blow molding, and pre-preg sheet lay up. A variety of plastic materials have been molded including: polypropylene, ABS, and polystyrene, as well as difficult process materials such as glass-reinforced nylon and polycarbonate.

The life of a spray metal tool depends on several factors, including the following:

1. The more complex the tool, and the greater the number of fine features, the shorter the tool life. Aluminum or brass inserts are usually used in spray metal tools where fine details and deep slots are required. This increases tool life, ensuring better-quality moldings.
2. Lower viscosity and lower processing temperatures of the injected plastic also provide longer tool life. Production numbers from spray metal tools

vary enormously. Some figures have been given as low as 25 to 250 for polypropylene, and as high as 1,800 for ABS. Thermal shock of the tool must be avoided. When operating the tool for the first time, partial shots should be used to gradually bring the tool up to normal operating temperature. Once the tool is operating, it is better to try to keep it running continuously. Using a spray metal tool for a few hours and then letting it cool down, before starting it again, will induce earlier tool failure.

3. The operator plays a large part in determining tool life. Spray metal tools need careful handling, as they are not as robust as aluminum or steel tools. Clamping pressures need to be reduced to avoid crushing the spray metal mold, unless a steel chase is used to withstand the clamping forces. The nozzle that injects the polymer also needs to be brought with care to the sprue of the mold.

A spray metal tool is generated using the following steps:

1. An RP&M master model is made, incorporating a suitable draft angle, as well as an allowance for shrinkage of the injection molded material.
2. The master is finished with the smoothness or texture required of the moldings. The sprayed shell will, fortunately or unfortunately, replicate virtually all features on the RP&M model, including defects.
3. The model is set up on a parting line within a chase, including the sprue and gates, so that the first half of the tool can be sprayed. Plasticine is often used to build this parting line.
4. The model and parting line must be coated with a release agent such as polyvinyl alcohol (PVA). If the RP&M master is difficult to remove because of inadequate draft angle, rough surfaces, or insufficient release agent, the shell may be damaged later.
5. Ejector pins and inserts must be positioned before spraying. Ejector pins are used for ejecting moldings from the tool, but they are also useful in removing the model from the shell. If too many ejector pins are added at this stage, it can be difficult to spray all over the shell.
6. The metal shell is sprayed to a thickness of about 0.08 inch (2 mm). This should be done in one session. Leaving the shell for a prolonged interval between coats allows oxidation to occur. This can lead to delamination of the sprayed shell.
7. A backing material is then used to support the thin shell. Epoxy resin is often employed. This is usually painted over the shell to ensure complete coverage. Additional epoxy can then be poured to the correct depth. Aluminum pellets, or fine powder, are often mixed with the epoxy to improve thermal conductivity and to reduce the cost of the tool, as epoxy is not inexpensive. Water cooling lines can be positioned within the epoxy at this stage.
8. Once the epoxy has set, the back of the tool can be machined flat.

The preceding eight steps yield only one half the tool. Producing the second half requires the following additional steps (see Figure 10-9 for an illustration of the main stages of producing a sprayed metal tool):

9. The chase is inverted and rests on the newly machined face. The second chase can then be located on top of the bottom section.
10. The parting line material is removed to expose the other side of the model.
11. Stages 4 to 8 are then repeated to complete the tool.
12. Removing the RP&M model is the most traumatic part of the operation. If this is not done carefully, part of the sprayed shell may come away with the master. The ejector pins should greatly help at this stage.
13. Depending upon the material to be processed, it may be necessary to brush plate the sprayed surfaces of the tool.

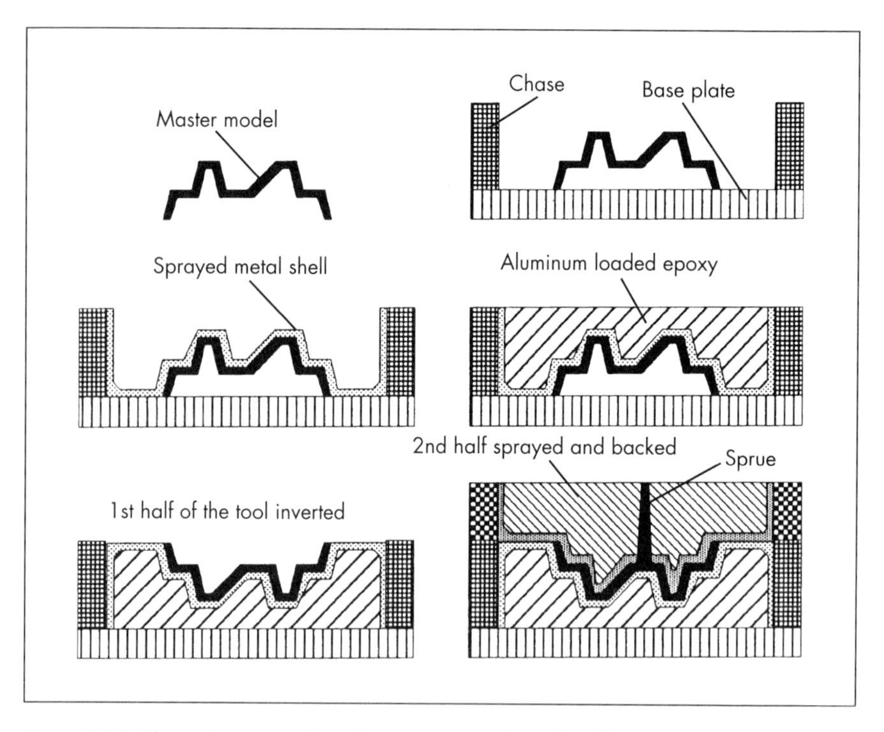

Figure 10-9. The main stages of producing a sprayed metal tool.

10.8 Chemically Bonded Ceramic Molds

This case study discusses the use of chemically bonded ceramic (CBC) as a tool material for injection molding thermoset plastic in the production of a large structural base for a new Pitney Bowes mailing machine.[14]

The original base of the mailing machine was made in aluminum as a die cast part. It measured about 27(l) × 20(w) × 4(h) inches (686 × 508 × 102 mm) and weighed about 10 lbs (4.5 kg). It was decided to redesign the part as a thermoset molding for several reasons, including reduced weight, smaller part count, and lower cost. To check the stiffness of the molded base, at least 10 parts were required in 22% glass-reinforced polyester.

Therefore, a prototype tooling technique was required in order to produce 25 to 50 moldings. Pitney Bowes had previous experience using CBC tools to mold reinforced plastic, but only for injection molded parts that were much smaller than the proposed base. The goals of this project were to produce a prototype tool in six weeks (versus 14 to 16 weeks), at one third the $100,000 cost of conventional aluminum tooling.

Pitney Bowes developed CBC molds for smaller parts in 1993, when teamed with Santin Engineering and CEMCOM Corporation. Santin provided expertise in pattern fabrication and mold design, while CEMCOM supplied knowledge in the use of CBC material. This time, because of the much larger sizes, 3D Systems built the SL master model on an SLA-500.

The part was designed on Pro/ENGINEER. The solid CAD model was also used for finite element analysis, which showed that some sections were not stiff enough. The pattern was too large even to fit on an SLA-500, so it was built as three pieces intentionally dovetailed together. The master included the shrinkage of the SL epoxy resin (0.06%), the CBC tooling material shrinkage (0.02%), and shrinkage of the final injection molded thermoset (0.1%). The model was built in the QuickCast build style, due to dewetting problems with an early experimental epoxy resin being tested at that time. However, building in QuickCast provided the added benefit of easier removal from the cast CBC, as the model was more able to flex than a solid SL part.

A steel mold frame was used to contain the CBC material and provide true alignment between the two mold halves. The parting line was machined in resin, based on data from the CAD model. Six steel inserts were used where there was concern over the aspect ratio of some features, or where the injected plastic might cause premature wear of the mold surface. Bronze rods 0.5 inch (12.7 mm) in diameter were located so that they could later be drilled to act as guides for the ejector pins used on both halves of the tool (40 ejector guides in all). The SL model was hand finished to eliminate any features that would hinder removal from the CBC mold halves.

CBC material is a calcium silicate-based castable material formed at room temperature. The specific material used in this mold was COMTEK 66, which incorporates fine chopped fibers to enhance fracture toughness and tensile strength. The physical and mechanical properties of CBC COMTEK 66 are listed in Table 10-1.

Table 10-1. Properties of COMTEK 66*

Properties	Imperial Units	S.I. Units
Compressive strength	50,000 psi	345 MPa
Flexural strength	6,000 psi	41 MPa
Elastic modulus	5.4 Mpsi	37 GPa
Hardness	R_b 65	.....
Density	225 lb/ft^3	3600 kg/m^3
Coefficient of thermal expansion	7.7×10^{-6}/°F	13.9×10^{-6}/°C
Thermal conductivity	17 BTU in/ft^2 hr °F	2.5 W/mK
Specific heat	0.19 BTU/lb °F	787 J/kg °C
Shrinkage	±0.0002 in/in	±0.02%
Maximum operating temperature	400° F	204° C

*Note: This material can operate at even higher temperatures, but with increased shrinkage.

The CBC material is vacuum mixed and then vacuum cast at room temperature. After 18 to 24 hours, it forms a gel with 25% of its final strength and 50% of its eventual stiffness. The stages of producing the two halves of the tool are similar to the spray metal procedure described earlier. When the two halves are cast, the RP&M master pattern is removed, and the CBC material is cured in an oven in the presence of moisture. The back surfaces of the two halves are subsequently machined flat. Finally, bronze bushings are drilled and reamed to accept the ejector pins.

Sufficient injection molded, 22% glass-reinforced polyester bases were produced from the tool and sent to Pitney Bowes for final machining and testing. The moldings passed all the necessary tests, including those for deflection, drop test, and cyclic testing. Among other benefits, this project proved that CBC tooling could produce functional prototypes in less than one third of the time, and at 40% of the cost of production tooling.

10.9 Suggestions for Further Research and Development

There are several current research projects studying different tooling techniques.

1. The European "MUST" project on simultaneous shot peening. This program is aimed at developing techniques for metal spraying of high-melting-point materials onto SL masters. If this can be perfected, it will then be possible to combine the simplicity of metal spraying with the higher strength and hardness of steel surfaces.
2. A European project to develop SL resins for building injection mold tools without any intermediate steps required. Research into the development of high temperature resins is being undertaken so that mold inserts can be designed in CAD, built in an SLA, and then used to directly produce injection molded products.
3. An additional European project to develop laminated sheet tools using laser cutting. Laminated sheets of steel have previously been clamped together and evaluated as blanking tools. This project is using the same principle to fabricate injection molding tools.
4. The use of concrete in Scandinavia and Germany to produce metal forming tools. This material is similar to the CBC ceramic described earlier, but the applications appear to be concentrating on sheet metal forming.

There are several other soft tooling R&D projects that have recently started or are about to begin very soon. These include:

1. More work on castable ceramics.
2. Metal sprayed electrodes for EDM.
3. Copper coated RP&M masters for EDM electrodes.
4. Cast ceramic electrodes for EDM.
5. Using ACES SL parts as tooling to directly produce wax patterns for limited quantity shell investment casting production.

The various opportunities involving soft tooling generated from RP&M master patterns still have a long way to go before they are fully developed. Many existing and future research projects at both corporations and universities will develop new techniques, enabling tools to be made more easily, in a shorter time, and at significantly lower cost.

References

1. Kimble, L., "A Materials Comparison for Rapid Prototyping Systems," Proceedings of the SME Rapid Prototyping and Manufacturing '93 Conference, Dearborn, MI, May 11-13, 1993.
2. Flint, R. and Ellis, D., "Secondary Tooling Using Rapid Prototyping," Proceedings of the SME Rapid Prototyping and Manufacturing '94 Conference, Dearborn, MI, April 26-28, 1994.

3. Roche, A.D., "MUST Process for Producing Prototype Steel Tooling," Proceedings of the 2nd European Conference on Rapid Prototyping & Manufacturing, ed. Dickens, P.M., Nottingham, England, 1993, pp. 143-156.
4. Muller, H., "EDM Electrodes made by Rapid Prototyping," EARP Newsletter No. 3, January 1994, Published by The Danish Technological Institute, Denmark.
5. Denton, K.R. and Jacobs, P.F., "QuickCast™ & Rapid Tooling: A Case History at Ford Motor Company," Proceedings of the 3rd European Conference on Rapid Prototyping & Manufacturing, ed. Dickens, P.M., Nottingham, England, 1994, pp. 53-71.
6. Schaer, L.S., "Spin Casting Assists Automotive Product Designers Developing Fully Functional Metal and Plastic Test Parts from SL Models," Proceedings of the SME Rapid Prototyping and Manufacturing '94 Conference, Dearborn, MI, April 26-28, 1994.
7. Ruder, A., Buchkremer, H.P., and Stover, D., "Wet Powder Pouring and Rapid Prototyping," Proceedings of the 1st European Conference on Rapid Prototyping, ed. Dickens, P.M., Nottingham, England, 1992, pp. 217-229.
8. Dickens, P.M. and Smith, P., "Stereolithography Tooling," Proceedings of the 1st European Conference on Rapid Prototyping, ed. Dickens, P.M., Nottingham, England, 1992, pp. 309-317.
9. Dickens, P.M. and Cobb, R.C., "Surface Finishing Techniques for Rapid Prototyping," Proceedings of the SME Rapid Prototyping and Manufacturing '93 Conference, Dearborn, MI, 1993.
10. Jacobs, P.F., "Fundamentals of Stereolithography," Proceedings of the Solid Freeform Fabrication Symposium, University of Texas, Austin, Texas, 1992, pp. 196-211.
11. Kowalczyk, B., "The Plaster Mold Casting Process as Utilized in Partnership with Rapid Prototyping Technologies," Proceedings of the SME Rapid Prototyping and Manufacturing '94 Conference, Dearborn, MI, April 26-28, 1994.
12. Mueller, T., "Recent Developments in the Use of Rapid Prototyping Techniques to Prototype Die Cast Parts," Proceedings of the SME Rapid Prototyping and Manufacturing '94 Conference, Dearborn, MI, April 26-28, 1994.
13. Martyniak, S.A., "Prototype and Limited Production, Spray Metal Tools for Injection and Blow Molding," Proceedings of the SME Rapid Prototyping and Manufacturing '93 Conference, Dearborn, MI, 1993.
14. Nagarsheth, V., Jacobs, P.F., Santin, D., Wise, S., and Yoder, J., "CBC Injection Mold Tooling for Thermoset Plastic Parts using Stereolithography Patterns," Proceedings of the SME Rapid Prototyping and Manufacturing '94 Conference, Dearborn, MI, April 26-28, 1994.

CHAPTER 11

Hard Tooling Applications of RP&M

"The world we have created with our current level of thinking has created problems which that kind of thinking will not solve. Therefore, imagination is now more valuable than knowledge."

Albert Einstein, 1938

11.1 Overview of the Rapid Tooling Process

Rapid tooling is a process that allows a tool for injection molding and die casting operations to be manufactured quickly and efficiently so the resultant part will be representative of production material. This definition means different things to different people. Production quantity for one company may be 1,000; for another it may be 500,000. Materials to be molded or formed differ not only from company to company, but from part to part as well. Several things have changed in recent years that enable us to create tooling based on more complex geometry. They involve various combinations of the following:

CAD - Solid Modeling	Steel Metallurgy
RP&M Technologies	Advances in Spray Metal Methods
QuickCast	New "Back Filling" Techniques
Investment Cast Tooling	Metal Vapor Deposition Processes

Table 11-1 illustrates the expected yield from various types of tools.

CAD Requirements

With the advent of more sophisticated solid modeling systems, part geometry can be expected to increase in design complexity. In the past, part complexity determined how, or if, the part could be manufactured. Ultimately, part design is

By Karl R. Denton, Rapid Prototyping Specialist, Valeo Climate Control, Rochester Hills, Michigan, and Sean B. O'Reilly, Technical Specialist, Ford Motor Company, Dearborn, Michigan.

Table 11-1. Expected Yield from Various Types of Tools

Conventional Tooling				
Number of Required Parts	Mold Material Type	Fabrication Cost $(000)	Fabrication Time (weeks)	Mold Life Parts Produced
1 to 30	Silicone	$5	2 to 3	30
30 to 1,000	Epoxy composite	$9	4 to 5	300
1,000 to 3,000	Kirksite (cast)	$25	12 to 14	1,500
1,000 to 3,000	Aluminum (cast)	$30	12 to 14	2,000
3,000 to 250,000	Steel (machined)	$60	16 to 40	250,000
Rapid Tooling				
Number of Required Parts	Mold Material Type	Fabrication Cost $(000)	Fabrication Time (weeks)	Mold Life Parts Produced
1 to 30	Silicone	$5	2 to 3	30
30 to 300	Epoxy composite	$9	4 to 5	300
300 to 1,400	Arc metal spray**	$25	6 to 7	1,000
1,400 to 15,000	Nickel vapor deposition*	$30	6 to 7	5,000
3,000 to 250,000	Steel (machined)	$60	16 to 18	250,000
3,000 to 25,000*	Steel (cast)	$15	4 to 6	250,000
*Mold life study is still in progress **Composite Assembly				

driven by styling and customer desires. It is left to the design engineer to figure out how to actually manufacture the part.

Because solid modeling has "come of age," engineers can now view the CAD solid model prior to prototyping. When the designer is satisfied, the part can be made using RP&M. At this point, there are still no design limitations for two reasons. First, CAD does not care what the part geometry looks like; second, the RP&M machine does not care what the part looks like.

Unfortunately, we may have created a part that is either extremely expensive, or simply impossible, to manufacture. As an example, consider the part shown in Figure 11-1. While good for a mathematical brain teaser, the Klein bottle would be impossible to manufacture. Yet, it exists because of advanced CAD solid modeling techniques and the implementation of RP&M.

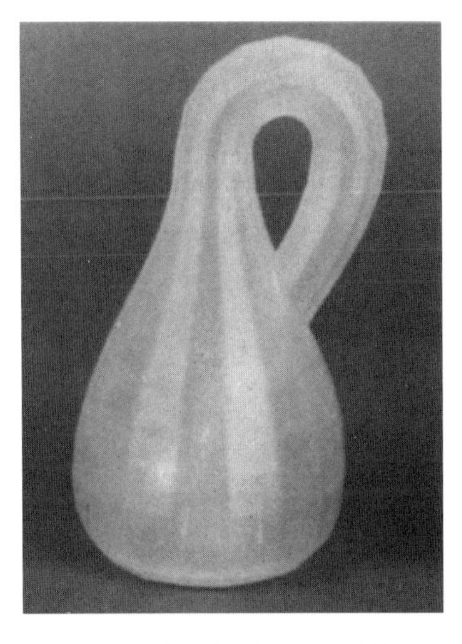

Figure 11-1. This Klein bottle would be impossible to manufacture without advanced CAD solid modeling techniques and the implementation of RP&M.

Tool Geometry Requirements

Solid modeling systems will now allow a tool to be generated based on the CAD model of the part. Unless strict rules are applied, the solid modeler may create tools that are incapable of producing parts. The modeling system still does not care what the geometry of the tool is, and will build it regardless of manufacturability issues. If you could envision a tool to produce the Klein bottle, what would it look like and, more importantly, how would it function?

Considerations of tool geometry must include the following.[1,2]

Parting lines—this is the most fundamental part of mold making. Parting line placement is determined by both part geometry and part appearance requirements.

Cooling lines—while parting lines are the most fundamental part of mold making, the routing of cooling lines may be the most important. Conventional methods of routing cooling lines are to drill in at various angles until passages are formed and the areas that are not used are then plugged, as illustrated in Figure 11-2. CAD systems will route cooling lines based on mold flow analysis, with no regard to machining limitations, as shown in Figure 11-3. This optimized routing can be incorporated into the tool in several ways.

Ejector pin holes—the ability to incorporate ejector pin holes in the solid model of the tool also allows the engineer or designer to "test" ejector placement for superior tool performance. Considerations may include placing ejectors at odd angles and using different sizes and materials.

Draft angles—another very important item in mold making. If no draft angle exists, the part may not be able to be removed from the mold. With excessive draft, the part will not be dimensionally accurate. Creating draft angles in solid modelers must be done carefully. Changes in part geometry by 1% have been known to create areas of concern (under/over cuts) that have been missed by the designers, thus building tools that must ultimately be ground to fit together.

Solid versus assembled hard tooling—this is the newest consideration in mold making. Because more complex cooling lines can be created in CAD, when in-

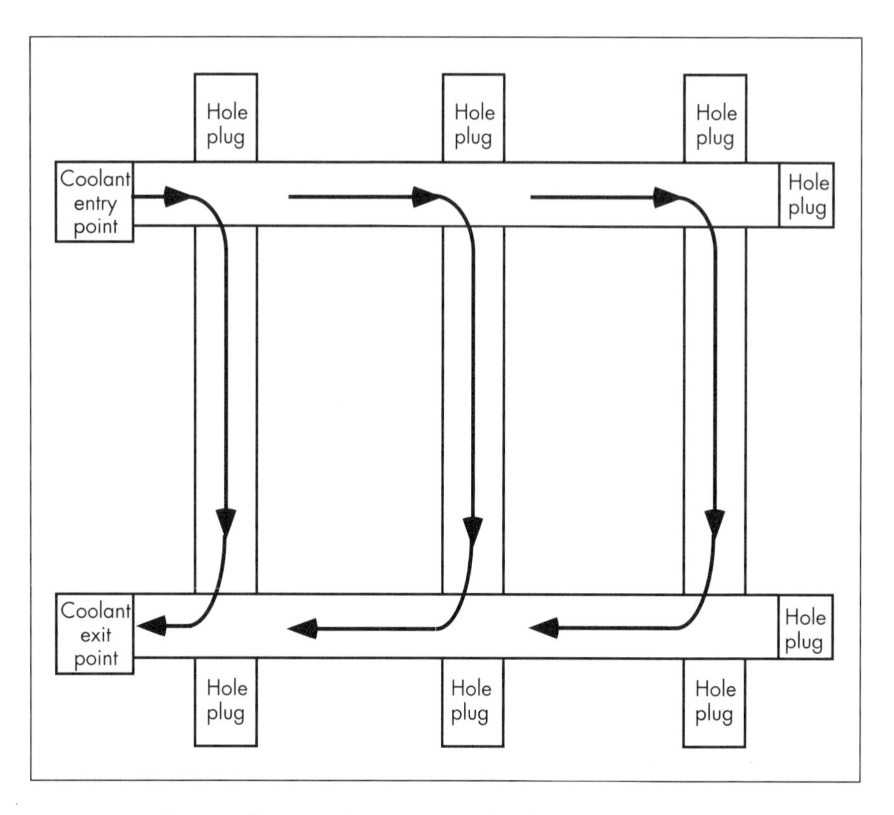

Figure 11-2. Schematic illustrating the conventional method of routing cooling lines.

vestment casting the tool, difficulties in removing the ceramic material from long, curved passages have surfaced. The thought behind assembled hard tooling is that a "subcore" is created within the main core that contains the cooling lines (Figure 11-4). The parting line for this would be the center of the cooling passages. The tool, after investment casting, would be assembled and a clamping plate would either be bolted or welded to the base of the core to seal the cooling passage.

Another consideration with regard to creating a tool via CAD solid modeling is mold base size. The size limitations of the RP&M system used will determine if the tool is an insert or both of the entire A&B plates. Items like part numbers and date stamps can usually be obtained off the shelf and placed into a slot or space prepped for the stamp. An exploded drawing of a complete tool is shown in Figure 11-5.

Using RP&M to Build "Fit & Finish" Parts

Although rapid tooling has the potential to be a great deal less expensive than conventional tooling, fit and finish parts should be built to verify part geometry.

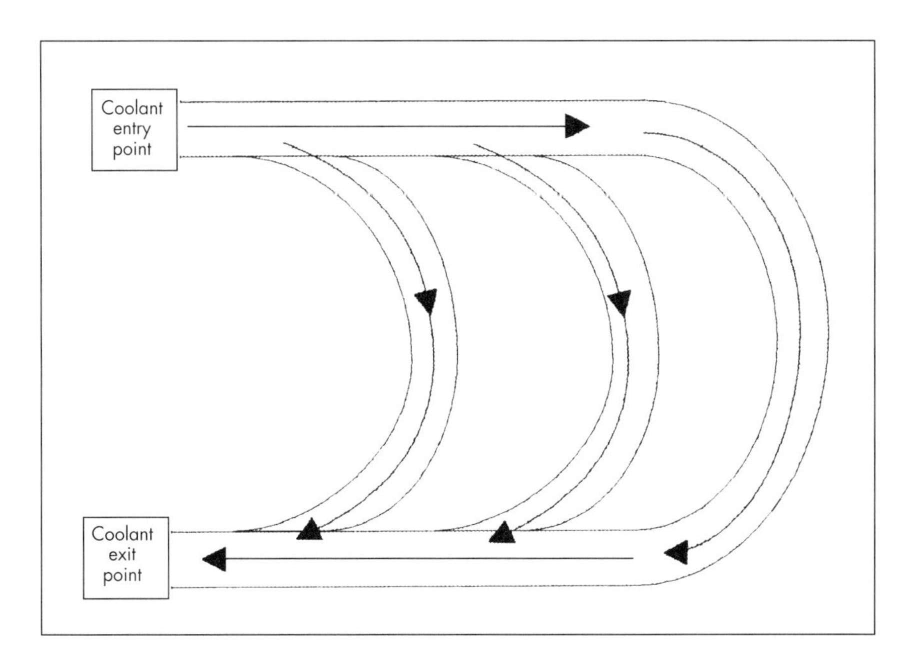

Figure 11-3. CAD systems route cooling lines based on mold flow analysis, with no regard to machining limitations.

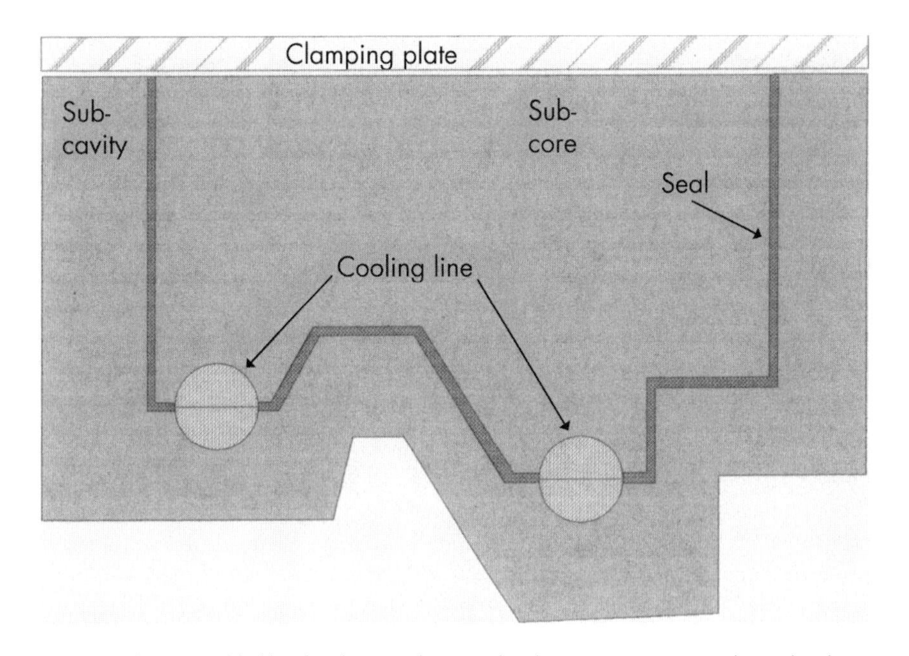

Figure 11-4. In assembled hard tooling, a subcore within the main core contains the cooling lines.

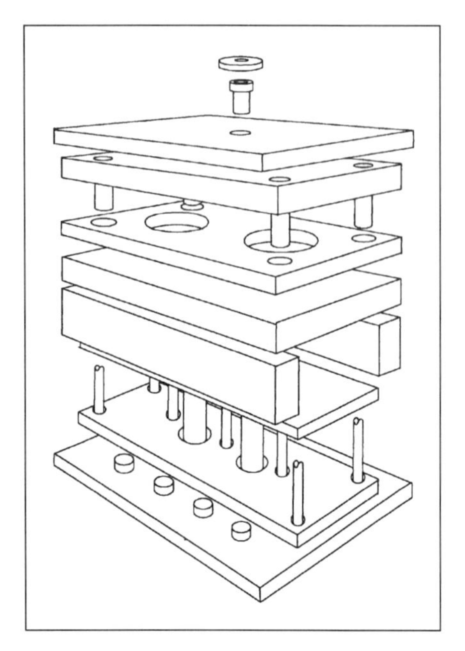

Figure 11-5. Exploded view of a complete tool created with RP&M.

When building the fit and finish part, remember to use only the RP&M shrinkage factor. This should be a "one-to-one" build of the final positive part that will ultimately be injection molded. If no other modifications are required, then the procedure continues. Otherwise, a second or third iteration may be needed.

Using RP&M to Produce Tools/Tool Inserts

To date, there are three RP&M technologies that have been used to create rapid tooling at Ford. They are SL, FDM, and SLS. The three RP&M methods have common concerns: size and material limitations, as well as shrinkage, accuracy, surface finish, and postprocessing issues.

Steel Selection. Prior to building the RP&M tool, the steel selection must be made. Considerations should include: (1) selection of an air hardening steel will result in very hard tools. In this case, the RP&M tool must be as close to net shape as possible. (2) If the final tool will be highly polished, the selection of a stainless steel may be the proper choice, e.g., P20. (3) Annealing after casting and then rehardening may result in tools that are dimensionally worthless. (4) Check and recheck the shrinkage factors of the steel selected.

Size. The largest build volume can be obtained using 3D Systems' SLA-500, the smallest build volume is obtained using the Stratasys machine, and the DTM machine will vary depending on the material used. It should be noted that, typically, the tooling is a great deal larger than the parts that will be created using the tools.

Materials. Depending on the technology and the desired outcome, the RP&M material can range from epoxy UV-curable photopolymer resin, as is used in the SLA-250/500, to investment casting wax or polycarbonate used on the SLS machine, or wax used on the FDM machine. Each will yield different results and different accuracy levels. To date, SL systems have been used to produce actual part molds at Ford Motor Company, while the other systems have been used to produce dunnage molds (typically these will be a great deal less accurate than part molds). DTM's sintered metal process shows some promise, although the build volume will still be smaller than that required for a great number of practical parts.

Shrinkage Factors. When producing rapid tooling, regardless of the technology, there are a number of shrinkage factors that must be included in the process for successful completion of a project. These shrinkage compensation factors include shrinkage of the final molded plastic material, shrinkage of the material that will be used as the RP&M pattern for casting, and shrinkage of the steel that will be poured. Shrinkage factors for the molded material and the RP&M build material are generally rather small and very controllable. The steel shrinkage factor will, however, require further investigation. As more tools are cast using the typical tool steels, we will begin to build a data base of information that, unfortunately, does not exist at this time.

Accuracy. This is the most important issue with regard to casting tools from RP&M machines. Accuracy of the part to be cast must be excellent and very repeatable. A reasonable definition of what accuracy is and how it relates to rapid tooling is simply that the resultant cast tooling should be as close to the CAD solid model as technology will allow. The ultimate tolerance goal is ±0.001 inch (±0.025 mm) up to 10 inches (254 mm), with a repeatability very close to 100%. This means that one should be able to build a part within a given tolerance, and build that part to the same tolerance every time. It should be noted that repeatability is, in the opinion of several people at Ford, the most important factor. The reason is . . . if the pattern is off the mark by 0.003 inch (0.076 mm) every time the part is built, we can compensate for that.

Postprocessing. This is simple; the tools produced from the RP&M technology should undergo as little postmachining as possible. Ideally, the tool should be finished in the cast state. Any issues regarding surface finish or other features should be applied at this time. If postprocessing causes a major shift in the pattern's dimensional stability prior to casting, it may be very difficult to regain accuracy after casting. Remember . . . investment cast the tool as close to net shape as possible. Following are some of the more important things to consider when casting from the three subject RP&M technologies.

Supports. The tool should be oriented in the vat to protect the most critical surfaces. This may mean the possibility of trapped volumes, longer build times, or other seemingly undesirable items in the build process. These considerations are minor compared to a critical surface that must have supports sanded or ground off, increasing the possibility of breaking through the skin and causing problems with the casting.

Surface Finish. Sintered parts should be designed for removing surface material after casting to obtain the desired surface quality. Work done to the part prior to casting should be kept to a minimum. In the case of QuickCast parts, no solvents (especially water) should be used. The RP&M pattern should be wiped down with a "soft" paper towel and postcured. As noted in Chapter 6, reagent grade-isopropyl alcohol (RG-IPA) may be used to clean the pattern prior to investment casting, but care should be taken to avoid getting RG-IPA inside the pattern.

With regard to stair-stepping issues, the most obvious answer is to build the RP&M patterns with the smallest possible layer thickness. This will minimize the stair-stepping, but will not eliminate it entirely. Again, part orientation is critical to successful building of the patterns. If this conflicts with the support issue previously discussed, the support issue must be addressed first. The stair-stepping artifacts can be worked on after the tool has been cast.

A note about the FDM process. Until now, very little has been said about the FDM process because an investment casting wax is used to form the RP&M part. This means that conventional methods can be used to treat the part once it has been built. All the foregoing precautions apply to the FDM process as well.

Investment Casting

Prior to arriving at this stage, the steel has been selected and the appropriate shrinkage factor has been applied to the RP&M pattern to be cast. The proper procedure for casting QuickCast (QC) parts was described in Chapters 6 and 7, but here are a few additional observations. The surface of the QC or SLS part can be treated (this may mean being coated with wax) to improve surface quality of the pattern, and ultimately the tool.

However, care must be taken not to affect the dimensional accuracy of the RP&M part. If at all possible, have the investment casting foundry develop a means to leave the cooling and ejector pin lines in the RP&M part. Drilling them out after casting will only reintroduce time and expense back into the project, as will be discussed in the Explorer wiper motor cover case history to follow.

Once a part design has been agreed upon, a tool is designed around that part using the CAD solid modeling process. Just as a "positive" part can be produced on an RP&M machine, so can the resultant "negative" tooling patterns. Once the tooling patterns are produced on the RP&M machine, they are sent to an investment casting foundry and prepped for the casting process. The RP&M patterns are gated, supported, and mounted to a tree assembly. Normally there may be as many as 100 parts on a tree, but because tools are much larger and heavier, only one tool half (core or cavity) will be attached at a time. A face coat of ceramic is applied to the pattern and entire tree assembly and allowed to dry. A thicker second coat of ceramic, as well as a stucco mixture, is then applied and also allowed to dry. This is repeated until the desired shell thickness is achieved.

After the final coat is dry, the assembly is placed in an oven and fired. This burns out the RP&M part inside the shell. The shell is heated close to the melting point of the steel that will be poured and placed in the pouring oven. The steel is melted and poured into the shell and allowed to harden. The assembly is later taken to a breakout area where the ceramic is removed and part inspection takes place. If all is well, the gates and pouring cup are removed and the steel castings are cleaned.

Machining and Fixturing the Mold

Depending on the type of steel and how close to "net shape" the RP&M pattern is, machining may be nothing more than tapping mounting holes and a light dusting of the tool surface. Or it can be several long days of surface grinding or hole drilling. It should be noted that when the completed casting is cooling, there will be several hardness levels throughout the part, unless the casting is allowed to cool in a temperature controlled environment. A study of cast steel hardness as a function of cooling rate is presently under way, but detailed information regarding hardness profiles is not yet available.

The comfort levels of all the parties involved in a rapid tooling project will have an important bearing on its success. The most important of these is the machine shop that will finish the tool and prep it for production of the desired part. The team picked to finish the part should first have a vested interest in the project's outcome, be skilled in working with a variety of tool steels, and, most importantly, have the desire and patience to work through initial problems to help develop a technology that will ultimately benefit the entire company.

Over the next several years, many rapid tooling projects will continue to push the technology, push the imagination of those using the technology, and push the RP&M industry. Until several successful projects have been completed by numerous companies that represent a wide variety of processes and part geometries, quite appropriately this will be considered an experimental tool making process. The case histories that follow describe the use of two different technologies to produce tools for various projects at Ford.

11.2 Case Histories

The automotive industry is constantly striving to find ways to produce final products with cost and time savings in the forefront of the designers', engineers', and manufacturers' minds. Parts constructed more rapidly and economically offer obvious savings. When generated in production materials, these parts allow several designs to be tested under real world conditions, enabling selection of the best possible design. The 1994 Ford Explorer™ "wiper motor cover" described in this section, although a simple part, illustrates the potential of today's technology. Simply stated, the tools created in this project were used to injection mold prototype quantities in production polypropylene material.

Wiper Motor Cover—Stereolithography/QuickCast

Ford Motor Company was a beta site for the QuickCast process during 1993. Recognizing the limitations of both spray metal and NVD tooling,[3] Ford decided to try QuickCast tooling. Once the QuickCast tooling process was selected, the task of learning how to actually accomplish the desired outcome was begun. The software selected to create the CAD solid model of the Explorer wiper motor cover was Parametric Technology Corporation's Pro/ENGINEER™ (Pro/E).

Pro/E was chosen based on the availability of the Pro/Mold software package. Pro/Mold allows the designer or engineer to create negative molds based on the positive of the desired final part geometry. As a starting point, a CAD positive solid model of the wiper motor cover was created in Pro/E and subsequently forwarded to Ford's SLA-250 for creation. As it turned out, the positive of the part allowed early detection of a packaging interference. A second iteration was created by altering the Pro/E solid model, and sending the new file to the SLA. This time, the positive of the wiper motor cover showed no interference, but required further changes so the final assembly could be more easily manufactured. Thus, a third iteration was generated in Pro/E.

When the design, engineering, and manufacturing staffs were in concurrence, mold halves were created by initially embedding the CAD solid model in a block-shaped volume called a workpiece in Pro/Mold. A parting line was then selected based on the manufacturer's requirements, and the two CAD solid model halves were then separated. Next, the cooling lines and ejector pin holes were positioned on the Pro/Mold solid models of the mold halves.

However, during a final meeting before actually preparing the STL file, the need for another design change became evident. This fourth iteration was made to the CAD solid model and was automatically incorporated into the mold halves. The STL files were then generated with the highest accuracy settings in Pro/E in order to obtain the best possible QuickCast pattern surface finish. A photograph of the final QuickCast mold insert patterns is shown in Figure 11-6. It is noteworthy that the classic SL sequence of design *visualization*, *verification*, multiple *iteration*, and final *optimization* was very evident during this project. Had traditional methods been used, significant, expensive, and time-consuming tooling rework would certainly have been required. In fact, it is quite possible that a new tool would have been necessary, adding many months to the delivery schedule.

QuickCast Patterns. Creating the SL patterns was fairly straightforward using 3D Systems' QuickCast build style described in detail in Chapters 6-9. Some items should be mentioned concerning the build process. Three separate shrinkage factors are needed, the final injection molded material (in this case polypropylene), stereolithography resin (in this case SL 5170), and finally the shrinkage factor of the A2 tool steel. The latter was the hardest to determine as there was little data to be found on shell investment cast shrinkage factors for A2 tool steel. The best available information suggested a value of about 2% in all directions, although the exact value may be even slightly higher. Ford is currently running samples to determine the shrinkage factors of various tool steels, as well as other data including hardness after shell investment casting.

Investment Casting. The QuickCast patterns were fully drained, cleaned, postcured, checked for pinholes, then shipped to Howmet Corporation, Whitehall, Michigan. They were rechecked for holes at Howmet, and those missed earlier were filled with investment casting wax. The original patterns included the ejector holes. However, because this was Howmet's first time shell investment casting

Figure 11-6. Final QuickCast mold insert pattern of Ford wiper motor cover.

A2 tool steel, they felt more comfortable filling the ejector holes with wax, leaving only a small locating dimple. The ejector holes would presumably be drilled after the steel had solidified and cooled.

Subsequent to the patterns being gated and all holes being filled, the QuickCast patterns were cleaned and prepped for dipping. This included an RG-IPA wipe and a final check for pinholes. The QuickCast patterns and attached gates were then invested in a vat of very fine "face coat" ceramic slurry. After drying, the process was repeated with subsequent layers sprinkled with refractory sand until the required ceramic shell thickness is achieved.

On completion of the shell, the patterns, gates, and ceramic shell assembly were then placed in a fully aspirated oven and fired at over 1,800° F (980° C) for one hour.

This firing cures and strengthens the ceramic shell. It also simultaneously burns out the QuickCast pattern, provided sufficient oxygen is present for the efficient conversion of the hydrocarbon-based resin to water vapor and carbon dioxide. After burnout was completed, the shells were checked for small cracks or fractures. It was determined that in some areas the pattern had stuck to the walls of the ceramic face coat, causing some surface imperfections. However, these were relatively minor and could be accounted for in the final polishing of the tool.

Next, the ceramic shells were preheated in an oven prior to pouring steel. Using an induction coil to melt steel ingots, Howmet poured the molten metal in a vacuum. After the steel was poured, the shells containing molten steel were placed in the open air to cool and harden. As an interesting and potentially significant sidelight, the low thermal mass and relatively high surface-to-volume ratio of the shell investment casting process resulted in very rapid metal cooling and, hence, extremely hard tool steel core and cavity pairs. Furthermore, the high hardness was not just at the surface, it extended through the entire thickness of the mold inserts. In fact, the resulting A2 tool steel was so hard that five carbide drill bits were rendered useless while attempting to drill the ejector holes. Ultimately, the holes were drilled with diamond tipped bits.

Quite obviously, in future QuickCast tooling projects, any and all registration holes and/or ejector holes will clearly not be filled with wax prior to casting. In fact, Figure 11-7 shows a hole accidentally created from overheating during attempts to cut off the extremely hard A2 tool steel gating.

Of potentially greater significance, however, is that the increased hardness of investment cast tool steel, relative to conventional billets of the same nominal material, may be of value in extending the abrasion resistance and, hence, the life of production tooling. Furthermore, it is quite possible that extended tool life can be achieved in this manner without the need for additional heat treatment steps. Heat treatment further increases tooling cost, extends tooling completion schedules, and can also lead to thermally induced distortion of the mold cavities, thereby requiring subsequent tooling rework and additional cost and time.

Figure 11-7. A hole accidently created from overheating during attempts to cut off extremely hard A2 tool steel gating.

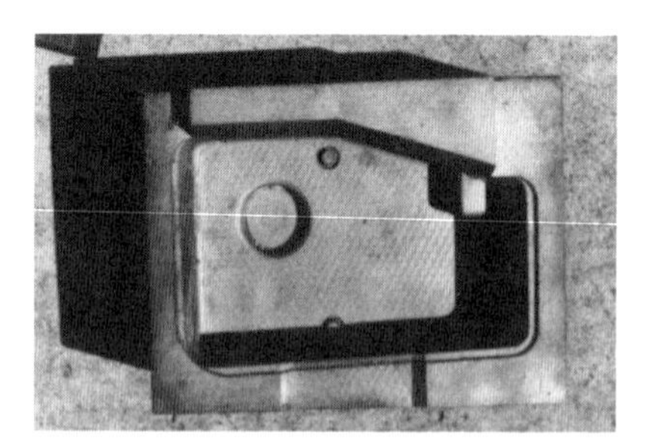

Figure 11-8. A cavity mold insert surface showing residual signs of QuickCast triangular structure.

Tool Preparation and Injection Molding. Upon completion of the investment casting portion of the project, the steel core and cavity inserts were forwarded to ATC Nymold Corporation, Brooklyn Heights, Ohio, for tooling preparation and injection molding of the wiper motor covers in polypropylene. Figure 11-8 is a photograph of the cavity mold insert surface, showing residual signs of the QuickCast triangular structure. This structure, present on the upfacing pattern surface, was later transferred to the tool steel insert during casting. Further signs of surface degradation are evident in the radii, also shown in Figure 11-8. It has been determined that during autoclave, the steam not only melted out the wax gating and softened the QuickCast pattern as intended, but either the steam or the alcohol used to clean the patterns apparently caused some sort of reaction with the still uncured ceramic shell material. This problem has subsequently been addressed at the foundry. Howmet now fully dries and cures the ceramic shell prior to autoclaving.

The hardness of the cast pattern is an issue that requires careful attention. ATC Nymold reports that the hardness level of the shell investment castings were about 48 Rockwell C, or a full 18 points higher than the normal value for conventional A2 tool steel. This hardness level, although certainly desirable for the final tool, actually caused some problems in tool preparation. Figure 11-9 is a photograph of the core and cavity investment cast A2 tool steel mold inserts prior to surface finishing. On the core half, a small indentation can be seen where one of the ejector holes was intended to be drilled, as noted earlier. The hole was only about 1/8 inch deep after breaking five carbide drill bits. After the holes were finally completed using diamond tipped drill bits, ATC Nymold reported that these extraordinary hardness levels extended through the entire thickness of the part. They also reported areas of exceptionally hard, brittle material they thought might be slag. However, these regions might be pockets of steel that were supercooled and therefore even harder than the remainder of the castings.

There are three possibilities for addressing the unusual problem of excessive tool steel hardness resulting from shell investment casting. The first, and perhaps simplest, is to investigate the resulting investment cast hardness of other candidate steel alloys, and then select the optimum choice. Here one would want a material hard enough to ensure long tool life, but not so hard as to impose special machining requirements on the finishing operations for the final tooling.

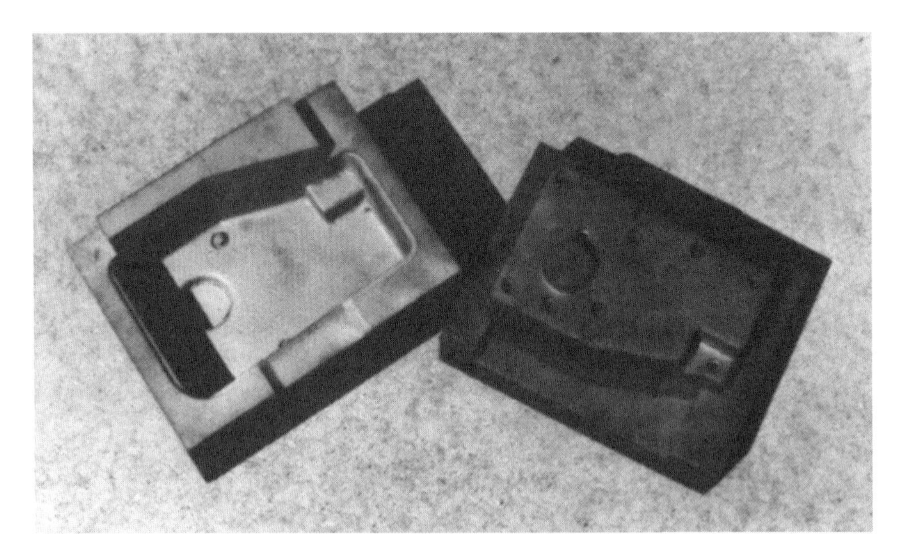

Figure 11-9. Core and cavity investment cast A2 tool steel mold inserts prior to surface finishing.

The second approach is to cool the casting in a programmable oven rather than the open air. In this way, the casting temperature can be reduced more slowly, and in a very controlled fashion, to ensure that the desired metallurgical properties are achieved.

Finally, the castings could be annealed after the ceramic shell is removed. However, this may prove to be the least desirable method as the annealing process itself could warp the inserts. Clearly, more work needs to be done to achieve optimal physical properties of the inserts and do it with the highest levels of accuracy, in the shortest possible time, and at the lowest cost.

Once the hardness issues were dealt with, the tool was fitted to the mold base. This also proved to be somewhat problematic because the two pieces were slightly different in size. The most probable explanation for the size difference was that the core and cavity were built in different orientations relative to the direction of recoater blade travel on the SLA-250.

A second possibility was that the shrinkage factors for the A2 tool steel might have been geometry dependent, thus not totally uniform. Again, further work is definitely needed with respect to the determination of accurate shrinkage factors for the shell investment casting of various candidate tool steels. Also, draft angles were not included in the original CAD model, the eventual QuickCast stereolithography patterns, nor in the resulting investment cast mold inserts. Rather, the draft angles were subsequently machined after the steel inserts were cast. As a result, the mold inserts had uneven wall thickness values. The obvious answer to this problem is to apply all draft angles, fillets, radiuses, etc., to the CAD design well before the inserts are actually investment cast.

When all these issues were resolved and corrected, the inserts were finally installed on their bases and placed in the injection molding machine. The resultant injection-molded polypropylene wiper motor covers were then used for water leakage testing. Although they would be going into an area of the vehicle not visible to the public, the surface finish prevented the modules from gaining release status. As seen in Figure 11-10, the surface quality of the final core and cavity halves required extensive machining and polishing. Figure 11-11 shows a photograph of the final injection-molded polypropylene Explorer wiper motor cover mounted on an actual wiper motor.

The potential advantages of QuickCast tooling through investment casting are evident relative to conventional machined steel tooling. Figure 11-12 shows the actual time and cost comparisons experienced by Ford on this project. Remarkably, this data includes all the time to solve the various problems associated with ascending the "learning curve" for the very first time. Figure 11-13 shows Ford's project plan for conventional tooling, while Figure 11-14 shows the actual results of the QuickCast tooling project.

Again, even with all the various learning curve delays, the time savings were noteworthy. Once the process is better defined, and additional information has been generated regarding the hardness as well as the shrinkage factors of various

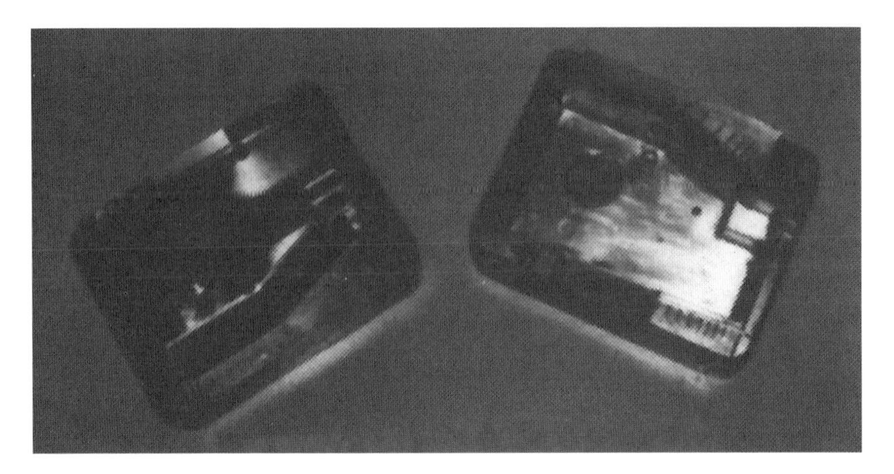

Figure 11-10. Final core and cavity halves that have been extensively machined and polished.

shell investment cast tool steel alloys, the time and cost savings relative to conventional tooling methods should be even more significant. Ford intends to monitor actual cost and time savings on at least two additional future projects so potential advantages of rapid tooling can be more accurately assessed.

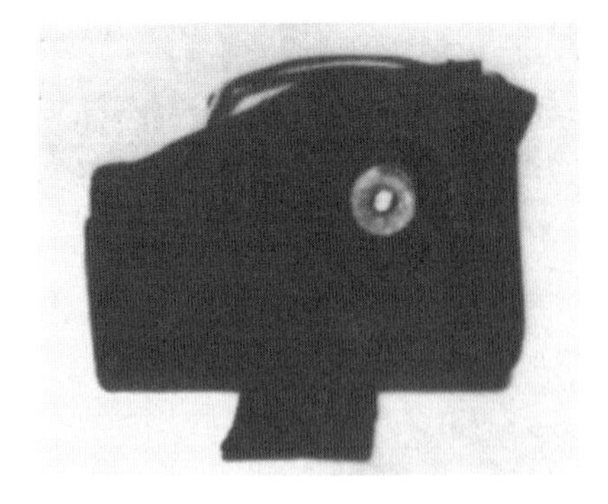

Figure 11-11. Final injection-molded polypropylene wiper motor cover mounted on an actual wiper motor.

Project Conclusions. With all the numbers laid out and the photographs, tables, and comparison figures in place, the question still comes to mind . . . was this project successful? The answer is most definitely yes! When the project started at Ford, the goal was to produce a part fabricated from production material, capable of being tested under realistic conditions. A secondary goal was using the resulting tooling for production run quantities. While the "first" rapid tool was not used for production, Ford does intend to use subsequent QuickCast tooling for actual production. Undoubtedly, the first QuickCast tool will ultimately become a conversation piece. Nonetheless, its value should not be understated. This project has already changed Ford's thinking about providing parts to our testing facility, as well as the way we will ultimately manufacture production run parts. From this first experience with investment cast tooling, the groundwork has been laid. Ford has authorized two additional QuickCast tooling projects, which are underway.

Cost and Timing Comparison
First part timing
6 weeks
17 weeks
Tooling cost
$18,000
$33,000
0
5
10
15
20
25
30
35
Cost in $000s—Time in weeks
Conventional mold
Investment cast mold inserts

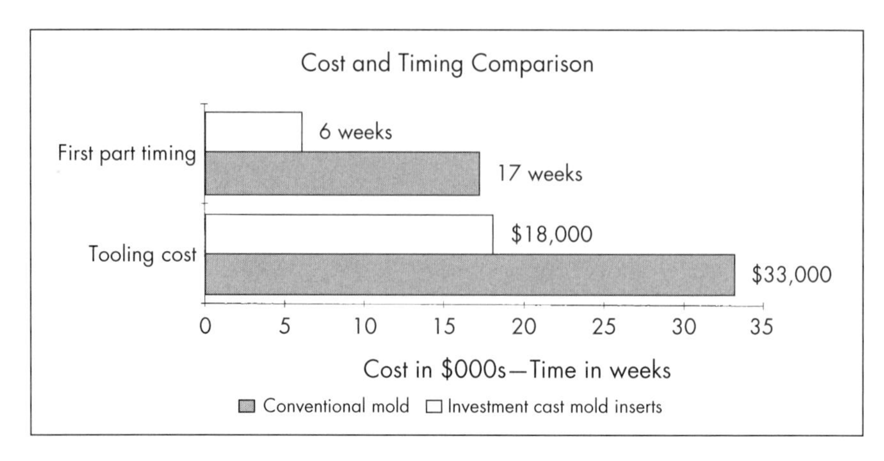

Figure 11-12. Data from Ford project showing time and cost comparisons between conventional mold and investment-cast mold inserts.

Timing Plan for Explorer Rear Wiper Motor Water Shield
Conventional Tooling
Event
May
June
July
August
September
October
November
Tooling drawings
Hard tooling
First prototypes
Testing
Durability (2 weeks)
Functional (1 week)
Torque backoff
Cost minimization
Parts to Trico
Parts to plant

Figure 11-13. Ford's project plan for conventional tooling.

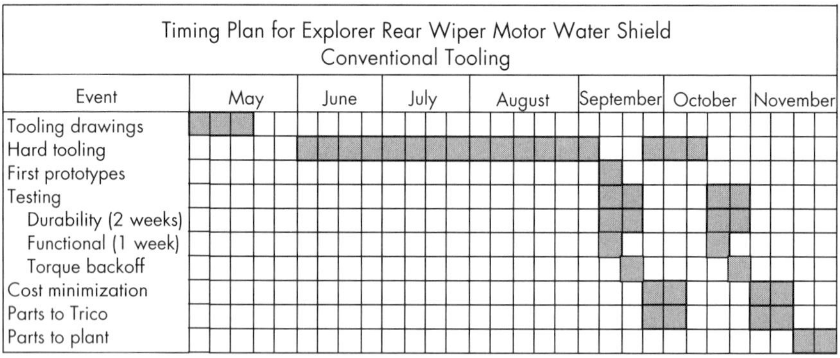

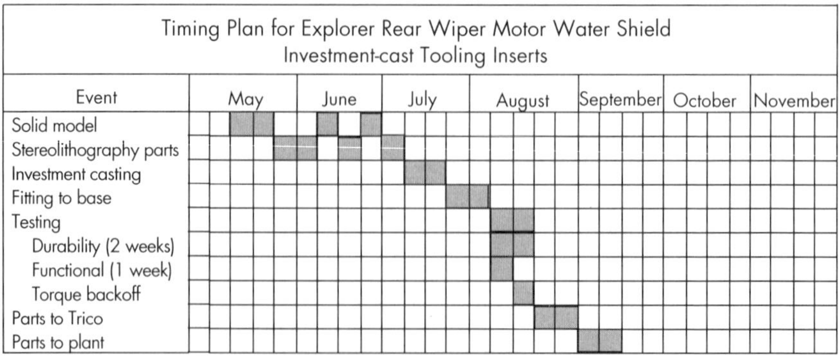

Timing Plan for Explorer Rear Wiper Motor Water Shield
Investment-cast Tooling Inserts
Event
May
June
July
August
September
October
November
Solid model
Stereolithography parts
Investment casting
Fitting to base
Testing
Durability (2 weeks)
Functional (1 week)
Torque backoff
Parts to Trico
Parts to plant

Figure 11-14. Actual results of QuickCast tooling project.

There are three areas where further research must be completed before we can consistently produce quality rapid hard tooling. First, the surface finish of QuickCast patterns, as well as the final investment castings, must be improved. 3D Systems has recently released an advanced version of QuickCast. Known as QuickCast 1.1, it directly addresses the issues of improved upfacing, downfacing, and vertical surface quality, as well as the elimination of pinholes during support removal. QuickCast 1.1 is discussed in Chapter 6.

Second is the matter of tool hardness. As noted, data is needed on the hardness of shell investment cast steel alloys as a function of cooling rate. Data on shrinkage, tensile properties, abrasion resistance, etc. are also needed to better understand the relationship between cast and cut tool life. Ford and 3D will be actively working with foundries to obtain such data. Assistance from other parties interested in developing this data base would certainly be welcome.

Third, we must establish realistic cost and time estimates for QuickCast tooling. Obviously, the project described in this paper is only the beginning. While we were able to produce these tools in a relatively short time, further reductions are definitely possible. However, because investment casting foundries have been used to the long lead times and large quantities of the aerospace industry, this may be viewed as typical. In the automotive industry, we do not produce great numbers of investment castings, but we might use a smaller number of investment cast tools to injection mold very large quantities of mass produced products.

In conclusion, we believe this project proved that QuickCast tooling can work. With the creative thinking and energy that has typified the RP&M industry, the wiper motor cover rapid tooling project can be considered the beginning of a very significant new application. As noted herein, much work remains to be done. However, we also believe that in the future, rapid hard tooling will have a major impact on manufacturing productivity.

Windstar™ Dunnage—Fused Deposition Modeling

Just as the parts that make up an automobile require injection molds (tools), so does the dunnage used to protect and hold components in place during shipping. During the course of production, the design of a part may change several times. It would not be economically sound to produce dunnage for every design change. Current tooling methods with respect to dunnage require 10 to 12 weeks. The project described here saved 50% of the time and expense required to produce machined tools in the same steel.

The Process

In some cases, the rack that the dunnage is attached to remains the same over the course of several car lines or platforms. It is therefore necessary to design only a new dunnage head that will interface with the new part, as shown in Figure 11-15. When it was determined that the dunnage design was satisfactory, the CAD (IGES) data was sent to Stratasys, Inc., to have the patterns constructed. The pattern material was ICW04, suitable for investment casting.

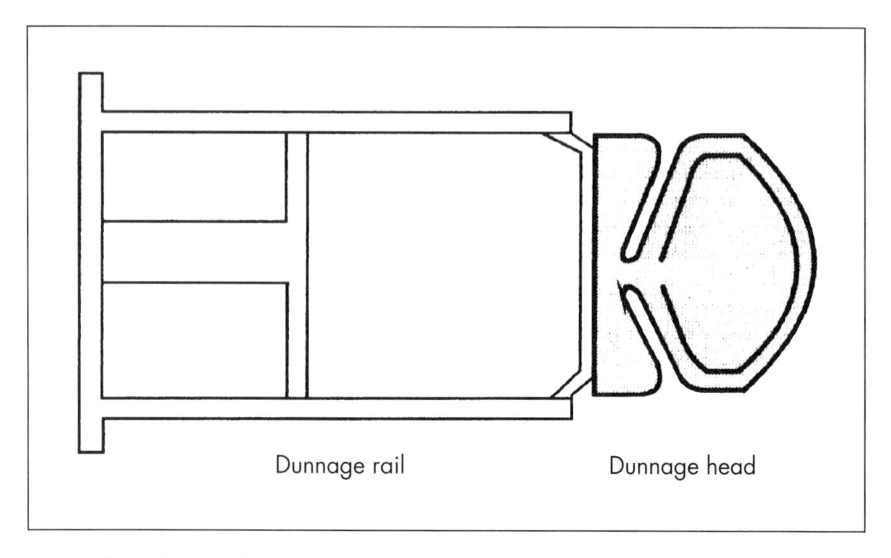

Figure 11-15. Design of a new dunnage head to interface with a new part.

It was determined at the beginning of the project that it would not be necessary to investment cast the entire thickness of the tool, but rather a "shell" that would be filled in later with a material called chemically bonded ceramic (CBC). Figures 11-16 and 11-17 show the FDM pattern from both the back and front views, prior to investment casting.

The next step was to have the wax tooling patterns investment cast. The foundry selected was Wisconsin Precision Cast, in East Troy, Wisconsin. The material was A2 tooling steel. Because the final investment cast tool would be back filled, several options were left open for cooling line placement. It was determined that the cooling required by the part geometry and material was such that a "half cooling line" cast into the back side of the tool and covered with a metal plate prior to back filling would be appropriate. When the investment casting process was complete, the gates and pouring cup were removed and the tools sent to J P Pattern in Butler, Wisconsin.

As with all of the current RP&M systems, the FDM process left signs of layering in the wax pattern. This was removed prior to investment casting and the steel tool needed only a polishing to provide the desired surface finish. J P Pattern machined in the cooling line entry and exit points to match the cast in lines, and sent the patterns to CEMCOM Corporation in Baltimore, Maryland. CEMCOM filled the back of the tools with CBC, providing the required strength to run production quantities of approximately 30,000 moldings. The CBC material exhibits high compression strength, minimal shrinkage, and high temperature characteristics.

Figures 11-16 and 11-17. At the beginning of the Windstar dunnage project, it was determined that it would be unnecessary to investment cast the tool's entire thickness. Instead, there would be a shell, filled in later with CBC. Figures 11-16 (above) and 11-17 (below) show the FDM pattern from both back and front prior to investment casting.

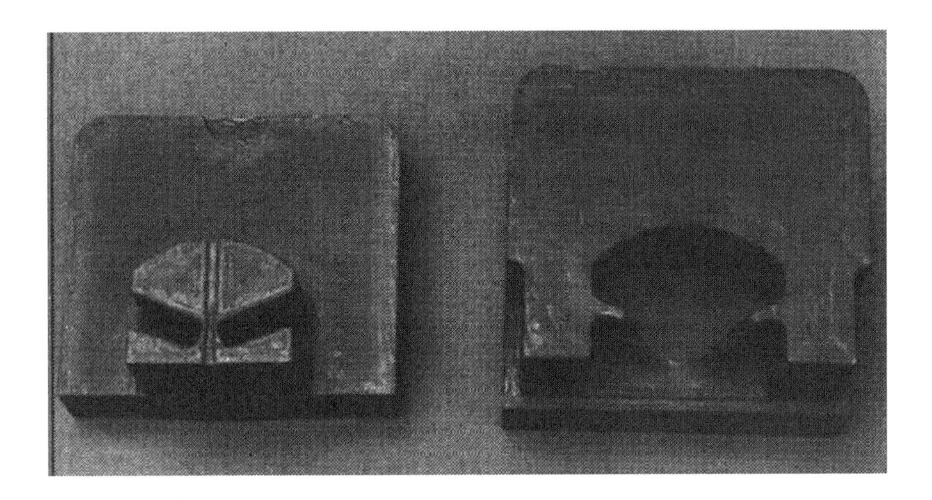

The final Windstar dunnage tool is shown in Figure 11-18. The tool was then sent to Creative Technologies, in Auburn Hills, Michigan, manufacturer of the injection molded part. Figure 11-19 shows the final injection molded dunnage.

Figure 11-18. The final Windstar dunnage tool.

Figure 11-19. The final injection molded dunnage.

Project Conclusion

Savings to Ford Motor Company on this project were 50% for both time and costs. This project is further proof that cast tooling is a valued option in the manufacturing process. There should be two things noted about this project. First, the accuracy issues discussed in the previous case history do not generally apply to this project. The required tolerances are much looser with dunnage than with part manufacturing. Second, surface finish did not have to meet customer standards since the dunnage would never be viewed by the public. Nonetheless, using the FDM process to produce tools for the Windstar dunnage proved to be a substantial savings to Ford Motor Company.

11.3 Other Considerations in Rapid Hard Tooling

Size and Weight

As stated earlier, tools are much larger than the parts they manufacture. Current RP&M technology is now reaching the limits of process capability in terms of size and weight. The size problem has two possible consequences, build the tool patterns in sections and then bond the RP&M patterns together prior to investment casting, or cast the tool in sections, weld the tool, and then rework the surface to remove signs of a "nit-line." The latter may prove a bit more difficult since some tools weigh many tons. Also, expecting a tool that has been literally welded together to last for the duration of a production run may be asking too much of the process.

This brings us to the weight problem. Investment casting foundries will have to develop the technology to handle the enormous amount of steel that some of the larger tools would contain. This is already being evaluated, as more and more companies look into the prospects of cast tooling. One other possibility, with regard to cutting the weight down on cast tooling, is to cast a shell, say 1 to 2 inches thick, and back fill it with a material that would withstand the rigors of the injection molding process. CEMCOM Corporation's development of the CBC process shows great promise for this application.[4]

Assembled or Whole Tooling. One answer to the size/weight problem may be to investment cast an insert as in the wiper motor cover case history. This would allow standard mold bases to be used, although the inserts would have to be machined to fit the mold base. This may require even more detail regarding the accuracy issues with RP&M investment cast tooling. If this is the route taken, attention must be paid to the steel used and the hardness level that might be obtained with that steel. Machining after the fact with the wrong steel could result in a project that was no more effective than conventional methods.

Surface Quality Considerations. Surface quality is an issue no matter what RP&M process is used. If the surface finish of the RP&M pattern is too smooth, the investment casting face coat may not stick. Then there is the texture of the face coat slurry. Rapid tooling technology may help push new research within the investment casting industry for still finer face coat materials. Current investment casting surface finishes obtainable for various alloys typically range from about 60 to 125 micro inches (1.5 to 3 microns) RMS.[5] For NET shape rapid tooling, the numbers ideally should be down around 20 to 40 micro inches (0.5 to 1 micron) RMS. This would provide the best "out of the shell" surface for final polishing.

Alternative Processes. Table 11-1 showed several different rapid tooling methods. All of these methods have a place in the manufacturing world. In one project at Ford, the platform team required 25 to 30 parts in production material. At the outset, it was determined that spray metal tooling would provide the best possible solution. The result was two dual cavity tools that were fully grained to a Ford specification. Total time to produce the tools was four weeks once we had the SL patterns, at a total project cost of $80,000. These tools produced the required number of parts and uncovered several problem areas for the engineers to work on before going to production tooling.

Another technology of interest is metal vapor deposition (MVD). Here an RP&M master is placed in a chamber and negatively charged to assist the deposition process. A metal vapor is passed through the chamber with a positive charge. As time progresses, metal is deposited on the RP&M part until the desired thickness is achieved. The metal coating is then separated from the master, forming the critical contour of half the mold. A box is constructed around the metal contour and back filled with an epoxy/aluminum mixture. The box is then flipped over and the process repeated to fabricate the second half of the tool.

Silicone RTV rubber is another popular soft tooling method. When using this procedure, a box is constructed and an RP&M master is suspended in the center. Either silicone or soft polyurethane is then poured around the master and allowed to cure. Once cured, the box is removed and the master is cut out. Silicone RTV molds are typically used when the part to be manufactured is some form of polyurethane or low-melt metal alloy. Several other methods of rapid tooling exist or are being developed every day.

11.4 Volumes, Tool Life, Schedule, and Economics

As new methods of rapid tooling are developed, four questions will occur. The answers will dictate the actual method used for a given project. First, will that method provide the desired number of parts? Second, can the parts be generated in the required time? Third, what is the desired final material? And fourth, what is the available budget?

In today's business world, leadership cannot be obtained by automatically using old, albeit tried and tested, methods of manufacturing. Leadership will be determined by those corporations that change and evolve with the times. The most dramatic method of adapting may be by changing the normal manufacturing processes and methods. Rapid tooling will yield several options and opportunities for the manufacturer today, and even more tomorrow. Companies must be willing to take the required risks on rapid tooling projects to remain competitive with the rest of the manufacturing world. If you don't do it, there is a good chance somebody else will!

Selection of the process to produce a part should be based on the following.

1. The number of parts required to satisfy the design requirement. A process should be selected that will yield at least the minimum desired quantity. If this is for prototyping, then a second question should be asked prior to method selection: what are the chances that we will need more than the initially requested amount?
2. Schedule. Is there enough time to realistically expect the desired outcome? The rapid tooling method chosen should be proven effective if the project will determine the success or failure of an entire program. If the method is expected to produce production quantities and has never done so in past projects, then an alternative or "backup" method should be selected. Let the natural progression of the manufacturing development process test and evaluate new rapid tooling processes, not the literal failure of a critical project.
3. Ask yourself if the part should be in a material that is production quality or only similar. If a polyurethane part will do for the time being, then it may be easier and cheaper to produce an RTV soft tool than any other method. On

the other hand, if production material is required, but only in small quantities, perhaps a spray metal method should be chosen.

4. The rapid tooling method selected should fall within budget requirements. Speed usually results in higher costs. If a project is in its early stages, the tooling options are numerous and will be generally less expensive than when the release of a product is just around the corner and design changes have kept tooling of any kind lagging the process time line. The closer one is to the release of a product, the more one should be prepared to spend for tooling, either rapid or conventional.

11.5 The Future of Rapid Tooling

As new rapid tooling methods are developed and implemented by companies that compete in the manufacturing world, those methods will help push the envelope of this evolving technology. We are consistently amazed at the tremendous amount of energy exerted by the RP&M industry today. Changes in technology produced by this energy challenge everybody connected with this industry to work hard at keeping up with advances taking place almost every day.

References

1. Weir, C., *Introduction to Injection Molding*, Society of Plastics Engineers, Brookfield, Connecticut, 1975.
2. Bryce, D., *Thermoplastic Troubleshooting for Injection Molders*, Society of Plastics Engineers, Brookfield, Connecticut, 1991.
3. Denton, K. and Jacobs, P., "QuickCast and Rapid Tooling: A Case History at Ford Motor Company," Proceedings of the SME Rapid Prototyping and Manufacturing '94 Conference, Dearborn, MI, April 26-28, 1994.
4. Nagarsheth, V., Jacobs, P., Santin, D., Wise, S., and Yoder, J., "CBC Injection Mold Tooling for Thermoset Plastic Parts using StereoLithography Patterns," Proceedings of the SME Rapid Prototyping and Manufacturing '94 Conference, Dearborn, MI, April 26-28, 1994.
5. Twarog, D., Barron, B., Warren, R., Helmer, J., Klemp, T., and Ray, E., *Handbook on the Investment Casting Process*, American Foundrymens' Society, Inc., Des Plaines, IL, 1993.

CHAPTER 12

Special Applications of RP&M

"Nothing remarkable was ever accomplished in a prosaic mood. The discoverers have found to be true more than was previously believed."

Henry David Thoreau, Cape Cod, 1850

Earlier chapters of this book have discussed in considerable detail a number of very important RP&M applications. These fall mainly into such categories as employing RP&M patterns to shell investment cast functional metal parts, discussed in Chapters 6 through 9, the use of various following processes enabling the development of soft tooling, described in Chapter 10, as well as methods aimed at utilizing RP&M for the generation of hard tooling, presented in Chapter 11.

However, these are not the only applications of interest to readers. This chapter describes four "special applications" that do not happen to fall neatly into any of the other categories, but are noteworthy nonetheless. While one might argue that other unique applications of RP&M are also presently available or at least currently under development, the author felt that these four were definitely worthy of inclusion.

All these specific applications share a number of characteristics. First, each was developed primarily by an end user, a university, or a service bureau, rather than by an RP&M system supplier. Second, none of these applications involve secondary processes. That is, they utilize the RP&M output directly. Third, they are extremely diverse in their scope. The first applies primarily to the automotive industry, the second would be appropriate for a broad range of industrial applications, the third would probably find a natural acceptance in the

By Paul F. Jacobs, Ph.D., Director of Research and Development, 3D Systems Corporation, Valencia, California.

aerospace sector, while the fourth is especially significant for various medical/surgical applications.

The four techniques have each been developed, tested, and applied by one or more individuals. Much description of the application, as well as the experimental data, test results, figures, references, etc., involved a direct contribution from these individuals. Therefore, rather than being given a simple reference within the text, they will be specifically cited as contributing authors at the beginning of each subsection.

12.1 Flow Visualization Utilizing RP&M Automotive Models

Dieter Steinhauser, Advanced CAD/CAM Engineering, Motor Sport Technology, Dr. Ing. h.c., F. Porsche AG, Weissach, Germany

Due to thermodynamic efficiency and endurance requirements, high-performance combustion engines require a constant flow of coolant through all cylinder units. In highly stressed areas of the cylinder head and crank case in particular, extensive testing is necessary to detect any sections insufficiently supplied with coolant. The thermodynamic consequences of uneven coolant distribution are slow burning of the fuel-air mixture and increased hydrocarbon emissions from "cold" cylinders, as well as "knocking" problems for "hot" cylinders.

Coolant distribution issues are becoming increasingly important since impending environmental restrictions will define new limits for automotive emissions. Due to the requirements of so-called "cold-start" emission test procedures, which are influenced by the duration of the cooling system warm-up phase, it is critical to optimize the coolant volume and flow distribution.

Existing methods generally involve experiments on a test bench. These tests make use of cast components. Unfortunately, the tooling and molds necessary for these castings are often available only at a very late stage of the project. Furthermore, various adaptations required for testing, such as runners for sensors and cutouts for optical systems, may falsify behavior of the real configuration in an operational engine. Beyond that, poor access to internal cavities can limit the integrity of test results by making it impossible to reach critical "hot spots," as well as casting doubt on the predictability of thermal transfer under extreme conditions.

More often than not, recommended alterations cannot be carried out because the tooling has progressed too far, and further changes would be expensive. As a result, various theoretical approaches, mainly based on finite element analysis (FEA), have been used for 3-dimensional flow computation. FEA techniques can be initiated much earlier than experimental "cut-and-try" methods, an important advantage.

However, accurate FEA computations demand high-quality finite element nets with a large number of cells. The time involved for the finite element meshing of a complete water jacket, including the cylinder heads and crank case, can easily exceed three months. The associated cost can also exceed $50,000. Also, the computational fluid dynamic programs used to visualize complex 3-dimensional flow patterns, and their velocity distributions, require experienced engineers capable of interpreting the results.

Alternative Approach Using RP&M Technologies

The goal of this test program was thc generation of a *transparent flow model*, including all the affected sections of the crank case and cylinder head. The specific subject was a cross-flow water jacket for a V6 high-performance racing engine. An accurate, closed, CAD surface representation of the parts served as the data base. The exterior geometry describes an envelope equipped with fittings for both fluid inflow and outflow.

Due to the large amount of data required to achieve acceptable flow field accuracy (the file exceeded 70 megabytes), a decision was made to convert the data from CAD to the RP&M system via the direct CAT-SLICE interface. CAT-SLICE enables the transfer of CATIA data directly into the SLC format, without any tessellation errors. The necessary support structures were created beforehand as CATIA entities, in collaboration with 3D Systems GmbH, Darmstadt, Germany. All components were produced on an SLA-250, operated by the European Technology Center at 3D Systems GmbH.

Part Size:	18.9 × 17.7 × 7.9 inches (480 × 450 × 200 mm)
Photopolymer:	Cibatool SL 5170
Build Style:	ACES
Turnaround Time:	Four weeks
Total Project Cost:	$23,000, including data conversion, part orientation, build time, postprocessing, and delivery

Although the epoxy resin SL 5170 had not been introduced to the market at the time this project was initiated, all parts met the required optical transparency levels specified by Ciba-Geigy. Subcomponent accuracy and surface finish were also excellent.

Experimental Results

Figure 12-1 shows the assembly of the ACES test parts. These components were equipped with more than 60 sensors to determine local flow temperature and pressure conditions. Additional information was provided through visualization of the coolant flow pattern. This was accomplished by accurately injecting very tiny air bubbles, and then recording their motion with a high speed video camera imaged and focused right through the ACES part, shown in Figure 12-2. The images were later analyzed on a frame-by-frame basis using a high-quality video cassette recorder (VCR). This proved to be a very useful tool to expose stagnation zones and, consequently, insufficiently cooled sections.[1]

Figure 12-1. Assembly of the ACES test parts.

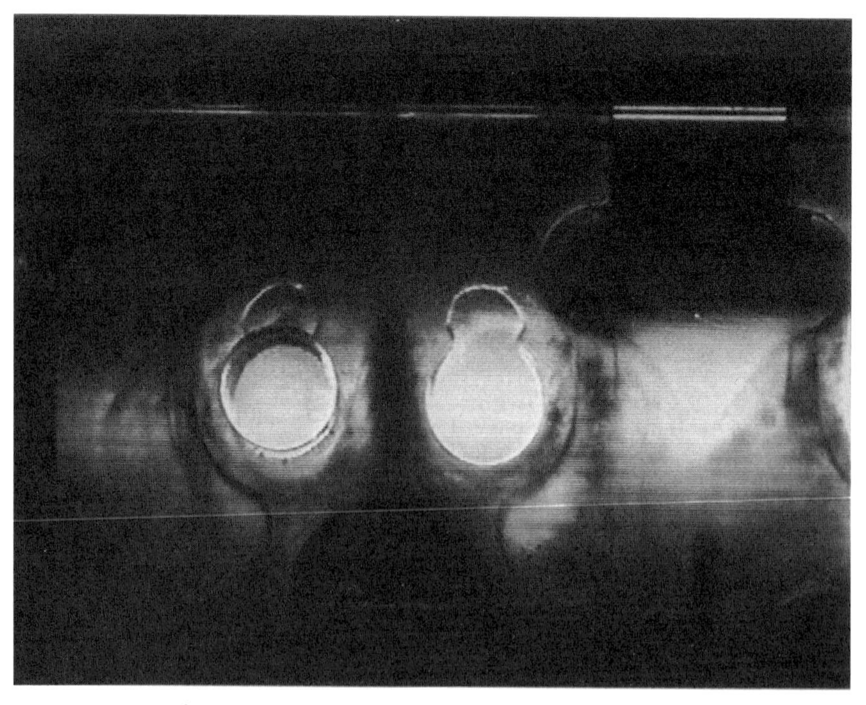

Figure 12-2. Visualization of the coolant flow pattern through the ACES part shown above was accomplished by injecting tiny air bubbles and recording their motion with a high-speed camera.

The experimental results showed an unacceptable coolant supply distribution through a number of cylinder units. Thus, critical sections were precisely removed, redesigned by CAD as required, and fabricated again in transparent SL 5170 resin. These parts, which were subsequently attached and sealed, enabled the testing of flow channel variations in a very attractive time frame. Typical lead times were about one week per iteration. Cost savings were also significant. A remarkable aspect of this program was that all necessary testing and design changes could be carried out before even a single dollar was spent on tooling.

Conclusions

The epoxy resin, SL 5170, coupled with the ACES build style, enables time-saving and cost-effective generation of *transparent models* to investigate the flow behavior of complete cooling systems in combustion engines. Accuracy and surface quality of the parts provided through stereolithography is more than sufficient for the intended application. Compared with previous methods, not only are considerable economic advantages apparent, the quality of the results is also improved.

Since RP&M technologies are now an integral part of our operation, positive results achieved in cooperation with competent partners will encourage F. Porsche AG to promote even more sophisticated applications.

12.2 Photoelastic Stress Analysis of RP&M Models

W. Steinchen, Ph.D., Professor and Director of the Laboratory of Photoelasticity and Holography, University of Kassel, Kassel, Germany; B. Kramer, Ph.D., Research Scientist, University of Kassel; G. Kupfer, Laboratory Engineer, University of Kassel.

The stresses and strains existing within a physical component can be determined, under the correct conditions, through the use of photoelastic testing.[2-4] This method is based on the temporary birefringence of a transparent material subjected to a specific load. A number of plastic materials exhibit birefringence, which can be illustrated by irradiating the test sample with polarized white or monochromatic light. The birefringence separates a single incident beam of polarized light into two beams oscillating perpendicular to one another. If the test material is both transparent and exhibits adequate birefringence, the directions of the twin refracted beams correspond to the directions of the principal stresses. The birefringence effects disappear when the load is removed.

The propagation speed of the two beams of light are slightly different due to small variations in the indices of refraction. These small differences in speed lead to corresponding path alterations. As a result, the differences in the principal stresses can be shown as lines of the same color, known as *isochromatics*. Fringes first appear at the points of highest stress, and as the load is increased, they move to the regions of lower stress. The fringes remain in the order in which they appear, and never cross over each other.[5]

Using ACES Models for Photoelastic Testing

To illustrate the use of photoelasticity as a test method in conjunction with SL, models of a connecting rod from a German automobile were built on both an SLA-250 and an SLA-500. The test models were fabricated using the new Ciba-Geigy epoxy resins SL 5170 and LMB 5353-1 (since commercially released and referred to hereafter as SL 5180). Figure 12-3 shows the models built using the ACES technique. These epoxy-based resins produce highly transparent parts, similar to test samples made from Araldite resin, which is commonly used for photoelastic testing.

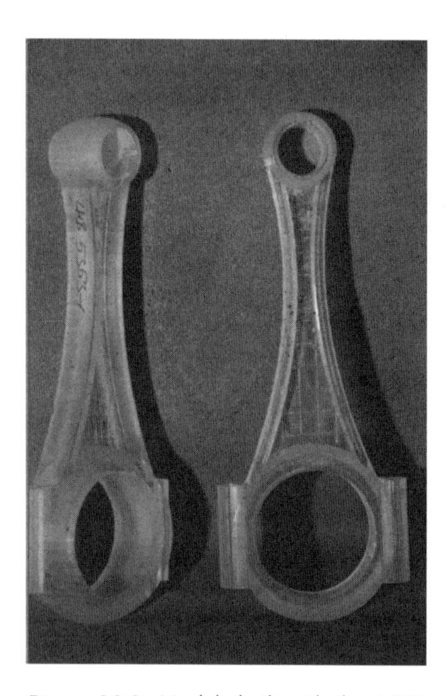

Figure 12-3. Models built with the ACES technique.

The fact that SL resins satisfy the conditions necessary for photoelastic testing was discovered during 1991 in the Laboratory of Photoelasticity and Holography at the University of Kassel.[6] For a given resin to be suitable as the basis of photoelastic stress analysis involving SL test models of production components, that resin must first satisfy the requirements placed on photopolymers as candidates for stereolithography. Such characteristics as adequate photospeed, sensitivity to the appropriate laser wavelength, good recoating properties, high green strength, low shrinkage, minimal creep, and the ability to produce accurate parts are all essential prerequisites.

Literally thousands of candidate resin formulations have been tested by the various commercial suppliers such as Ciba-Geigy, Allied Signal, and DuPont. To date, only about 20 resins have been commercialized for stereolithography. This is a testimony to the difficulty of simultaneously achieving all these goals.

However, to be of value for photoelastic testing, the resin must also possess the following properties:

- optical birefringence;
- optical transparency;
- fringe order proportional to the applied force; and
- an invariant photoelastic coefficient.

A schematic of the photoelastic test setup is shown in Figure 12-4. Fortunately, the equipment required is both relatively simple and inexpensive. For example, the irradiance source can be a standard incandescent light bulb, and polarization

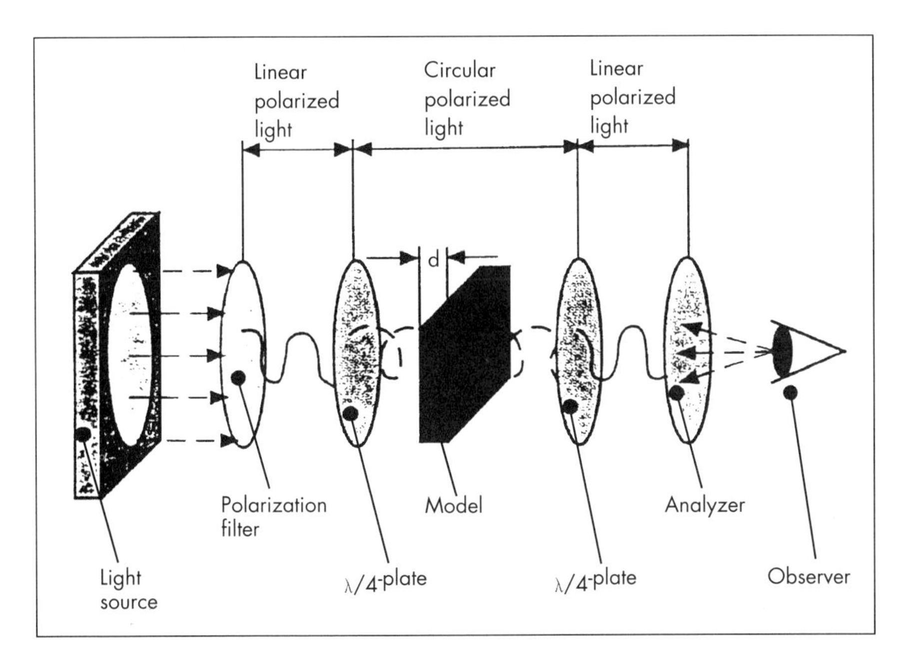

Figure 12-4. Schematic of the photoelastic test setup.

filters and quarter-wave plates are available from numerous optical component suppliers.

A light wave passing through the first polarization filter becomes linearly polarized. If a birefringent model under load is irradiated with linearly polarized light, both isochromatics and dark lines are obtained. These dark lines, known as *isoclinics*, also show the principal directions of stress. Unfortunately, for some applications the isoclinics interfere with the isochromatic fringe pattern. The isoclinics can be eliminated through the use of circularly polarized light, obtained with two quarter-wavelength plates.

The principal stress difference is determined from the basic photoelastic equation:

$$\sigma_1 - \sigma_2 = \frac{SN}{d} \tag{12-1}$$

where σ_1 and σ_2 are the principal stresses in the model (Newton/mm^2), S is the photoelastic coefficient of the material (Newton/mm × fringe order), N is the isochromatic fringe order, and d is the thickness of the model section being studied (mm).

The photoelastic coefficient, S, of a given material is determined by calibration. The material in question is subjected to a known load. In a compression test, force is applied across opposite sides of a cylindrical disk, as shown schematically in Figure 12-5. In this case, the diametral compression force F results in principal stresses σ_1 and σ_2. The photoelastic coefficient, S, is then determined through the relation

$$S = \frac{(\sigma_1 - \sigma_2)\,d}{N} = \frac{8F}{\pi DN} \tag{12-2}$$

where D is the diameter of the disk. Thus equation (12-2) indicates that if the photoelastic coefficient is truly a "constant of the material," then $N = (8/\pi DS)\,F = kF$, and the isochromatic fringe order N would be linearly proportional to the applied force F.

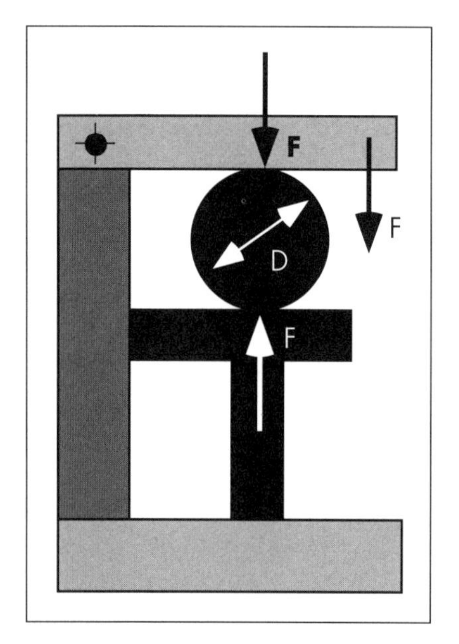

Figure 12-5. In a compression test, force is applied across opposite sides of a cylindrical disk.

Circular disks generated using various SL resins and build styles were supplied to the Laboratory of Photoelasticity and Holography at the University of Kassel by 3D Systems and Ciba-Geigy. These disks were then calibrated using the compression test method referred to in Reference 7. Figure 12-6 shows S over a range of applied loads for the Ciba-Geigy acrylate resins used in an SLA-250. Calibrations were completed involving circular disk samples fabricated using the STAR-WEAVE build style in resins SL 5081-1, SL 5134, SL 5143, and SL 5149.

Figure 12-7 shows the results of similar calibrations performed on circular disks produced using the ACES build style with the more recently released Ciba-Geigy epoxy resins SL 5170 and SL 5180. As a point of comparison, the photoelastic coefficient of Araldite B is also plotted. All of the acrylate and epoxy-based photopolymers demonstrate a linear dependence of the isochromatic fringe order N versus the applied calibration force F, in good agreement with equation (12-2). When quarter-wave plates are not used, isoclinics appear in the photoelastic images of the calibration disks. The isoclinics enable a complete evaluation of the stress distribution. To determine the isoclinics, the load is decreased nearly to zero. The isochromatics then disappear, but the isoclinics remain.

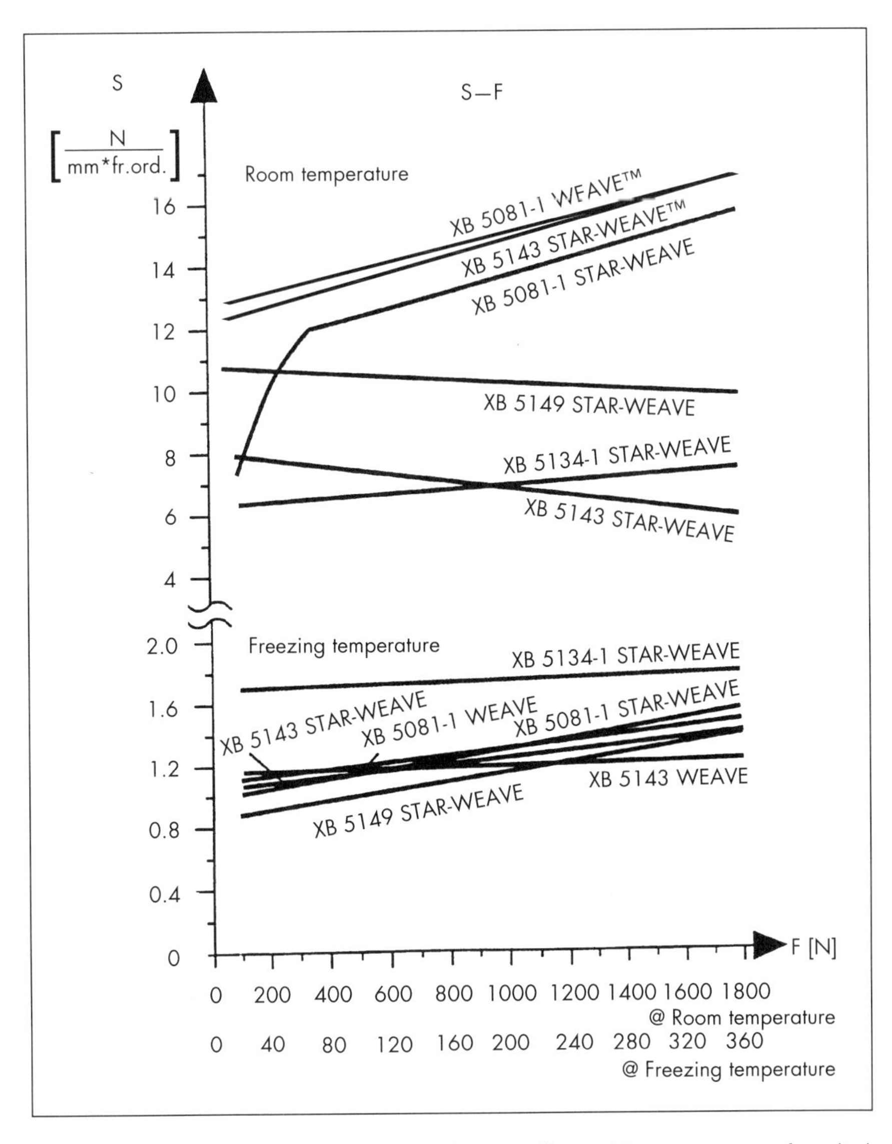

Figure 12-6. The drawing depicts the photoelastic coefficient (S) over a range of applied loads for Ciba-Geigy acrylate resins used in an SLA-250.

Residual stresses are normally introduced into plastic through rolling and extrusion. However, when SL components are generated, residual stresses are a function of the particular build style. The residual stresses in ACES-generated SL parts are very small.

An additional important benefit of using SL generated models for photoelastic stress analysis lies in the ability to "freeze" the strains and stresses by warming

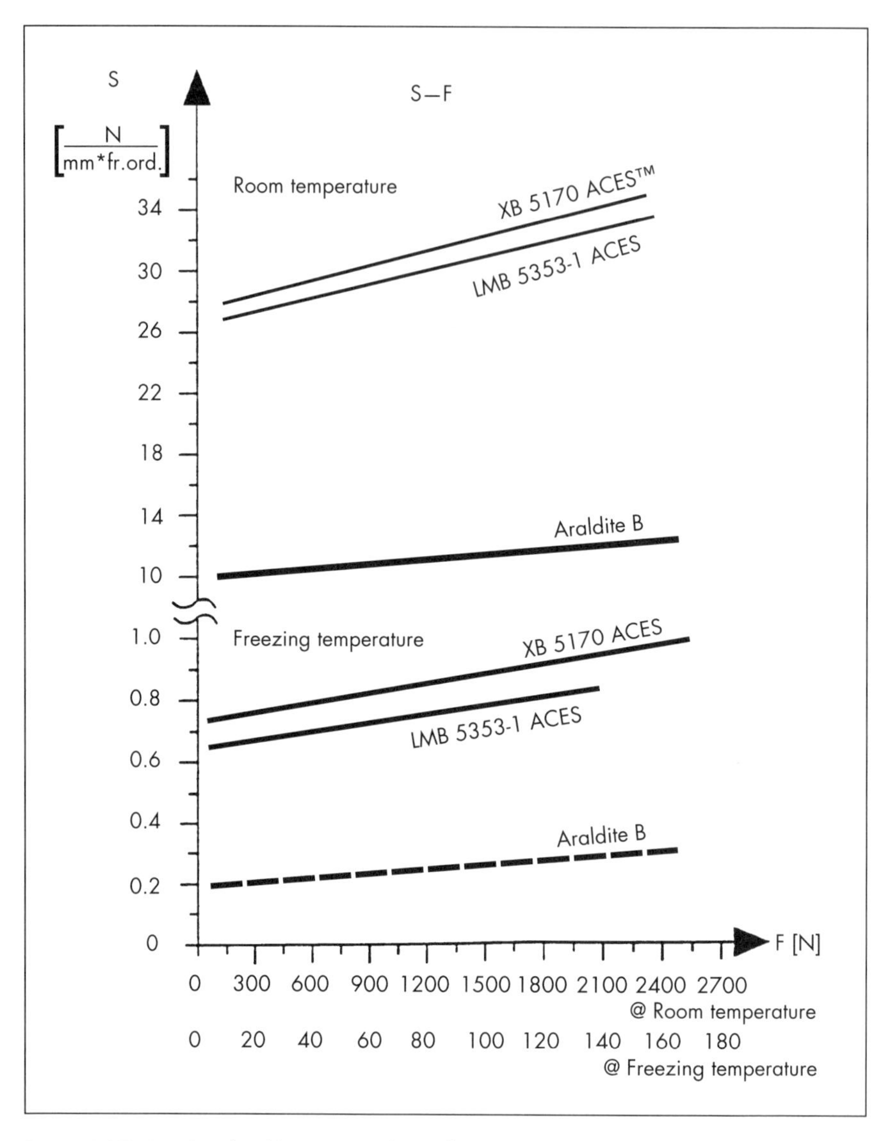

Figure 12-7. Results of calibrations performed on circular disks produced using the ACES build style with the more recently released Ciba-Geigy epoxy resins SL 5170 and SL 5180.

the loaded model to a level above the resin "glass transition temperature," T_g, and then gradually cooling the model back to room temperature.[8]

The glass transition temperature of the epoxy resins SL 5170 and SL 5180 is about 176° F (80° C). A thorough relaxation of the applied stresses and strains can be achieved at 212° F (100° C). At this point, the model is allowed to return to room temperature, after which the applied loads can be removed. The result is then a photoelastic test part where the previously-applied stress distribution is "frozen-in." This enables one to study the stress distribution within the part without the need to continue the application of any loading whatsoever. As an example, Figure 12-8 shows the frozen stress distribution for a model of an automotive connecting rod viewed in circularly polarized light. The twin views are for the whole 3-dimensional SL model (left), and from a thin quasi-2-dimensional slice through the model (right).

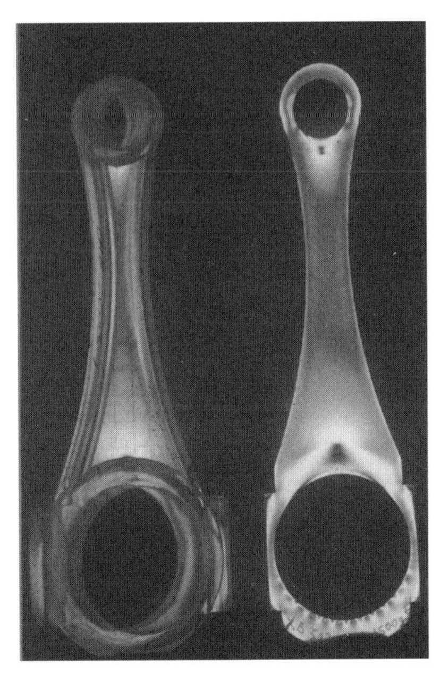

Figure 12-8. The frozen stress distribution for a model of an automotive connecting rod.

On the basis of a thin slice taken from the center of the frozen stress case, a complete evaluation of both the isochromatic and isoclinic fringe patterns can be realized. The stress distribution surrounding the gudgeon hole at the top of the connecting rod was evaluated in detail. The results show that both SL 5170 and SL 5180 resins are definitely suitable for photoelastic investigations of complex 3-dimensional objects.

Photoelastic Analysis of Prototypes

The foregoing techniques can be transferred to functional metal components by using the results from the photoelastic analysis of SL models, in conjunction with fundamental similarity laws. Using these relationships, it is then possible to make reasonably accurate predictions regarding the actual stress distributions anticipated in functional prototypes.

If we designate conditions within the test *model* with a subscript M, and those within the functional *prototype* with a subscript P, then the basic similarity principle takes the form

$$\frac{\sigma_P}{\sigma_M} = \frac{F_P}{F_M}\left(\frac{L_M}{L_P}\right)^2 \tag{12-3}$$

where σ, F, and L are the stress, applied force, and characteristic length, of either the prototype or the model respectively.

Thus, while the forces on the SL resin model cannot be as large as those that will be applied to a functional metal component, stress similarity can nonetheless be achieved through proper scaling. As an example, if the force on the metal prototype would normally be 25 times the greatest load the model can withstand before failure, then by scaling the model to 1/5 the linear dimensions of the prototype, the stress levels measured in the model should accurately reflect those anticipated in the functional component.

Since scale factors are easily applied to SL parts, and smaller parts build more quickly, stress similarity is both simply and rapidly achieved. As a result, an optimization of the automotive connecting rod component through the use of photoelastic stress analysis methods, applied to a scaled ACES model, is definitely feasible.

In contrast to Araldite or the epoxy resins, creep in the acrylate resins presents a significant problem for the application of photoelastic methods. The stress distribution in the models is no longer solely the result of applied forces, but is also dependent upon the stresses induced during curing and shrinking of the acrylate photopolymer resins. Furthermore, the relaxation of these stresses, in the form of creep, is also time-dependent. As a consequence, it is difficult to accurately interpret the results of photoelastic analysis when applied to models made from acrylate resins.

However, this is not the case with the epoxy resins. Here, the internally induced stresses, resulting from resin shrinkage subsequent to adhesion to the previous layer, are very much lower. As noted in Chapter 2, creep is over an order of magnitude smaller for parts built from epoxy resins than that found with acrylate models. As a result, the stress distributions within epoxy resin models, as measured by photoelastic methods, far more accurately reflect the effects of the applied loads.

A major benefit of building SL models for photoelastic stress analysis with epoxy resins SL 5170 or SL 5180 is that no tooling is needed to mold the object to be tested. While Araldite possesses excellent photoelastic properties, it is not a photopolymer and, hence, must be molded to form complex shapes. Considerable time and money are necessary to CNC machine the requisite tooling, which has been a significant impediment to photoelastic analysis. As a result, increased utilization of FEA methods began to supplant photoelastic techniques over the past decade.

However, because of the rapid advances occurring in SL during the past five years, it is now possible to build complex geometric models with high accuracy, and to do this without any tooling at all. If these models are built in ACES using the recently released epoxy resins, the potential for rapid, low-cost, accurate, photoelastic stress analysis is greatly enhanced. The results presented here have been restricted to static photoelastic stress analysis. In the near future, ACES models should also make it possible to perform dynamic photoelastic stress analysis of complex geometries. These techniques should provide still further benefits to the product engineer.

12.3 Dynamic Testing of RP&M Models

William Dornfeld, Ph.D., Allied Signal Engines, Stratford, Connecticut. (Based on a paper presented at The International Gas Turbine and Aeroengine Congress, The Hague, Netherlands, 1994[9], courtesy of the American Society of Mechanical Engineers).

Model testing has long been used to validate, augment, or replace structural analysis. The extent of model testing depends on the schedule, budget, and skills available, as well as the complexity of the structure and the consequences of the results. In particular, plastic models have seen significant use in attempting to understand the static and dynamic behavior of many different industrial structures. Today's ability to rapidly and economically generate scaled RP&M models directly from a solid CAD data base has opened a new chapter for model testing.

Specifically, SL models are readily scaled, quickly produced, and some SL photopolymers provide final parts with mechanical properties well suited for dynamic testing. The use of these models makes it possible to perform tests in the initial stages of product development. This accelerated testing can uncover problems earlier, reduce costs, and shrink the overall development cycle.

To a considerable degree, structural modeling has been driven by the availability of appropriate modeling materials. The Westinghouse Research Laboratories are credited with pioneering the use of plastic models in the design of turbo machinery in the early 1950s. During the 1960s and 1970s, vacuum-formed vinyl acetate sheets were increasingly used for dynamic testing involving 3/8 scale models of automobile structures.[10] Aluminum-filled epoxy (ModelTech™) was also utilized to test structures as diverse as landing gear and pressure vessels.

Since the 1980s, experimental modeling has seen reduced usage. This is most likely due to the enhanced capability and economic advantages of FEA methods. However, the recent advances in RP&M technology present modeling opportunities not previously possible. The ability to proceed from a CAD model, to a scaled test model, and then directly to static and dynamic testing of that model, presents potentially significant time and cost benefits that deserve serious consideration.

Why Model?

Experimental modeling is performed for three basic reasons: (1) to validate analysis, (2) to augment analysis, or (3) to replace analysis. The approach selected is influenced by the time available, the cost and difficulty of either the analysis or the test, similarity to previous experience, and the required confidence level of the results. Experimental models can have a number of characteristics that make them attractive compared to testing actual functional components. Table 12-1 lists several of these features.

The use of scale models for dynamic testing has been a standard practice in the development of gas turbines, especially for smaller engines. Figure 12-9 shows a 10:1 aluminum scale model of a turbine blade mounted on an electromagnetic shaker. These models were used through the mid 1970s with brittle lacquer coat-

Table 12-1. Advantages of experimental models.

Features	Examples
1. Enlarge small things	Turbine blades
2. Reduce large things	Wind tunnel aircraft and automotive models
3. Utilize special properties	Photoelastic testing of transparent models
4. Simplify testing	Fewer massive fixtures, smaller test loads
5. Results available earlier	Easily machined materials, and now RP&M
6. Functional similarity	Fluid flow structures
7. Accelerate test development	Dress rehearsal for testing the real part
8. Geometric verification	Mock-ups

Figure 12-9. This 10:1 scaled aluminum turbine blade has frequencies about one-tenth the actual engine component.

ings and strain gages to establish resonant frequencies, mode shapes, and stress distributions. Note that the blade attachment fir-tree was replaced with a rectangular section that accommodated the shaker mounting requirements, with known blade-attachment boundary conditions.

From a dynamics standpoint, the resonant frequencies, f_M, of the models are related to the component frequencies, f_C, by the following expression.[11]

$$f_M = f_C \; [L_C/L_M] \; \{ (Y_M/\rho_M)^{1/2}/(Y_C/\rho_C)^{1/2} \} \tag{12-3}$$

where Y is the Young's modulus, or modulus of elasticity, L is a characteristic length, and ρ is the density. The materials are assumed to be (1) isotropic, (2) elastic, and (3) homogeneous. Equation (12-3) indicates that a proportional half-scale model will have resonances at twice the frequencies of the original, and that 1:1 scale models made from different materials will exhibit resonant frequencies determined by the square root of the elasticity/density ratios of those materials.

Note that the modulus, Y, has units of force per unit area, or F/L^2, and the density, ρ, has units of mass per unit volume, or M/L^3. Since $F = Ma = ML/T^2$, then the quantity $(Y/\rho)^{1/2}$ has units of $[(ML/T^2)/(L^2)/(M/L^3)]^{1/2} = [L^2/T^2]^{1/2} = L/T$, which has the units of velocity, and is in fact equal to the speed of propagation of transverse waves in an elastic solid.[12] Since $(Y/\rho)^{1/2}$ for steel and aluminum are nearly equal, the 10:1 scale aluminum turbine blade shown in Figure 12-9 has frequencies that are about one-tenth those of the actual engine component.

Why Stereolithographic Models?

Of the various RP&M processes available today, SL is very attractive for experimental model generation. The several benefits of SL models are:

1. They can be fabricated directly from solid CAD models, the trend in current aerospace design.
2. They are easily scaled in size.
3. They can be built quickly.
4. The process is generic, requiring little or no part-specific tooling.
5. The resulting materials possess properties that are suitable for dynamic mechanical testing.
6. The process accurately builds internal cavities.
7. The model geometry can easily be modified for various attachments, or to accommodate special boundary conditions.

The parts can also be built as QuickCast patterns to shell investment cast a scaled test model from different materials, such as aluminum or steel, or directly as an ACES model for photoelastic analysis.

Of these benefits, the one directly relevant to this discussion is number 5. The remainder of this section will cover test results demonstrating that SL polymers possess properties suitable for use in dynamic model testing.

The Basic Cantilever Beam Test

Table 12-2 lists the room temperature properties of some relevant materials, including steel and aluminum, the traditional plastic modeling materials[13], as well as two Ciba-Geigy acrylate resins, SL 5131 and SL 5154, appropriate for use in the SLA-500 system.[14]

Table 12-2. Room Temperature Properties of Various Materials

Material	Y (10^6 psi)	ρ (lb/in.^{3})	[Y/ρ] (10^6)	$[Y/\rho]^{1/2}$ (10^3)
Steel	29	0.28	104	10.2
Aluminum	10	0.10	100	10.0
PVC-A Vinyl	0.41	0.048	8.5	2.9
PE Epoxy	0.44	0.042	10.4	3.2
ModelTech™	0.60	0.050	11.9	3.5
SL 5131	0.58	0.044	13.2	3.6
SL 5154	0.23	0.042	5.5	2.3

From these properties, we would expect that an SL 5131 model would have its resonances at 3.6/10.0, or 36%, of the resonant frequencies of a corresponding aluminum part of the same size, while a model made from SL 5154 would have frequencies at 23% of the aluminum component values. To confirm this, three 0.25 inch × 1 inch × 5 inch (6.35 × 25.4 × 127 mm) cantilever beams were fabricated. One was made from aluminum, one from SL 5131 and one from SL 5154. Each test sample was clamped in a vise and impact excited. Vibration response was measured with an ultra light (0.14 gram) accelerometer located at the tip of each cantilever beam.

Spectral response functions (amplitude versus frequency) for the three beams are shown in Figures 12-10, 12-11, and 12-12. The frequency behavior is summarized in Table 12-3, which lists the frequency ratios for the first four or five bending modes of each beam relative to their fundamental bending frequency. Also given are the theoretical ratios for a uniform cantilever beam.

The SL 5131 beam has a first bending mode frequency at 115/300 = 38% of the aluminum beam, and the SL 5154 beam's frequency is at 75/300 = 25%. The Mode 1 frequency ratios for these SL models agree with the $[Y/\rho]^{1/2}$ ratios because the precise elastic moduli for both resins were calculated from the measured Mode 1 frequencies and densities. This was done because the values of Y determined in this manner are often more accurate than those established from the slope of the elastic portion of a conventionally measured stress versus strain plot. However, once the precise value of the moduli has been determined, the consistency of all the higher modal frequency ratios for the SL generated samples

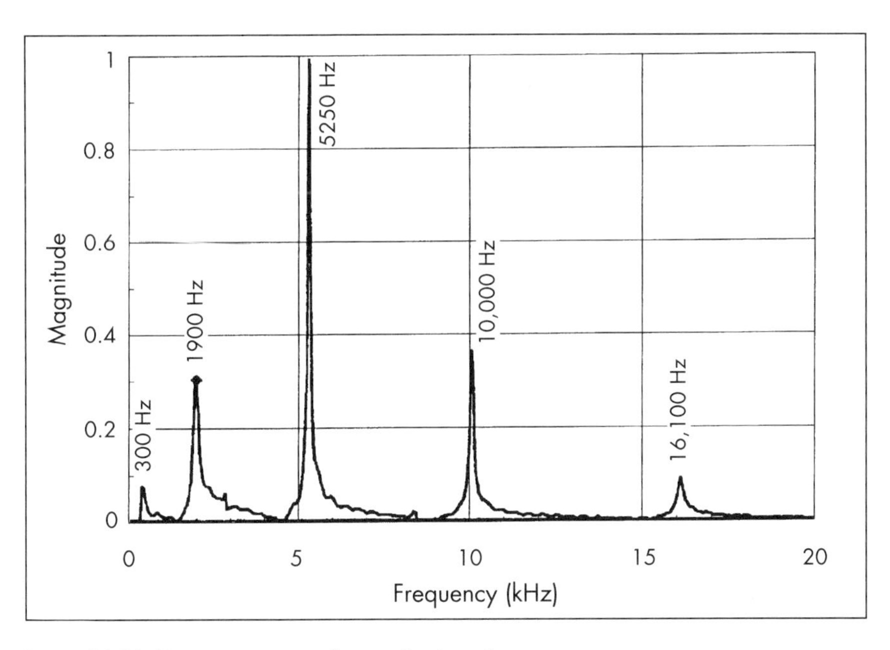

Figure 12-10. Frequency spectra for an aluminum beam.

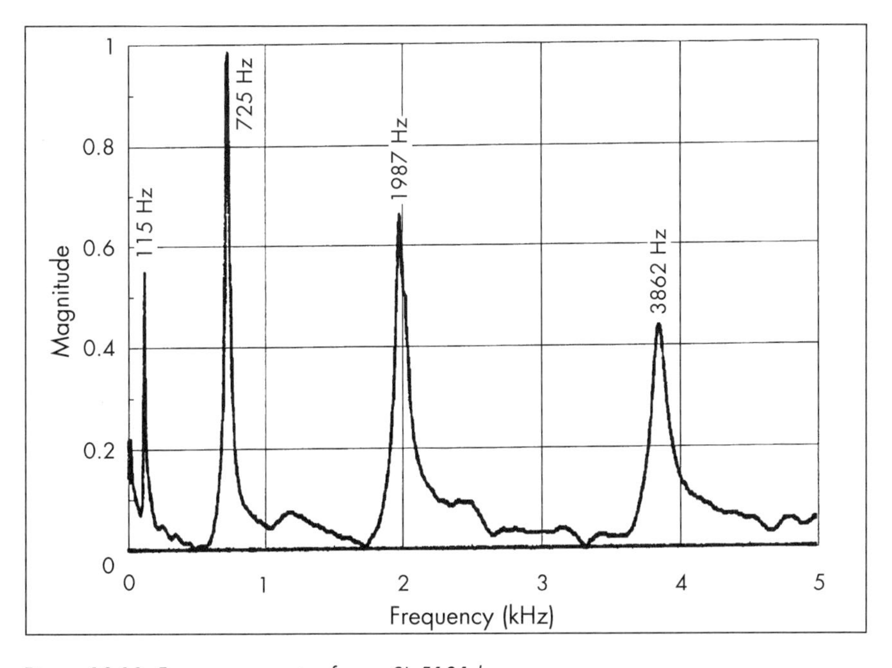

Figure 12-11. Frequency spectra for an SL 5131 beam.

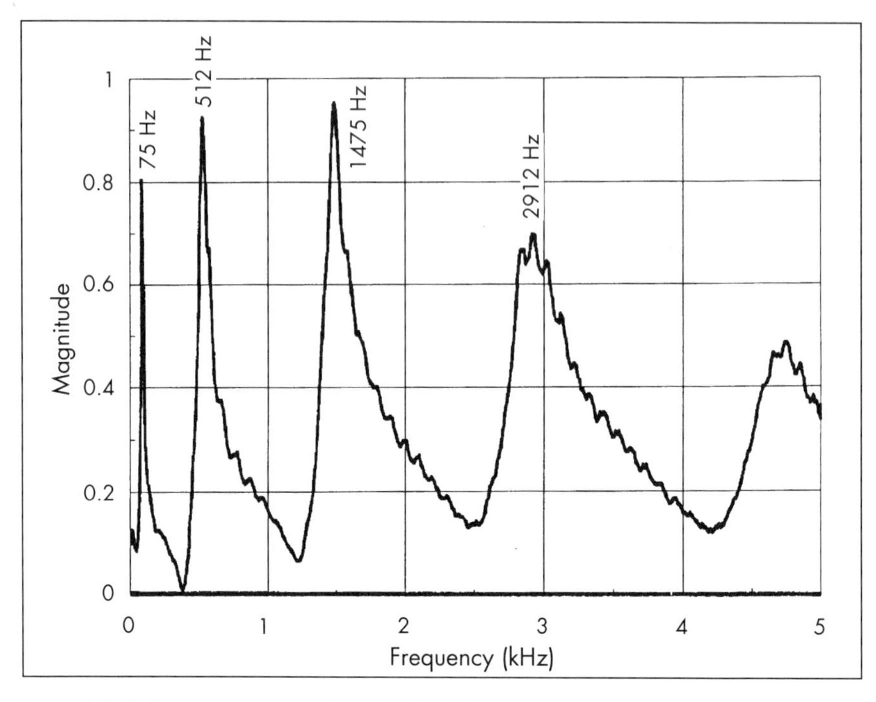

Figure 12-12. Frequency spectra for an SL 5154 beam.

Table 12-3. Frequency Ratios for Beam Bending Modes

Modes	Frequency Ratio (Theoretical)	Aluminum (Measured)	SL 5131 (Measured)	SL 5154 (Measured)
1	1.0	1.0 (300 Hz)	1.0 (115 Hz)	1.0 (75 Hz)
2	6.4	6.3	6.3	6.8
3	17.5	17.5	17.3	19.7
4	34.4	33.3	33.6	38.8
5	56.8	53.7		

demonstrates that indeed the model cantilever beams are behaving dynamically according to their $(Y/\rho)^{1/2}$ ratios, analogous to the uniform metal beam.

Furthermore, two observations regarding damping can be drawn from the spectra shown in Figures 12-10 to 12-12. First, the aluminum beam and the hard SL 5131 beam are both lightly damped at about 2–3%. Second, the softer SL 5154 resin has approximately 9–10 % damping, and a corresponding broadening of the higher frequency modes is apparent. Although SL 5154 is much less brittle than SL 5131

(10% elongation versus 2–3% respectively), the tradeoff is the greater damping. Higher damping makes accurate frequency measurements more challenging, but not impractical.

In a follow-on test, the free length of the aluminum beam was increased to 5.48 inches (139 mm) through reclamping, and the beam was excited again. This length yielded modal frequencies that were very nearly double those of the SL 5131 model. The resulting frequency spectrum of the modified aluminum beam is shown in Figure 12-13, as an overlay on the SL 5131 beam spectrum of Figure 12-11. The excellent frequency correlation is evident.

Inherent in all RP&M processes is the generation of the part through the adhesion of multiple thin layers. To determine the influence of build orientation, and hence lamination, on the dynamic behavior of SL models, two cantilever beams were built at the same time in the same resin. One beam was built lying flat while the other was built standing on edge. Frequency testing showed no significant effect of build orientation.

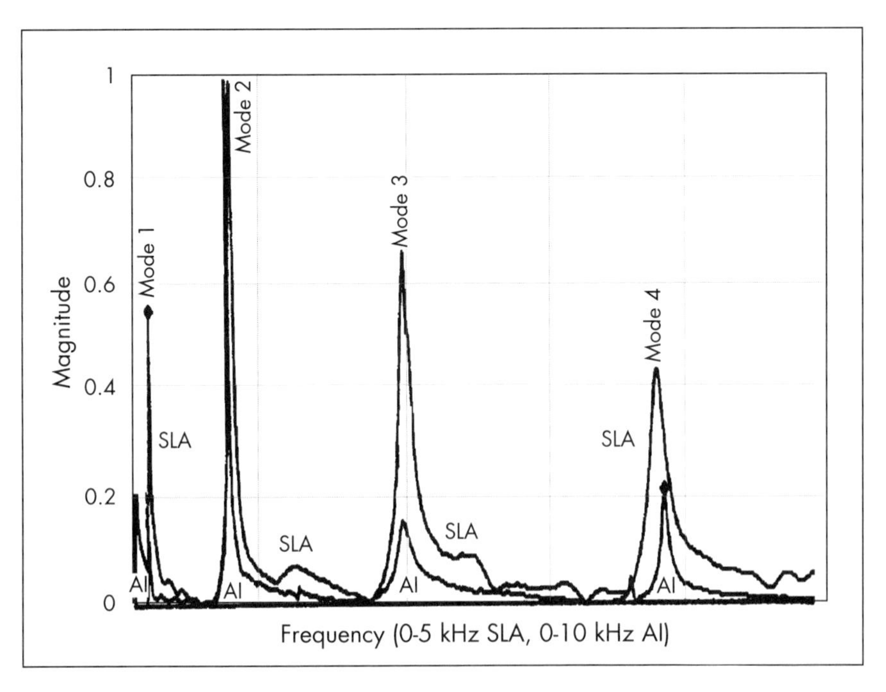

Figure 12-13. Frequency spectrum of a modified aluminum beam overlayed with that from an SL 5131 beam.

Testing a Turbine Blade

To put the procedure to a practical test, models of an Allied Signal turbine blade were built using SL 5131 resin at both 1:1 scale and 3:1 scale. The CAD geometry and FEA grid model of the turbine blade are shown in Figure 12-14.

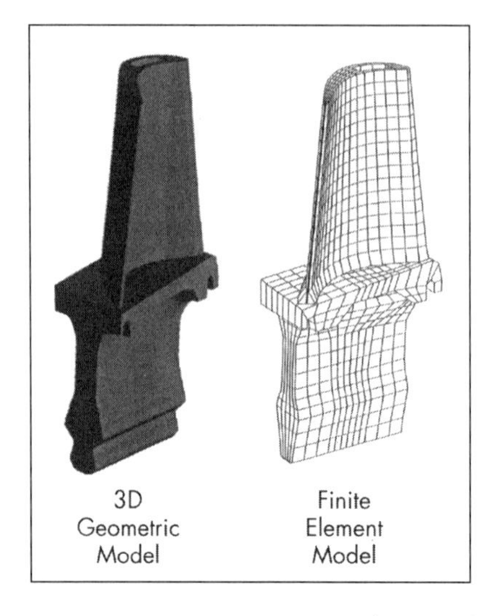

Figure 12-14. CAD geometry and FEA grid models of Allied Signal turbine blade.

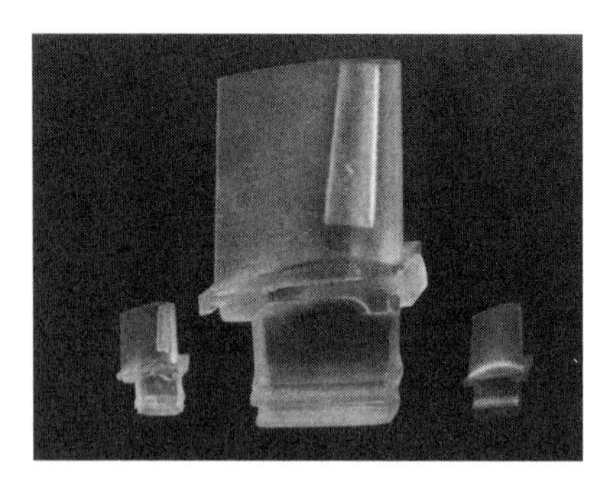

Figure 12-15. The metal blade and two SL models.

The STL file was generated directly from the CAD model. The metal blade, as well as the two SL models, can be seen in Figure 12-15.

The 3:1 scale model was built for two reasons: first, because of the very small size of the actual turbine blade and, second, to verify the size-scaling properties of these SL models. Resonant frequencies of all three test specimens were determined using either electromagnetic or piezoelectric exciters, as well as a combination of ultra light accelerometers and holography. Table 12-4 lists the resonant frequencies of the first five vibration modes of the actual blade, and of both sizes of the SL 5131 models, along with the FEA predicted frequencies.

These results are displayed in Figure 12-16, where the SL model frequencies are plotted against the actual metal blade frequencies. Also shown are the "best-fit" scaling lines for each measurement set. Note that the FEA model predicts frequencies from 3.6% lower to as much as 19.4% higher than the actual blade, possibly due to the blade not having been tip ground. The 1:1 SL model frequencies lie along a 35% scale line.

Based upon the 101.4×10^6 elasticity/density ratio measured for the actual high-temperature steel alloy turbine blade, the SL model should have frequencies at 3.6/10.07 or 35.7% of the metal blade. This compares extremely well with the test results. Also, the 3:1 blade should have frequencies at 11.9% and, in fact, the measured frequencies lie very close to the 12% scale line. The agreement between the theoretical scaling law and the experimental measurements is excellent.

Finally, as a further check of dynamic testing fidelity using SL models, holographic testing was also performed on the steel alloy turbine blade and the 1:1 SL model. The holographic images for three different vibration modes are shown in Figure 12-17. This figure also shows the deflection contours predicted by finite element analysis for each of the three modes. The good correlation involving the corresponding mode shapes for the 1:1 SL 5131 model further supports the validity of performing dynamic tests using stereolithography models.

Table 12-4. Resonant Frequencies of Blade Vibration Modes

Modes	Blade Hz	FEA Model Hz	1:1 SL 5131 Hz	3:1 SL 5131 Hz
1	2,470	2,948	1,080	338
2	8,470	8,169	2,990	963
3	9,320	10,745	3,520	1,138
4	15,160	16,124	4,380	1,850
5	18,560	19,905	6,800	2,325

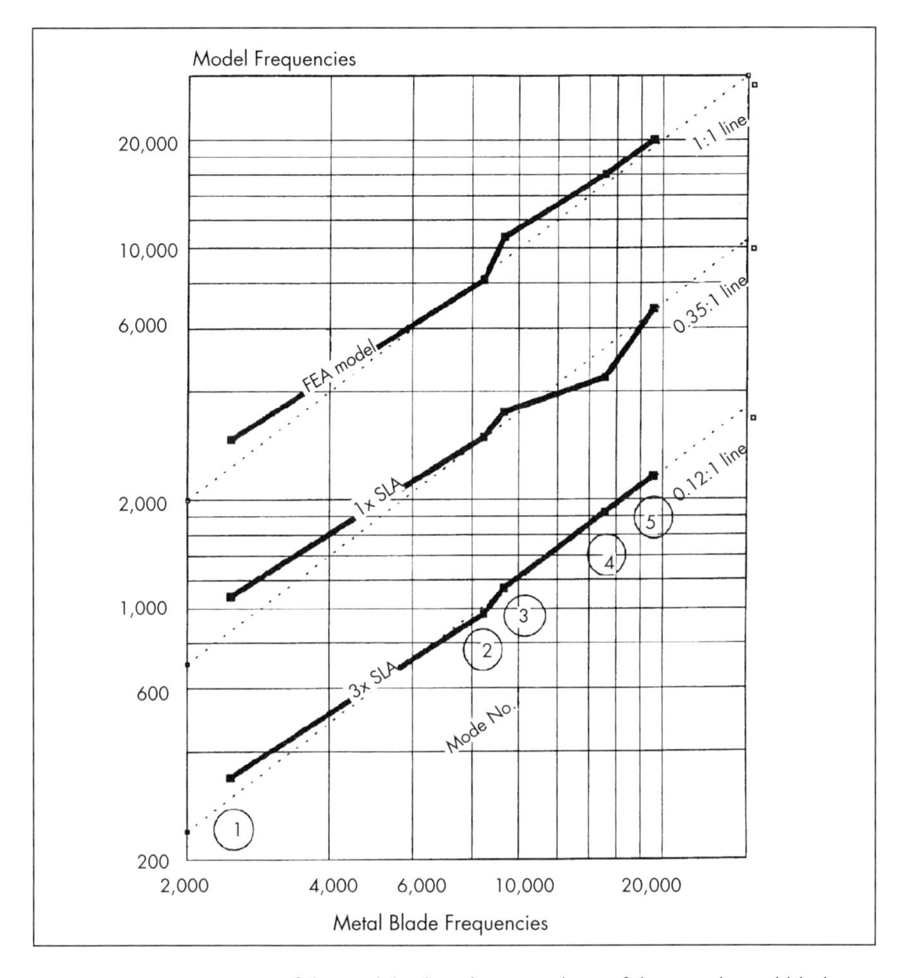

Figure 12-16. Frequencies of the models plotted against those of the actual metal blade.

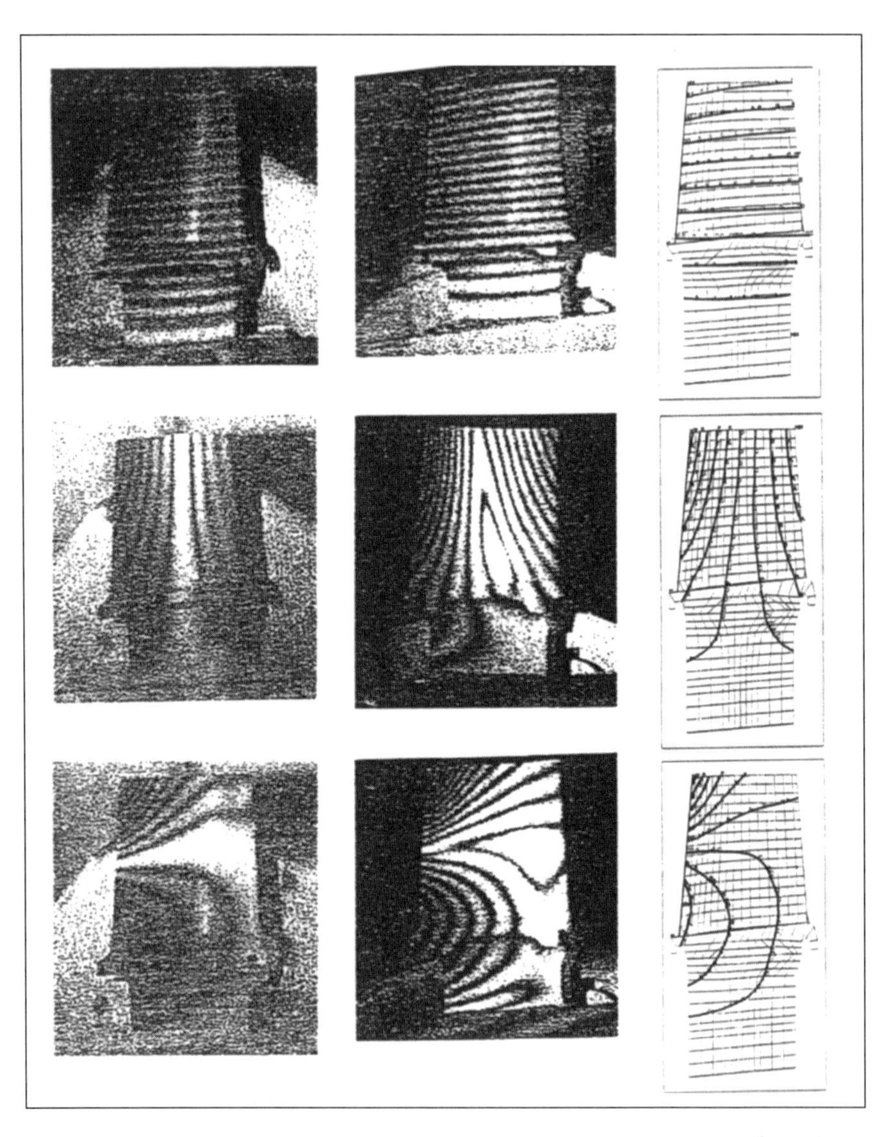

Figure 12-17. In a check of dynamic testing fidelity, holographic testing was performed on a steel alloy turbine and a 1:1 SL model. Here, we see holographic images of three vibration modes. At right are deflection contours predicted by finite element analysis for each mode.

In a concurrent engineering environment, dynamic testing of RP&M models offers the possibility of determining the dynamic characteristics of new designs shortly after the solid CAD model of the component has become available. The challenge lies in identifying those situations where it is cost effective to use this method to either augment or replace FEA techniques.

12.4 Medical Models Using RP&M Techniques

B. Swaelens, Materialise NV, Heverlee, Belgium; F. Noorman van der Dussen, Eeuwfeestkliniek, Antwerpen, Belgium; W. Vancraen, Materialise NV, Heverlee, Belgium.

In the early days of stereolithography, design visualization and verification were the most important applications for models. Today, SL models are also useful as functional prototypes or as a first step in generating either soft tooling, as discussed in Chapter 10, or hard tooling, as noted in Chapter 11. Each of these processes is moving in the direction of manufacturing and production.

The same story is now being repeated in the field of medical imaging. Specifically, computer assisted tomography (CT) and magnetic resonance imaging (MRI) scanning systems provide high resolution images of the internal structure of the human body. During the past few years, these scanning techniques, along with the associated software and hardware, have undergone substantial development. As a result, it is now possible to represent 3-dimensional data as shaded images, virtual reality, and holograms. However, the ultimate 3-dimensional representation is still an actual physical model.

The use of medical models is also evolving from visualization and verification toward simulation (prototyping) and, eventually, prosthetic production. In various industrial applications, the new RP&M techniques have legitimately introduced the term "rapid" into the prototyping and manufacturing environments. However, in many medical applications, 3-dimensional visualization is rather new, prototyping was previously almost nonexistent, and individual prosthetic production was both very difficult and expensive.

Three potentially important applications of RP&M in medicine are:

1. *Visualization*

A physical model can serve as a hard copy of the data set, providing both visual and tactile documentation for diagnosis, therapy planning, and didactic purposes. The model facilitates better communication between surgical team members, between radiologists and surgeons, and between physicians and patients. Radiologists are often skillful in interpreting 2-dimensional scans and shaded images as a result of extensive training and experience. However, surgeons are typically not used to performing this type of interpretation, and would generally prefer an actual physical model over an image.

2. *Simulation*

Physical models can be very useful in planning complex surgery, which may involve simulation performed on the model. For instance, in facial surgery, life-size models of the patient's skull are used to plan osteotomies (displacements of bone segments) prior to actual surgery, by rehearsing on the model itself.

3. *Prosthesis Generation*

When an accurate physical model of the existing structure is available, the design of prosthetic devices is much easier, and fixation plates can be bent prior to surgery. The model can serve as a "negative" mold from which the implant is manually generated before surgery, or it can serve as a "positive" master for generating soft tooling.

Mirror images can also be generated readily to achieve symmetry in facial or breast reconstruction. The ultimate step will be achieved when the implants can be made directly in an SLA using biocompatible materials. Animal tests have shown that some acrylic resins do indeed possess a high degree of biocompatibility. However, at present, no resin specifically formulated for SL has been certified for use within the body.

Generating Medical Models

Since RP&M parts are built in a layer-by-layer manner, they are well suited to the layer-by-layer information coming from medical scanners. Unfortunately, despite this natural connection between the information generated by the scanners and the fundamental build procedure of all RP&M systems, it has not proven to be a simple matter to provide a good interface. The RP&M machines and their associated software were developed to produce parts from CAD information, not from CT or MRI images.

This is the reason why the earliest attempts to build medical models on an SLA still made use of a CAD system. Contours, introduced in a CAD system, are used to generate surface models. For structures such as a portion of a bone, this is possible, but for more complicated structures with multiple cavities, such as a skull, this becomes an almost impossible task if one wishes to obtain a faithful representation of the original.

In order to overcome the huge manual effort required, automatic systems were developed, mostly starting from existing 3-dimensional representation software. Indeed, much work had previously been done to enable the display of 3-dimensional CT data in the form of shaded images. A standard approach was to calculate triangles that would represent the surface of the scanned entity (e.g., a bone). Then, these triangles only needed to be converted into the STL format.

Unfortunately, the major problem with this approach is the huge amount of triangles that need to be processed. Cases involving more than 1 million triangles were not unusual. The consequences of this approach were very long slice and part preparation times. To overcome these shortcomings, special purpose software was developed to interface from both CT and MRI scanners directly to the RP&M systems, as shown in Figure 12-18.

The CT-Modeller System

The x-ray tube and the detector of a CT scanner are firmly coupled on a rotating arm. This arm turns 360° around the long axis of the patient. This enables the CT scanner to make measurements through the patient from different vantage points. A computer then utilizes the measurement data to generate an artificial image of a cross-section through the patient. The resulting CT image is created from many pixels. Each pixel is assigned a numerical value representing the amount of x-radiation absorbed at that pixel location. These absorption values are repre-

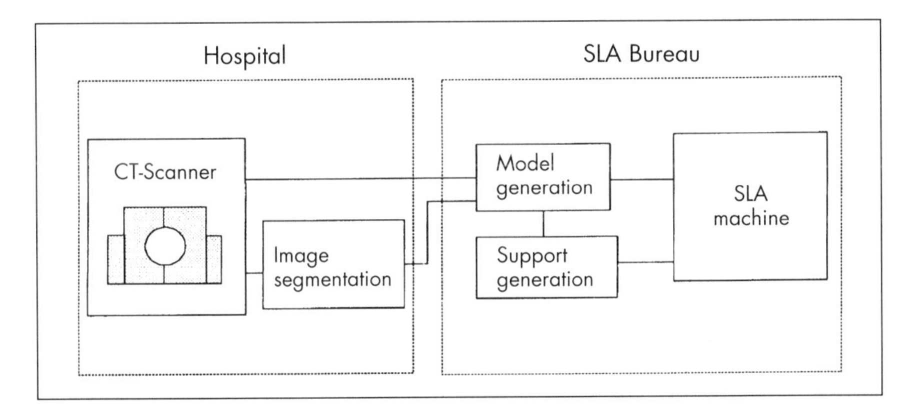

Figure 12-18. This special software was developed to help overcome the problems of too many triangles present during attempts to build medical models.

sented by an arbitrary scale from -1000 to +1000 (i.e., Houndsfield units). The scanner is calibrated such that -1000 corresponds to air, 0 to water, and +1000 to cortical bone. A 3-dimensional data base is then constructed by connecting adjacent CT scans. A stack of 2D slices then forms the 3-dimensional data set, where each pixel now represents a tiny volume element or voxel, equal in size to the pixel area times the slice thickness.

Spiral scan is achieved by slowly transporting the patient through the CT gantry while the scanner arm is synchronously rotated about the long axis of the patient. The result is a multitude of 360° rotations. The continuous rotation of the detector and the x-ray tube, coupled with the uniform translation of the patient, results in a spiral or helical path of the x-ray focal point relative to the patient. Spiral CT requires special data processing software since the direct reconstruction of images over any 360° segment would result in artifacts due to the patient transport. These artifacts are the result of the pitch of the helix (i.e., the "gaps" between successive spiral segments). Measurements obtained in planar geometry have to be synthesized from the volume data set. With spiral CT systems, an entire 3-dimensional data set is obtained in about 30 seconds. The reconstruction of flat images can be performed afterward.

MRI is based upon the phenomenon that nuclei with an odd atomic mass possess spin and, hence, a magnetic moment. When such nuclei are placed in an externally applied magnetic field and irradiated with low-frequency radio waves, the magnetic properties of the nuclei enable the generation of extremely detailed cross-sectional images. The radio waves used are optimized for various types of body tissue. As a result, it is possible to differentiate between various organs. In general, CT is most suitable in differentiating hard from soft tissues, such as bone from muscle. However, MRI is better at distinguishing between different types of soft tissue, such as the liver from a kidney.

The first aspect of model creation is determining which portions of the image are bone and which are not. This is referred to as *segmentation*. The standard segmentation technique involves "thresholding." All the pixels that have a gray value higher than some preselected threshold are assumed to be bone. Unfortunately, a major problem with this approach is the so-called "partial volume effect." This effect produces image artifacts that result from a violation of the assumption that each voxel within the patient is occupied by only one type of tissue.

Since the voxels are finite in size, it is quite possible that the edge of a bone lies part way into a voxel. The result would then be a measured value that is a volume-weighted mean for the two different tissue types. This is the case for voxels overlapping the interface between two different tissue types, or for voxels containing objects thinner than some of the voxel dimensions.

By taking into account the partial volume effect, the resolution of the images can be improved dramatically. A linear interpolation is performed between pixels, which greatly improves the contour relative to the original object, as shown in Figure 12-19. This interpolation gives reliable results when only two different tissue types are present in the neighborhood of the threshold value. For CT image segmentation between cortical bone and soft tissue, or between soft tissue and air, this is a good assumption.

Other artifacts can arise from metallic implants or prostheses. For instance, dental fillings cause severe artifacts in scans through the jaw and teeth. These are the result of extreme x-ray absorption through the metal fillings. Thus, practically no photons reach the detector, and information is missing. The only practical means of dealing with these artifacts is manual editing.

All these image interpretation issues are combined in a stand-alone software package called MIMICS™, recently developed by Materialise NV. This software was intended for use by physicians and surgeons. The MIMICS software utilizes some of the latest and most advanced image segmentation algorithms to enable the RP&M generation of physical, 3-dimensional medical models from either CT or MRI data.

Medical Model Production

After completion of the medical procedures, contour images are transferred to the RP&M site. Here the CT-Modeller™ software package combines all the technological aspects. Both the model and its support structure are automatically generated using this software. The first step in this operation involves the 3-dimensional interpolation process. CT scans are typically taken with a pitch of 0.06 or 0.08 inches (1.5 or 2 mm), and the pixel size is commonly about 0.02 to 0.04 inches (0.5 to 1 mm). Consequently, if simple thresholding is employed, a model is obtained that involves "stairstepping" in the X and Y directions as well as in the Z direction.

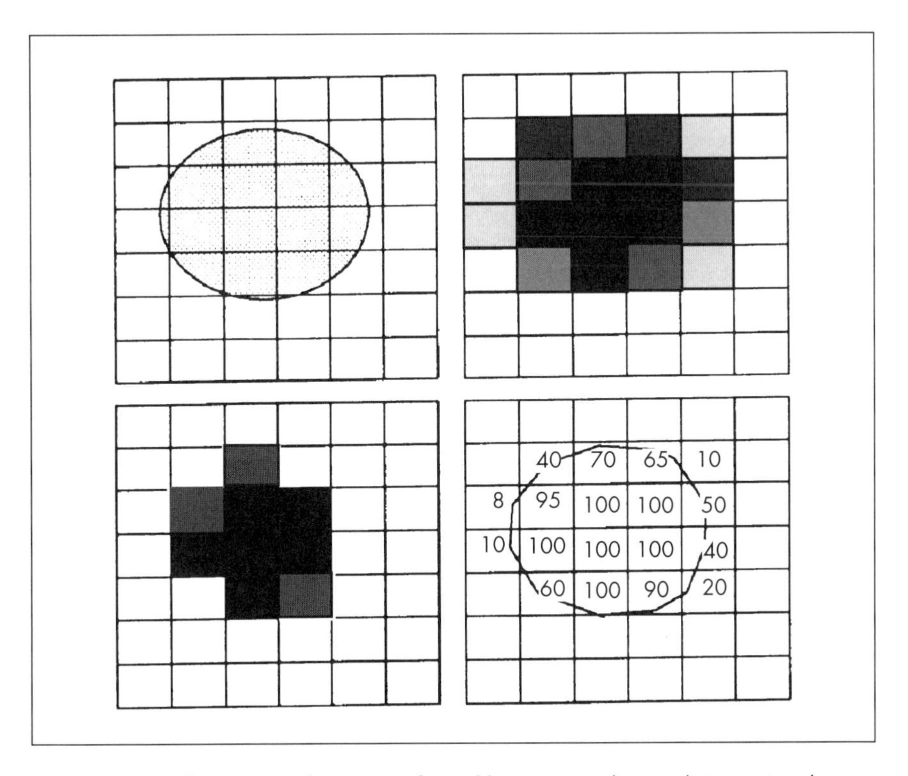

Figure 12-19. A linear interpolation is performed between pixels, greatly improving the contour relative to the original object.

Hence, a good interpolation scheme is very important. Because of the morphological forms of medical models, we know in advance, at least roughly, what they will look like. Simply stated, human organs do not have sharp corners or irregular stairstep-like shapes. This knowledge is useful in generating the software. Specifically, if the contours established in the CT scans are faithful approximations of the cross-sections, then the cubic interpolation method utilized by the CT-Modeller will provide a more accurate model. In short, models with stairstepping effects will always hide some details, and they will provide less than optimal visual and tactile information.

The advantages of direct layer interfacing are obvious; the most accurate directions of the input data (i.e., the scanning plane) are produced on the most accurate directions of the SLA. A disadvantage is the lack of a surface model. This implies that the orientation of the part in the SLA should be the same as the orientation used during patient scanning. Fortunately, because of the morphology of the human skeletal system, this is not really a significant disadvantage. In most cases, the CT scanning plane turns out to be the best plane in which to build the model.

Another implication is that the standard techniques for producing support structures cannot be used. Automatic support generation starting from STL files is not possible since these files do not exist. Furthermore, manual generation of supports is also impossible due to the lack of a CAD model. Therefore, automatic support generation software was developed to provide the required build supports starting from contour information. At this time, CSUP™ (Contour SUPports) is the only commercially available automatic support generator able to perform this operation. The CSUP software is also used for automatic support generation in technical applications when interfacing directly from a CAD system to layers (e.g., CATIA to SLC).

The demands on model surface finish depend upon the application. For some models, optical transparency is very important. An example would be the ability to see the mandibular channel in the jaw. However, for other applications, transparency is not desirable because it may provide misleading visual effects. In these cases, sandblasting or even spray painting the model may be helpful. Also, the removal of support structures is best done before postcuring. This is the case because polishing the model can hide small details or "landmarks" which are often used for both identification and measurement.

Case Study of Surgical Preplanning

The first skeletal surgical intervention in a patient with Parry Romburg disease, shown in Figure 12-20, involved the correction of the orbito-naso-frontal structure. The planning of the hard tissue reconstruction for such a syndrome, with cranio-orbital involvement, was done in the Eeuwfeest Clinic in Antwerp, Belgium, using an SL model, shown in Figure 12-21.

The most important aspects of presurgical planning involving SL models are not so much where the osteotomy lines should be located. The major advantages involve planning the displacements of bone segments while having a complete overview as shown in Figure 12-21, and while also being able to measure all the necessary displacements in advance, as shown in Figure 12-22.

By means of a bicoronal incision, the surgeon was able to approach the frontal orbital region. After lifting the frontal bone flap, the frontal bar is outlined and cut. The frontal bar is then rotated 180° and fixed with titanium plates and screws. Also, the orbital framework is osteotomised and repositioned in the most symmetric location.

The displacements are measured in advance on the SL model (Figure 12-22). In this way, time-consuming "fitting and chipping" is avoided during the operation because the surgeon already knows exactly the shape and dimensions involved before the surgery even starts. The stabilization of the bony parts is very accurate, and the surgeon can thereby proceed directly to the fixation procedure.

Physical attractiveness has much to do with a symmetric facial appearance about the midline as well as parallelism of the facial planes. This patient with Romberg disease was classified as having "fair appearance" one year postopera-

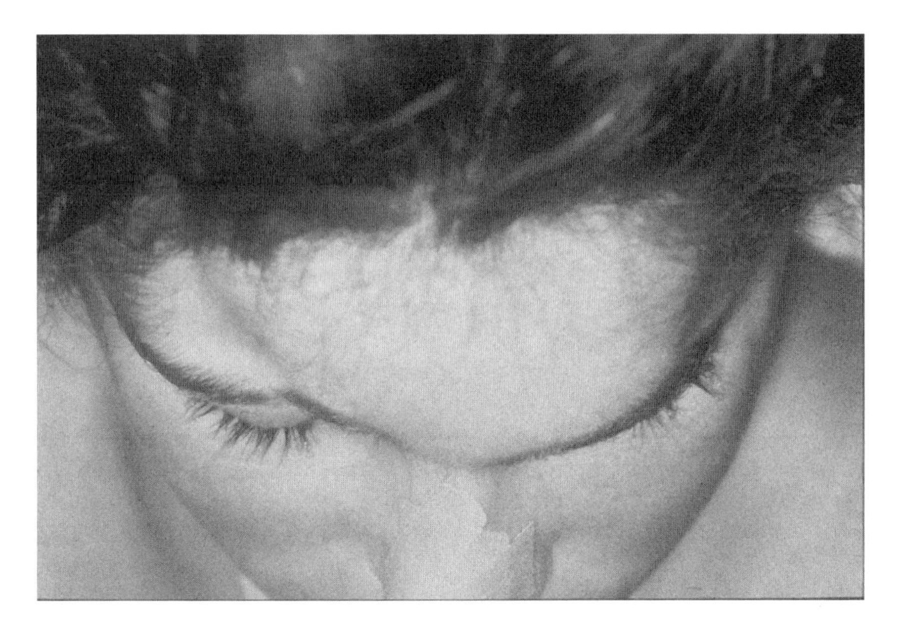

Figure 12-20. The first skeletal surgical intervention in a patient with Parry Romburg disease, shown above, involved correcting the orbito-naso-frontal structure.

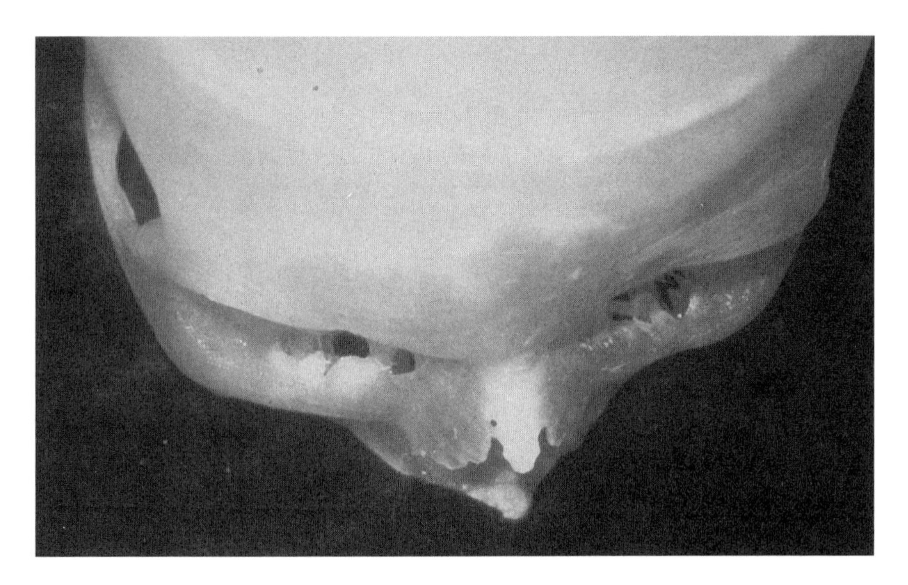

Figure 12-21. Planning for the hard tissue reconstruction for the Parry Romburg patient was done in Belgium, with an SL model.

Figure 12-22. A major advantage of presurgical planning with SL models is the ability to measure all necessary displacements in advance.

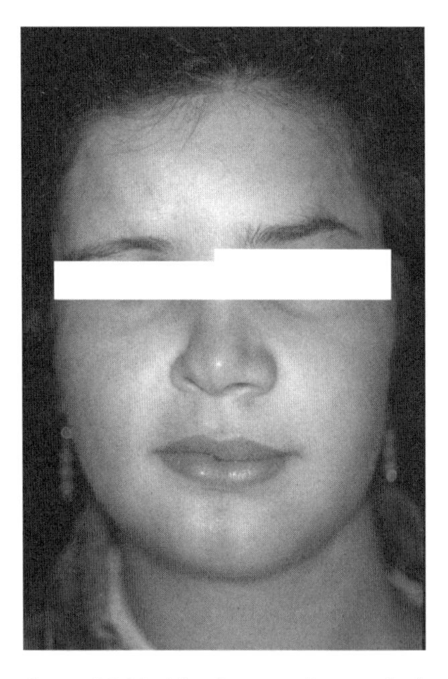

Figure 12-23. After the procedure in which the SL models were involved, the fronto-orbital region was perfectly symmetric.

tively. The soft tissue evaluations, involving the differences between the right and left sides of the face, was within the 5% deficiency rate typical of "normal people." However, there was still an obliquity at the third level and chin. The nasal tip deviation still remains to be restored. The fronto-orbital region is now perfectly symmetric, as shown in Figure 12-23.

Conclusions

The generation and use of medical models directly from CT or MRI scan data is a new and interesting application of RP&M. The recently developed CT-Modeller is a complete medical modelling system, allowing every RP&M user to interface with CT scanners and produce accurate models of internal bone structures. Furthermore, the medical models are extremely useful as a visualization tool for surgical rehearsal as well as in preparing implants. Currently, RP&M-generated medical models are mostly used in maxillo-facial, cranio-facial and oral surgery. However, they have also seen limited use in conjunction with spinal, hand, foot, and hip surgery as well.

Nonetheless, these medical applications are rather new. Surgical preplanning with accurate RP&M-generated medical models of internal structure only becomes truly interesting when surgeons regularly utilize them prior to opening the patient. The potential for saving time, saving expense, and improving the surgical result is evident.

References

1. Steinhauser, D., "Flow Analysis of Complex Configurations: Another Successful Application of RP&M Techniques," presented at the 1994 European

Stereolithography Users Group Meeting, Strasbourg, France, November 7, 1994.
2. Wolf, H., *Spannungsoptik*, Second Edition, Springer-Verlag, Berlin, 1976.
3. Dally, J. and Riley, W., *Experimental Stress Analysis*, McGraw-Hill, New York, 1978, pp. 339-454.
4. Steinchen, W., Kramer, B., and Kupfer, G., "Photoelasticity Cuts Part-Development Costs," *Photonics Spectra*, May 1994, pp. 157-162.
5. Delnegro, J. and Gargiulo, E., "Stereolithography for Photoelastic Stress Analysis," Proceedings of the Fourth International Conference on Rapid Prototyping, University of Dayton, Dayton, Ohio, June 14-17, 1993, pp. 7-14.
6. Steinchen, W., "Neue Werkstoffe für Spannungsoptische Untersuchungen," Proceedings of the 1991 European Stereolithography Users Group Meeting, Darmstadt, Germany, October 1991.
7. Kramer, B., "Kalibrierung der Spannungsoptischen Konstanten für die Fotopolymere XB 5081-1, XB 5143, XB 5134-1 and Somos 3100 des Stereolithografie-Verfahrens," Ph.D. Thesis, University of Kassel, 1992.
8. Cernosek, J., "Three-Dimensional Photoelasticity by Stress Freezing," *Journal of Experimental Mechanics*, Vol. 20, No. 12, December 1980, pp. 417-426.
9. Dornfeld, W., "Direct Dynamic Testing of Stereolithographic Models," presented at the International Gas Turbine and Aeroengine Congress and Exposition, The Hague, Netherlands, June 13-16, 1994.
10. Elliott, W., "Plastic Models for Dynamic Structural Analysis," Society of Automotive Engineers Transactions, Vol. 80, 1971, paper number 710262.
11. Langhaar, H., *Dimensional Analysis and Theory of Models*, John Wiley, New York, 1960, Chapter 6.
12. Slater, J. and Frank, N., *Introduction to Theoretical Physics*, McGraw-Hill, New York, 1933, Chapter XVI, pp. 176-179.
13. Redner, A., "Model Approach in Structural Design," American Society of Civil Engineers, National Structural Engineering Meeting, Cincinnati, Ohio, April 22-26, 1974, preprint number 2205.
14. Smith-Moritz, G., "Rapid Prototyping Materials," *Rapid Prototyping Report*, Volume 2, Number 11, November 1992.

CHAPTER 13

RP&M Service Bureaus

"He that will not apply new remedies must expect new evils."

Francis Bacon

Service bureaus have played an important part in developing the RP&M market. These entrepreneurial companies have been among the leaders in pursuing new applications for the various RP&M technologies. In addition, service bureaus have introduced many larger companies to RP&M through educational efforts and by building parts for these organizations effectively and economically.

The first RP&M service bureaus were formed in the late 1980s. As RP&M became an increasingly common method for creating new models, a need was established for an outside resource that would allow companies to benefit from RP&M without purchasing and operating the equipment in-house.

The first stereolithography machine, the SLA-1, was installed at Baxter Health Care in Illinois, in a division that acted as a service bureau for both in-company and outside clients. Other service bureaus appeared soon after, some as divisions of existing corporations, but more as start-up companies that bought RP&M equipment and then opened their doors for business.

In the early years, when the SL process was being developed and tested, many service bureaus established additional capabilities, such as CAD services and silicone RTV rubber moldmaking to supplement their businesses. After SL had become a fully accepted process, these additional capabilities strengthened the position of the bureaus, allowing them to offer a valuable mix of services.

13.1 How Service Bureaus Work

All currently commercial rapid prototyping processes can be obtained through service bureaus, including SL, SLS, SGC, LOM, and FDM. Most service bureaus specialize in one of these processes, although some offer two or more methods. Many also have alliances with other bureaus that allow them to contract out for additional services.

By Frost Prioleau, President, Plynetics Corporation, San Leandro, California.

Most RP&M service bureaus handle CAD modeling, and will transform blueprints, 2D IGES files, or 3D IGES data into solid CAD models in order to create the STL files required for the RP&M processes. Some service bureaus can also accept special formats such as computer axial tomography (CAT-SCAN) and magnetic resonance imaging (MRI) medical data. As CNC machining complements many of the RP&M processes, some service bureaus offer this capability as well.

In addition, many service bureaus offer secondary processes, including part finishing and painting, silicone rubber moldmaking, polyurethane part making, epoxy toolmaking for wax patterns, epoxy toolmaking for injected molded parts, and spray metal toolmaking for injection molded parts.

13.2 Reasons for Using Service Bureaus

There are a number of reasons why a company, government agency, or educational institution might choose to utilize a service bureau to handle their RP&M requirements. Today many organizations are adopting the credo, "unless we are going to be the best at something, we should contract it out," thereby defining their core competencies.

Expertise in Secondary Processes

To leverage the benefits of RP&M, the technology often must be coupled with secondary processes. Service bureaus provide added value by serving as advisors and consultants on the best process or combination of processes. Because service bureaus deal with a wide variety of end-use applications, budgetary requirements and time schedules, they often have expertise a single company may not possess in-house.

Multiple Technologies

Each of the RP&M technologies has its own specific strengths, and a company may have a wide variety of RP&M needs. For one application, a very tough part may be needed; for another, the need may be for extreme accuracy. A company can use multiple service bureaus to obtain the best technology for each application, rather than "putting all its eggs in one basket" by purchasing a specific RP&M technology.

Occasional Prototyping Requirements

The prototyping requirements of some companies are simply too small to justify purchasing the equipment. If a machine is used less than half the available time, it is probably not worth the investment to the company. Owning the machinery typically requires a continuing investment in the equipment, upgrades, consumables, and technical expertise. Other companies may have a high annual requirement for prototyping, but those requirements might occur in peaks and valleys. For example, a company might have an urgent need for prototyping five weeks out of every quarter. During that time, the company's requirements are so high that 10 machines are required. During the remainder of the year, the company may have very minimal requirements for prototypes. Using a service bureau, or multiple service bureaus, gives the company the flexibility to handle a highly variable work load.

Expense of the Equipment

Many companies needing RP&M are small operations that either do not have the capital equipment budget to purchase the equipment outright or the ongoing budget to support the continuing investment in the equipment, upgrades, consumables, and technical expertise described above. Some of the RP&M machines are very expensive to own and to operate.

For companies that do have the resources to purchase and maintain the equipment, it often makes sense to engage a service bureau first to test the technology as well as the results. By doing so, companies can ensure that the process being evaluated is the correct one for the specific applications most commonly required. In addition to testing the technical capabilities of RP&M processes, organizations considering the purchase of RP&M equipment often find it useful to test market the technologies with in-house users to gain a better understanding of their level of demand and technical requirements.

Rapidly Changing Technology

RP&M continues to evolve rapidly, as better, faster, less expensive, and more accurate processes are developed and refined. Many companies feel that at this stage in the development of RP&M, they would rather maintain the flexibility to use today's best technology, rather than being locked into what may later prove to be an outdated system.

Hazardous Materials

Some of the RP&M processes require the use of hazardous materials, such as the resins used for SL and SGC. These materials need to be handled and disposed of properly, and some companies have restrictions that relate to the handling of such materials.

13.3 The Mechanics of Using Service Bureaus

Most companies will come to a service bureau after an initial design of the basic geometry has been developed. This design does not have to be complete or even very detailed. In many cases, it is helpful to consult early in the design process. The company will generally fax a sketch of the parts or visit a service bureau to discuss the specific application in greater detail.

Identifying Part Application

The first step is to identify the application of the part(s) to be produced. For example, the part might be needed for an industrial, engineering, or manufacturing design review. Or the part might possibly be required for functional testing as a mold master or as a metal part. The application of the part will determine the specific requirements, such as the need for accuracy, a specific surface finish, mechanical properties, or thermal stability. Size, quantity, and cost are also factors that influence the RP&M approach.

For example, a part intended for an industrial design review must have a high-quality surface finish and appearance, but mechanical properties would not be as important. For functional testing, mechanical properties and thermal stability are important, but surface finish may not be critical. If the part is being used as a mold master, accuracy and surface finish are absolutely critical.

Once the most important requirements have been determined, the service bureau will be able to advise the company on the part-building process best suited for the particular application and geometry being considered, and the approach that also provides the most cost effective result for the required quantity.

For example, at Plynetics we find SL to be the best process for parts that include fine detail and applications in which parts will be used for mold masters. We find SLS to be best for functional applications, in which part toughness and/or thermal stability are required.

Part Quantities

RP&M is often the best method for producing small quantities of parts. For higher quantities, it may be most economical to use a combination of processes, creating a master through RP&M, and using a secondary process to create the balance. For example, the production process may entail producing a master through SL, and subsequently using this master to create a rubber mold to cast functional polyurethane parts. The expertise to determine the best production process for a particular project is one of the most important value-added capabilities provided by service bureaus.

Transferring the Part

Once a service bureau has been selected, the part design needs to be transferred to their site. This design can be as basic as a rough sketch or as complete as a solid CAD data base. Most service bureaus can accept a range of data in almost any format. The more complete the data, however, the lower the cost to the company, and the faster the turnaround. CAD data can be transferred electronically via a direct modem line or via the Internet. Alternatively, the data can be sent on a tape or computer diskette.

After the service bureau receives or creates the CAD data, initial parts from the RP&M system are often ready within 1 to 7 days. Longer lead time is generally required for larger parts, or parts requiring a high degree of postfinishing. If a secondary process is involved, the service bureau will generally coordinate all the required steps. Consequently, the operator at the service bureau makes several decisions regarding building the part, including build orientation, layer thickness, and the amount and type of finishing. All these items affect the final quality of the part.

Service bureaus often encourage customers to specify critical dimensions to be inspected or to supply mating components against which the parts can be checked. Once the parts have been completed, they are generally sent to the customer via overnight courier.

13.4 Cost and Time Requirements

The costs of using service bureaus are primarily based on the amount of time required to complete the job. In some cases, material costs also become a significant factor, but typically these costs are absorbed into machine rates. The factors that combine to determine what a project will cost for completion at a service bureau are listed below.

CAD Time

If completed CAD data or STL files are supplied to the service bureau, then data development and translation cost is typically zero. However, if blueprints, 2-dimensional CAD data, or 3-dimensional wireframe CAD data are supplied, the service bureau must spend time to create the 3-dimensional solid or the fully surfaced CAD database from which the STL file will be extracted. The time required to accomplish this task depends on the complexity of the design, and can take as little as 1 hour to as much as 40 hours or more. Some service bureaus also charge separately for the engineering time required to set up a part to be built on an RP&M machine. This time might include orienting, positioning, and scaling the STL file, generating support structures, and setting machine building parameters.

Build Time

The amount of time required to produce a part (or group of parts) on the RP&M machine is typically the main cost driver for a project. Build time is determined primarily by the size of the part. Depending upon the RP&M system being used, part volume, part height, bounding box volume, or a combination of these attributes typically determine build time. In extreme cases, very intricate detail and complex support structures also can add significantly to build time. However, RP&M processes differ from conventional subtractive processes in that complexity is not typically a primary factor in determining build time.

Finishing Time

For some applications, the RP&M models can be used right off the machine, with little or no finishing required. Other applications, such as mold masters or appearance models, require hand finishing in order to achieve the necessary surface finish and/or appearance. The estimated finishing time is also driven by size and complexity, as well as the level of finish required.

Secondary Process Costs

Finally, project cost may also be affected by additional processes required to bring the project to completion. Such additional processes may include machining certain features to achieve tight tolerances, making silicone rubber molds and casting polyurethane prototypes, sending patterns out for investment casting, or one of many other processes.

13.5 Factors in Choosing a Service Bureau

Appropriate Process/Technology

Each RP&M process has its own set of strengths and weaknesses when compared to other prototyping or manufacturing processes. It is important for users of RP&M services to understand the requirements of their applications and to compare these requirements to the relative capabilities of the available RP&M technologies.

Coordinating Secondary Processes

The best prototyping solution is often achieved by combining an RP&M process with complementary secondary processes such as urethane casting or investment casting. In such cases, close coordination with secondary processes is often vital to the success of the overall project. The service bureau should have the ability to handle secondary processes either in-house or through closely managed relationships with outside suppliers such as foundries, spray metal tool shops, platers, etc. Users of RP&M services can help ensure a successful project by choosing a service bureau with significant experience in the specific secondary process to be employed.

Quality

The quality of parts built with RP&M processes is affected by many factors, including layer thickness, build orientation, build parameters, equipment condition/calibration, materials used, finishing techniques, and more. A detailed discussion of these factors and how they affect part quality is beyond the scope of this chapter. Indeed, this information is beyond the scope of knowledge that one must attain to be a successful user of RP&M service bureaus.

Users of RP&M service bureaus depend upon a bureau's expertise in the technology or technologies offered. In order to ensure that the parts are of acceptable quality, the user should determine that the service bureau:

- understands the capabilities and limitations of the RP&M process being proposed;
- has the ability, experience, and expertise to get the most out of the RP&M process;
- understands the requirements of the application; and

- is committed to delivering parts that meet the requirements of the application.

Quality can mean different things to different users of RP&M technologies. Only by understanding the application for which RP&M is to be used can a service bureau reliably deliver quality parts.

CAD Compatibility

The service bureau selected should have the ability to easily receive and make the best use of any available CAD data. If the user has the ability to output STL files, then almost all service bureaus will be able to directly read this data and build parts with little engineering effort. However, if the user cannot output STL files, it is important to select an RP&M service bureau that can most quickly, reliably and economically generate STL files from the data that is available.

Many RP&M service bureaus offer in-house CAD modeling services so that they can accept designs in the form of blueprints, 2-dimensional CAD files, or 3-dimensional wireframe CAD files.

Competitive Pricing and Turnaround Time

Pricing and turnaround time can vary among RP&M service bureaus based on the technologies used, capacity available, and methods for determining costs. Because many RP&M projects are on tight schedules where time is not often allocated for receiving multiple bids, many users of RP&M service bureaus choose to establish close relationships with one or two service bureaus to do the bulk of their RP&M work. In these relationships, standard rates and turnaround times can be established early so that time is not wasted in a paperwork-quoting cycle when parts are urgently needed.

References

It is always helpful to speak with other customers of a service bureau to determine their level of satisfaction with the service provided. Most service bureaus will provide references upon request.

13.6 Case Study—Farallon Computing

One example of a company using RP&M through a service bureau with great success is Farallon Computing, a leading supplier of innovative, plug-and-play networking products for personal computers. When Farallon designed its innovative EtherWave™ adapter, a primary concern was how to speed the product development cycle and deliver the product to market as quickly as possible. Working with our company, Plynetics, Inc., an RP&M service bureau based in San Leandro, California, Farallon used stereolithography to test and refine the design of the EtherWave without losing valuable production time. This time- and cost-efficient production method allowed Farallon to succeed in its goal of creating a product that combines a contemporary eye-catching design with top-of-the-line innovative networking technology.

The product has met with great success. The Farallon EtherWave won *MacUser* awards for Hardware Product of the Year and Best New Network Hardware for 1993. It also won five stars from *MacWorld*, five diamonds from *MacWeek*, four and a half mice from *MacUser*, and a 4.0 rating and Product of the Week designation from *PC Week*. The product is on the "Best Seller" list for both Merisel and Ingram/Micro.

Combining Form With Function

Farallon's EtherWave adapter offers a number of distinct advantages over its competitors. The patented design simplifies the electrical connection between computers. The adapter uses standard unshielded twisted pair cable, which is less expensive and more flexible than coaxial wire. The fact that the adapter can be "daisy-chained" makes it easier to use for computer networking.

To ensure that the functional advantages of the product were enhanced by its design, Farallon hired Lunar Design of Palo Alto, California, to develop a unique look for the product. The innovative stingray design was selected for both its sleek looks and its functionality. A custom color, a muted gray with a green undertone, was developed to further distinguish the Farallon product from its competitors.

From Concept to Reality

Once the design of the part had been completed, Plynetics was sent a tape cassette containing the Pro/ENGINEER CAD data, and was then able to use this data to interface directly with its stereolithography process. No additional paperwork or documentation was required.

Plynetics used the CAD data to produce a set of SL parts in one week. The initial SL parts were test fit to various components, and changes were made to the CAD data in order to incorporate slightly larger electrical components. Using the revised data, a second set of SL parts was generated and subsequently used to create silicone RTV molds. These molds then produced 60 sets of polyurethane parts, the first fabricated just two and a half weeks after Plynetics first received the Pro/ENGINEER files.

These initial polyurethane parts were used for marketing applications, including publicity photos and trade show demonstrations. In addition, 50 parts were sent to beta sites for testing. Simultaneously, Farallon conducted mechanical tests on the part.

Louis Ornelas, design engineer for Farallon, cites speed of production and the ability to handle many processes concurrently as the key benefits of using stereolithography. "The fact that we were able to create beta units and do a mechanical fit up front was very important," said Ornelas. "We started the process of final tooling while we were running tests on the urethane parts, and we were able to incorporate important refinements into the tooling."

According to Ornelas, Farallon's typical production cycle involved generating mechanical drawings, and then going straight to tooling. If there were errors in the production tool, more time and money would have to be spent to remedy the problem and rework the tool. "By producing models first, we reduce our risk," says Ornelas. "It's now Farallon's policy to model everything before we go into production."

An Efficient Solution

By employing stereolithography to create parts for use in the product development and testing stage, Farallon was able to take advantage of concurrent engineering, reducing the cost and time required to bring the EtherWave Adapter to market. The result is a unique, award-winning product that reflects the benefits of extensive testing and improvements.

13.7 Directory of Worldwide RP&M Service Bureaus

(Courtesy of Denton & Company, Dearborn Heights, Michigan)

UNITED STATES

ALABAMA
Fastec
Terry Vance
1020 Winchester Road
Huntsville, AL 35811
(205) 852-0500
SL

ARIZONA
Adams Mold and Engineering
Joe Adams
8641 N. 79th Avenue
Peoria, AZ 85345
(602) 486-2019
FDM

Phoenix Analysis & Design Technology, Inc.
Rey Chu
4500 South Lakeshore Drive
Suite 348
Tempe, AZ 85282
(602) 820-6449
SL

CALIFORNIA
3D Cam, Inc.
Gary Vassighi
9139 Lurline Avenue
Chatsworth, CA 91311
(818) 773-8777
SL

3D Systems, Inc.
Technology Center
Max Gerdts
26081 Avenue Hall
Valencia, CA 91355
(805) 295-5600
SL

Arrk Creative Network
Kevin Robertson
8880 Rehco Road
San Diego, CA 92121
(619) 552-1587
SL

Helisys, Inc.
Michael Feygin
2750 Oregon Court
Building M-10
Torrance, CA 90503
(310) 782-1949
LOM

Hughes Aircraft Company
Matt Wilson
2000 E. El Segundo Blvd.
El Segundo, CA 90245
(310) 616-8623
SL

Laser-Tech Engineering
Randall Wood
971 Via Rodeo
Placentia, CA 92670
(714) 577-3940
SL

Marbeth Industries, Inc.
Ralph Young
5217 Industry Avenue
Pico Rivera, CA 90660
(310) 948-2618
FDM

Plynetics
Frost Prioleau
627 McCormick Street
San Leandro, CA 94577
(510) 613-8300
SL/SLS

Precision Measurement Laboratories
Robert Chmelka
201 W. Beach Avenue
Inglewood, CA 90302
(310) 671-4345
LOM

Pure Fluid Magic
Norma Jean Kardos
Bill Evans
1800 De La Cruz Blvd.
Santa Clara, CA 95050
(408) 748-1222
SL

Rapid ProtoCad
M.G. Risenberg
1095 Mountain View Blvd.
Suite G
Walnut Creek, CA 94596
(510) 938-0499
LOM

Satellite Models
Kelly Hand
950 Rengstorff, Suite C
Mountain View, CA 94043
(415) 903-3540
SL

Scicon Technologies
Scott Turner
26027 Huntington Lane, Unit D
Valencia, CA 91355
(805) 295-8630
SL

Solid Concepts
Joe Allison
28231 Avenue Crocker, Unit 10
Valencia, CA 91355
(805) 257-9300
SL

Tri-Tech Precision, Inc.
Ernie Husted
1577 N. Harmony Circle
Anaheim, CA 92807-6003
(714) 970-1363
FDM

COLORADO

Commercial Pattern, Inc.
Bill Shaver
800 E. 73rd Avenue, Unit #9
Denver, CO 80229
(303) 288-2373
SL

Protogenic, Inc.
Steve Stewart
2820 Wilderness Place, D
Boulder, CO 80301
(303) 442-4604
SL/SGC

Rapid Prototyping Corp.
Rod Ward
Mark Kelley
1840 Industrial Circle, Suite 200
Longmont, CO 80501
(303) 684-0088
SL

DELAWARE

DuPont Somos Solid Imaging Material Group
Daniel J. Mickish
2 Penn's Way, Suite 401
New Castle, DE 14720
(302) 328-5435
SL

FLORIDA

Cutler Technologies, Ltd.
Jim Harrison
19501 NE 10th Avenue, Suite 300
North Miami Beach, FL 33179
(305) 653-9098
SL

Magna Manufacturing
Greg Medla
85 Hill Avenue
P.O. Box 279
Fort Walton Beach, FL 32549-0279
(904) 243-1112
SL

GEORGIA

Compression Engineering
Frank Bowker
2475 Meadowbrook Parkway
Suite H
Duluth, GA 30136
(404) 495-8085
SLS

Lockheed Aeronautical Systems Company
Mark Skeehan
86 S. Cobb Drive
Marietta, GA 30063
(404) 494-1246
SL

ILLINOIS

Arrow Pattern & Foundry
9725 S. Industrial Drive
Bridgeview, IL 60455
(708) 598-0300
LOM

Baxter Healthcare
Terry Kreplin
Route 120 & Wilson Road, RLP-30
Round Lake, IL 60073-0490
(708) 270-4067
SL/SGC

Clinkenbeard and Associates, Inc.
Ronald Gustafson
577 Graple Street
Rockford, IL 61109-2001
(815) 226-0291
LOM

Designcraft
Judy Jensen
70 S. Lively Blvd.
Elk Grove Village, IL 60007
(708) 593-2640
SL

Digital Prototype Services
Jiten Shah
1950 North Washington
Naperville, IL 60566
(708) 505-0750
LOM

Laser Modeling, Inc.
Frank McRae
1143 Tower Rd.
Schaumburg, IL 60173
(708) 884-0662
SL

Micro-Cut Engineering
Chuck Benning
1534 Burgundy Parkway
Streamwood, IL 60107
(708) 837-7001
SL

Proto Tech Engineering, Inc.
Kevin Kennedy
663 Executive Drive
Willowbrook, IL 60521
(708) 920-0370
SL

Prototype Express, Inc.
Dave Flynn
Thomas Mueller
2301 Hammond Drive
Schaumburg, IL 60173
(708) 925-9900
SL

Rock Valley College
Stan McCord
3301 N. Mulford Road
Rockford, IL 61111
(815) 654-4345
SL

Uptrend Models, Inc.
Tony Fuderer
221 North Hemlock
Wood Dale, IL 60191
(708) 766-6070
SL

INDIANA

Compression Engineering
Darrell Pufahl
7752 Moller Road
Indianapolis, IN 46268
(317) 228-2200
SL/SLS/FDM

Metro Plastics Technologies
Jim Warthon
9175 E. 146th Street
P.O. Box 1208
Nobleville, IN 46060
(317) 776-0860
SL

IOWA

Eldora Plastics
Jamie Robertson
1814 21st Street
P.O. Box 127
Eldora, IA 50627
(515) 858-2634
LOM

KANSAS

Pittsburg State University
David Lomsbek
School of Technology and
Applied Science
103B Whitesitt Hall
Pittsburg, KS 66762
(316) 235-4830
LOM/FDM

Prototype Specialists, Inc.
Lyle Britt
219 S. Water
Derby, KS 67037
(316) 788-8975
SL

KENTUCKY

Accelerated Technologies, Inc.
Tim Perkins
1780 Anderson Blvd.
Hebron, KY 41048
(606) 586-9300
SL/FDM

University of Kentucky
R.J. Robinson
Center for Robotics & Manufacturing Systems
Lexington, KY 40506
(606) 257-6262
SL

MASSACHUSETTS

Mor-Tech
Charles Wood
P.O. Box 215
126 Industrial Drive
East Longmeadow, MA 01028
(413) 525-6321
SLS

Santin Engineering, Inc.
Drew Santin
One Lakeland Park Drive
West Peabody, MA 01960
(508) 535-5511
SL

MICHIGAN

3-Dimensional Services
Alan Peterson
2547 Product Drive
Rochester Hills, MI 48309
(810) 852-1333
SL/LOM

Auburn Engineering
Rob Schweikhart
2917 Waterview Drive
Rochester Hills, MI 48309
(810) 852-6250
SL

Cascade Engineering
Mark Bonnema
5141 36th St., SE
Grand Rapids, MI 49512
(616) 975-4800
SL

D & C Molding
Frank McDonald
13455 Stamford Court
Livonia, MI 48150
(313) 261-2300
SLS

Eagle Design & Technology, Inc.
Bruce Okkema
2437 84th Avenue
Zeeland, MI 49464
(616) 748-1022
SL

Eurotech Design, Inc.
Kevin T. Kerrigan
32751 Edward
Madison Heights, MI 48071
(810) 588-0000
SGC/LOM

Invenio Design and Development
Mert Wreford
5715 East 13 Mile Road
Warren, MI 48092
(810) 979-0890
LOM

Laserform, Inc.
David Tait
Al Dewitt
1124 Centre Road
Auburn Hills, MI 48326
(810) 373-4400
SL

Line Precision, Inc.
Jerry Miller
32371 West Eight Mile Road
Livonia, MI 48152
(313) 474-5280
LOM

Mack Industries, Inc.
Merlin Warner
2696 American Drive
Troy, MI 48083-4618
(810) 588-1742
SL/LOM

McKenna GST, Inc.
David P. Haydon
2904 Bond Street
Rochester Hills, MI 48309
(810) 852-4930
LOM

Nationwide Design, Inc.
George Ackerman
6605 Burroughs
Sterling Heights, MI 48314
(810) 254-5493
LOM

PEC Technical Center Midwest
Mike Jezdimir
1811 Vanderbilt Road
Kalamazoo, MI 49002
(616) 327-3023
SL

Pinnacle Technologies
Robert Hamood
16153 Common Road
Roseville, MI 48066
(810) 779-8807
SL

Select Manufacturing Services
Ross Gates
1210 E. Barrey Avenue
Muskegon, MI 49444
(616) 733-1133
SL

Stature Prototyping
Ernie Guinn
20201 Hoover Road
Detroit, MI 48205
(313) 839-8245
SLS/SGC

MINNESOTA

Alliant Technical Systems
Gary Klingelhutz
600 Second Street NE
Hopkins, MN 55436
(612) 931-6760
SL

Craft Pattern & Model
Dennis Hagen
1410 County Road 90
Maple Plain, MN 55359
(612) 479-1969
SL/LOM

ENTEC Services, Inc.
Brad Fox
2500 Niagara Lane
Minneapolis, MN 55447
(612) 884-8500
SL

General Pattern Company
Bob Grainger
3075 84th Lane, NE
Blaine, MN 55449
(612) 780-3518
SL/SGC

Proto Tech Services
Tim Thellin
7145 Shady Oak Road
Eden Prairie, MN 55344
(612) 943-1513
FDM

MISSOURI

Solutions in 3D
Chris Franks
13705 Shoreline Court
St. Louis, MO 63045
(314) 291-2224
SL

NEW JERSEY

Ethicon, Inc.
Michael Bregen
Route 22
Somerville, NJ 08876
(908) 218-2247
SL

Laser Prototypes, Inc.
Todd Grimm
3155 Route 10
Denville, NJ 07834
(201) 361-7666
SL

Merck & Company, Inc.
Arthur Lifshey
P.O. Box 100, WSIE-55
Whitehouse Station, NJ 08889
(908) 423-5274
SL

SiCAM Corporation
Peter Sayki
One Harvard Way, Suite One
Somerville, NJ 08876
(908) 685-2211
SL

NEW YORK

Cadview Technologies
Corporation
100 Doxee Drive
Freeport, NY 11520
(516) 223-1575
LOM

Design Prototyping Technologies
8 Adler Drive
East Syracuse, NY 13057
(315) 434-1869
SL

Eastman Kodak Company
Douglas Van Putte
Manufacturing Research &
Engineering
Kodak Park Building 604
Rochester, NY 14652-4102
(716) 477-1492
SL/SLS

Solid Imaging Ltd.
Kim Phillips
731 Park Avenue
Rochester, NY 14607
(716) 473-0990
SL/SLS/LOM

NORTH CAROLINA

Accent on Design
Greg Thune
4912 Wallace Neel Road
Charlotte, NC 28208
(704) 398-0887
SL/LOM

Joel Wittkamp Design
Joel Wittkamp
1101-E Aviation Parkway
Morrisville, NC 27560
(919) 469-9908
SL

OHIO

Ashland Chemical, Inc.
Kelley Kerns
5200 Blazer Parkway
Columbus, OH 93216
(614) 889-3698

Astro Model Development
Mark Horner
35280 Lakeland Blvd.
Eastlake, OH 44095
(216) 946-8171
SL/LOM

Bastech Engineering Services
Ben Staub
3500 Stop Eight Road
Dayton, OH 45414
(513) 890-9292
SL

Consolidated Technologies
Robert Burgess
425 Fairway Drive
Springboro, OH 45066
(513) 748-2052

Laser Reproductions
Paul Bordner
6400 E. Main St.
Reynoldsburg, OH 43068
(614) 759-2200
SL

LM Industries
Frank Leyshon
620 Wall Avenue, Apt. B
Cambridge, OH 43725
(614) 432-2969
FDM

Morris Technologies, Inc.
Greg Morris
4480 W. Lake Forrest Drive
Suite 308
Cincinnati, OH 45242
(513) 733-1611
SL

Toledo Molding & Die
Mel Harbaugh
4 East Lasskey Road
P.O. Box 6760
Toledo, OH 43612
(419) 476-0581
SGC

PENNSYLVANIA

ALCOA-Aluminum Company of America
Paul Fussel
Alcoa Technical Center
100 Technical Drive
Alcoa Center, PA 15069-0001
(412) 337-2721
SL

ProtoCAM
Ray C. Biery
3848 Cherryville Road
Northampton, PA 18067
(610) 261-9010
SL

The Jade Corporation
Michael Dougherty
1120 Industrial Highway
Southampton, PA 18966
(215) 947-3333
SLS

TENNESSEE

Certified Industries
Lincoln Highway - 412 West
123 Certified Drive
Hohenwald, TN 38462
(615) 796-3202
LOM

TEXAS

Accelerated Technologies, Inc.
Mike Durham
12919 Dessau Road
Austin, TX 78754
(512) 990-7199
SLS

PI Components Corporation
Bob Behrend
350 Loop 290 South
Brenham, TX 77833
(409) 830-5400
SL

UTAH

ENTEC Services, Inc. West
Ken Miller
2391 South 1560 West
Woods Cross, UT 84087
(801) 296-2500
SL

Lone Peak Engineering, Inc.
Curtis Griffin
12660 S. Fort Street (950E)
Draper, UT 84020
(801) 553-1732
LOM

Rapid Prototyping, Inc.
Mark Jones
6146 S. 350 West #B
Salt Lake City, UT 84107
(801) 268-4206
FDM

Solid Design & Analysis
Greg Goin
Alan Seamons
165 Wright Brothers Drive
Salt Lake City, UT 84116
(801) 328-8220
SL

VIRGINIA

Hub Pattern Corporation
John Cloeter
2113 Salem Avenue SW
Roanoke, VA 34016
(703) 342-3505
FDM

WASHINGTON

Triquest Precision Plastics
Vance Prather
3000 Lewis & Clark Highway
Vancouver, WA 98661
(360) 690-0142
SL

WISCONSIN

J.P. Pattern
Pat Jaquish
5038 North 125th Street
Butler, WI 53007
(414) 781-2040
FDM

Light Sculpting, Inc.
Efreum Fudim
4465 N. Oakland Avenue
Milwaukee, WI 53211
(414) 964-5737

Manitowoc Prototypes
Jim Vander Linden
4211 Clipper Drive
Manitowoc, WI 54220
(414) 682-8825
SL

Phillips Plastics
Jim Klos
1201 Haney Road
Hudson, WI 54016
(715) 386-4320
SL

AUSTRALIA

Queensland Manufacturing Institute
Scott Loose
P.O. Box 4012
Eight Mile Plains
Australia Q 4113
61-72-34-0459
SL

South Australian Centre for Manufacturing
Jeff Groves
853 Port Road
Wood Ville, South Australia
61-8-300-1500
SLS/LOM

AUSTRIA

Laserform H. Prihoda u. Mitges. Prihoda
Missindorfstrasse 21
Vienna, Austria A-1140
43-1-9854545
SL

Plot
Weisshappel
Salztorgasse 8
Vienna, Austria A-1010
43-1-5330807
SL

BELGIUM

CRIF
T. Dormal
Parc Scientifique Bois Saint Jean
Hil2
Seraing, Belgium B-4102
32-41-618700
SL/FDM

Materialise, N.V.
Bart Swaelens
Wilfried Vancraen
Kapeldreef 60
Heverlee, Belgium B-3001
32-16-270364
SL

BRAZIL

Digicon S.A.
Alexandre Censi
Distrito Industrial C.P. 131
CEP 94000-970
Gravatai, RS, Brazil
(051) 489-1333
SL

CANADA

ARC Digital Prototyping
George Thorpe
190, 3553-31st Street NW
Calgary, Alberta
Canada T2L 2K7
(403) 297-7546
SGC

Factotum Plastics Technologies, Inc.
Andy Juhasz
5800 Amber Dr., Suite 200
Mississauga, Ontario
Canada L4W 4J4
(416) 602-7986
SL

ENGLAND

Advanced Technology Center
Warwick Manufacturing Group
David Wimpenny
Lee Styger
University of Warwick
Coventry, England CV4 7AL
44-020-352-3687
SL/LOM

ARRK Europe Limited
Peter Rawson
Unit 11, Commercial Way
Abbey Road
London, England NW10 7XF
44-181-961-6366
SL

AMSYS
D. Griffith
The Industry Centre
Hylton Riverside West
Sunderland, England SR5 3XB
44-191-515-3333
SLS

Defence Research Agency
Paul Dummer
WX9f StereoLithography
407 Building
DRA Pyestock
Farnborough, England GU14 OLS
44-1252-374446/374409
SL

Formation Engineering Services
Tim Plunkett
Spinnaker House
Hempsted Gloucester, England
GL2 6JA
44-1452-380336
SL

IMI Rapid Prototyping
Tony Macis
Holdford Road, Witton
Birmingham, England B6 7ES
44-121-344-4429
SL

Laser Integrated Prototypes
142-144 Station Road, Mach
Cambridgeshire, England PE15 8NH
44-1354-50789
SL/LOM

Laser Line Ltd.
Andrew May
Beaumont Close, Banbury
Oxon, England OX16 7TQ
44-295-267755
FDM

Quo-Tec Limited
Norman Waterman
The Dray House, School Lane
Amersham Bucks, England HP7 0ET
049-443-2277

Rolls-Royce PLC
Richard Rogers
CRDF
P.O. Box 3
Filton, Bristol
England BS12 7QE
44-117-979-6199
SL

Rover Group
Graham Tromans
Rover Group Ltd., Bldg. 41
Fletchamsted Highway
Canley, Coventry
England CV4 9DB
44-1203-875726
SL

Styles Rapid Prototyping
Gordon Styles
Unit BT50/19A Dukesway
Teesside Industrial Estate
Thornaby Cleveland
England TS17 9LT
44-1642-769-930
SL

The University of Nottingham
Dept. of Manufacturing Engineering
Dr. Philip Dickens
University Park
Nottingham, England NG7 2RD
44-115-951-4063
SL

Umak Limited
Simon Graham
BSA Business Park
Armoury Road
Birmingham, England BI1 2RQ
44-21-766-8844
LOM

FINLAND

Electrolux Rapid Development
Olli Nyrhila
Aholanite 17, FIN-21290
Rusko, Finland
358-21-819-600
SL

Oulu Institute of Technology
Esa Niinisalo
Kofkantie 1
Oulu. Finland FIN-900250
358-81-537-0481
SL

FRANCE

Ateliers Cini, S.A.
Yannick Seeleuthner
107-109 Boulevard Tolstoi
Tomblaine, France F-54510
33-83-18-13-13
SL

Bel 3D
Eric Gross
Parc Technologique, 9000
Belfort, France
33-84-21-39-93
SL

Centre de Transfert de Technologie du Mans
Vincent Reymond
J. Grevin
Technopole Universite
20 Rue Thales de Milet
72000 Le Mans, France
33-43-39-46-23
SGC

Cotraitance Mediterranee Aeronautique
Francois Nonnenmacher
Chemin de la Roubine
Mandeliueu, France F-06210
33-93-90-36-36
SL

Laser 3D
Claude Medard
6, Allee Pelletier-Doisy
F. 54603 Villers-Les-Nancy, France
338-361-4476
SL

Laservision Systems
Christophe Massan
125 rue du Vieux Pont de Sevres
Boulonge Billancourt, France F-92100
33-47-61-01-90
SL

Modelage Mecanique de Montreuil
Daniel Defouloy
94 bis rue Marceau
Montreuil, France F-93100
33-48-58-70-00
SL

Multistation S.A.
Yannick Loisance
BP 15-35801 Dinard Cedex
France
33-99-46-11-11
FDM

Parangon, S.A.
BP 39-Z.A. de Langelin, Edern
Briec-de-l'Odet, France 29510
33-98-59-3815
LOM

Plastida
Herve de Borde
Usine de la Riviere
Dange St. Romain
France F-860000
33-49-86-41-42
SL

GERMANY

3D-Schilling
Schilling
Frankenhauser Strasse 64a
Sondershausen, Germany 99706
49-3632-652250
SL

Ascam
Michael Junghanb
73614 Schorndorf
Wiensenstrasse, 33, Germany
49-71-81-211
FDM

Bavarian Laser Centre
Dr. Peter Hoffmann
Heberstrasse, 2
Erlangen, Germany 91058
49-9131-858342
SLS/LOM

Bertrandt Ingenieurburo GmbH
Bichler
Reichenback Strasse 26
Stuttgart, Germany 70372
49-711-550090
SL

Borgware GmbH
Eggert
Hauptstrasse 8
Haigerloch, Germany 72401
49-7474-6980
SL

Faust Sonderbearbeitungen, GmbH
Herbert Faust
Heilbronnerstrasse 6
Stuttgart-Echterdingen
Germany D-7022
49-711-796021
SL

FKM GmbH
Frank Henckel
Breidenstein-Goldbergstrasse
Biedenkopf, Germany 35216
496-461-8405
SLS

Fockele & Schwarze
Stereolithographictechnik
Matthias Fockele
Dieter Schwarze
Alter Kirchweg 34
Borchen-Alfen
Germany D-33178
49-251-39-19-25
SL

Fraunhofer Institut
Matthias Greul
Lesumer Heerstrasse 36
W-2820 Bremen 77
Germany
49-241-63830
FDM

Grunewald & Partner:
Rapid Prototyping
Universitatsstrasse 142
Bochum, Germany 44799
49-234-97-06-50
SL

Invenio Design & Development
Markstrasse 2-4
65428 Russelsheim, Germany
49-6142-899-210
LOM

Jos. Schneider Feinwerktechnik
GmbH & Co. KG
Mr. Wynands
Enderstrasse 94
01277 Dresden, Germany
49-351-2329-250
SGC

Kegelmann Technik GmbH
Stephan Kegelmann
Burgermeister-Mahr-Strasse 6
Obertshausen, Germany 63179
49-6104-709905
SL

Laser Bearbeitongs-und
Beartongszeutrum (LBBZ-NRW)
Wolfgang Mathar
Hans-Dieter Plum
Steinbachstrasse 15
Aachen, Germany 52704
49-241-8906-458
SL/SLS

Metafot GmbH
Neumann
Harkotstrasse 3
Velbert, Germany 42551
49-2051-20010
SL

Misslbeck Modell-und Formenbau
GmbH
Lehmeyer
Hebbelstrasse 65
Ingolstadt, Germany 85055
49-841-95300
SL

Modell Technik Rapid
Prototyping GmbH
Volker Kuehne
Herrenmuehlenstrasse 32
Eisenach, Germany D-99817
49-3691-29050
SL

Mucheyer Engineering
Norbert Mucheyer
Waxensteinstrasse 20
Schongau, Germany 86956
49-8861-7966
SL

M. Weihbrecht GmbH
Weihbrecht
Frankenstrasse 1
Wolpertshausen, Germany 74549
49-7904-1081
SL

Robert Hofmann GmbH
Hofmann
An der Zeil 6
Lichtenfels, Germany 96215
49-9571-76650
SL

Schneider Prototyping GmbH
Chr. Kappler
Postfach 24 63, Ringstrasse 132
Bad Kreuznach, Germany 55543
49-671-67726
SGC

SLM-Modellbau GmbH
Neugebauer
Lange Strasse 20
Oebisfelde, Germany 39646
49-39002-3887
SL

Stemo-Tec Design GmbH
Sushardt
Europaplatz 14
Castrop-Rauxel, Germany 44575
49-2305-33003
SL

HONG KONG

Asia InfoSciences Corporation,
Ltd.
Louis Lui
Room 4A02, 4/F, 78 Tat Chee Ave.
HKPC Building
Kowloon, Hong Kong
852-788-5083
SL

ISRAEL

Conceptland, Ltd.
Friedman
852
Kfar Shmariahu
Israel 46910
972-9-580388
SL

ITALY

EOS Image SRL
M. Orio
Via Breve 22
25086 Rezzato (BS), Italy
39-30-259-2570
SGC

Stelit SRL
Ignazio Pomini
Via Degasperi, 16/3
Trento, Italy I-38100
39-461-93-02-04
SL

Technimold SRL
Gianfranco Galter
Via Greto di Cornigliano 6R
Genova, Italy 16152
39-10-650-8640
FDM

Tekno Line, Italia
Stefane Grigiante
Via Canelli 54
Torino, Italy 10127
39-11-66-38-865
SL/SGC/LOM/FDM

JAPAN

D-MEC Ltd.
Eijiro Tegami
2-11-24 Tsukiji, Chuo-Ku
Tokyo, Japan
81-355-65-66-61
SL

Sado Giken
Y. Sawatari
Futasuya, 1-273, Kitamoto-shi
Saitama, Japan
81-485-92-50-80
SL

Tokuda Kogyo
H. Kaneko
Shimokirihimegaoka 1-54
Kani-shi
Gifu, Japan
81-574-62-51-21
SL

INCS
Sinjiro Yamata
1-1544-10 Kanjin Minami
Funabashi-shi, Chiba
Japan 273
81-474-31-33-65
SL

Nissin-DTM Company Ltd.
Sheichi Terashima
Kyoto Research Park SCB3 1,
Chuodoji Awata-cho
Shimogyo-ku, Kyoto
Japan 600
81-753-15-98-00
SLS

Nakashima Propeller
M. Miyake
Jodokitakata 688-1
Okayama-shi, Okayama
Japan 700-91
81-862-79-51-11
SL

Nagai
T. Terao
Wakaehigashi-machi 3-2-18
Higashiosaka-shi, Osaka
Japan 578
81-672-18-85-1
SL

Marubeni Hytech Corporation
Toru Watanabe
4-20-22 Koishikawa
Bunky-ku, Tokyo
Japan 112
813-38174931
FDM

KOREA

Komotech, Inc.
I. Kim
423-5, Togok-Dong
Kangnam-ku, Seoul
Korea
82-257-556-83
SL

NETHERLANDS

Modelling and Imaging Center, Nederland
E. Advocatt
L. Leidsedwarsstrasse 210
1017 NR
Amsterdam, Netherlands
31-20-420-79-23
SGC

Philips Components
B. Slaats
Building S.A.N.P.
P.O. Box 218, 5600 M.B.
Eindhoven, Netherlands
31-40-73-61-01
SL

NORTHERN IRELAND

Laser Prototypes Europe Ltd.
Tom Walls
Manley House, Cargan Crescent
Duncrue Road
Belfast, Northern Ireland
44-232-77-98-77
SL

The Automotive Design Centre
Dermot Sterne
Northern Ireland Technology Centre
Cloreen Park, Malone Road
Belfast, Northern Ireland
44-232-33-54-24
SL

NORWAY

Sintef Production Engineering
Nils-Aksel Ruud
Richard Birkelandsvei 2b
7034 Trondheim, Norway
47-73-597125
SGC

SINGAPORE

Accelerated Technologies, Singapore Pte Ltd.
Michael Saram
14 Science Park Drive
#2-04 Maxwell Singapore Science Park
Singapore 0522
65-776-8322
SL

Innomation Systems & Technology
Gregory Lee
Block 3014, Ubi Road #03-282
Singapore, Singapore 1440
65-746-6698
FDM

Champion Supplies Ltd.
32, Ang Mo Kio Industrial Park
Sing Industrial Complex #01-08
Singapore, Singapore 2056
65-481-7000
LOM

SPAIN

Ilarak
Jaime Munoz
Poligono Industrialdea 29
20-160 Lasarte, Spain
34-43-366-043
FDM

SWEDEN

Electrolux Rapid Development
Bo Svensson
Huskvarna, Sweden S-56181
46-36-14-63-58
SL

IVF
Urban Harrtsson
Argongatan 30
S-431 5s Molndal, Sweden
46-31-706-6000
SLS/LOM

Protech
Evald Ottosson
Rudbecks Vag 52, S-191-05
Sollentuna, Sweden
46-8-625-00-48
FDM

VTC-Seeden YSTAD
Anders Tufvesson
Dragongatan 47
S-271 44 YSTAD, Sweden
46-411-747-32
FDM

SWITZERLAND

Agil, A.G.
Otto Laubacher
Stansstaderstrasse 104
Stans, Switzerland CH-6370
41-61-614961
SL

H. Weidmann Ltd.
Jochen Ziegler
Rapperswil, Switzerland CH-8640
41-55-21-41-1
SGC

Lithopol AG
Wafler
Schachenafee 29
Aarau, Switzerland CH-5000
41-64-2696560
SL

Loosli FORMTECH AG
Aerni
Derendigenstrasse 60
Biberist, Switzerland CH-4562
41-65-425455
SL

Proform Ltd.
P. Bernhard
Rute de Fribourg 42
Marly, Switzerland CH-1723
41-37-46-43-83
SL

Von Allmen AG
Von Allmen
Witzbergstrasse 23
8330 Pfaffikon, Switzerland
41-1-950-4358
SGC

TAIWAN

CADTech Corporation
Jimmy Huang
189 San Tou 3rd Road
Kaoshung, Taiwan
886-733-590-18
SL

Erintech
Charles Jarvis
4A, 4th Floor, #250 Section 4
Chung Hsiao East Road
Taipei, Taiwan
886-277-258-55
LOM

Tatung Company
Jack J.K. Hauang
22 Chungshan N. Road
3rd Section
Taipei, Taiwan
China 104
866-2-592-5252
FDM

URUGUAY

Robtec SA
Ariel Litjtenstein
Zona Franca de Montevideo
Ruta 8 KM 17500 Montevideo
CP-12200
Uruguay
598-298-4928
SL

CHAPTER 14

Laminated Object Manufacturing™

"There's more than one way to skin a cat."

Anonymous

14.1 Company History

Helisys, Inc. was founded in 1985 with a charter to provide state-of-the-art, 3-dimensional modeling capability for a wide range of industrial applications. Early R&D activities were financed in large part by government grants, resulting in the development of the Laminated Object Manufacturing (LOM) process. Since the 1991 introduction of its LOM machines, Helisys has experienced significant growth and is among the leading manufacturers of RP&M systems in the world. LOM machines are installed and running 24 hours a day at corporate and industrial sites spanning the globe.

The LOM process, covered in a patent issued to Helisys in 1988, was one of the first to demonstrate a method to fabricate complex 3-D objects with virtually no human intervention. The patent has 80 claims[1] and covers a number of process, apparatus, and material variations, including powders, photoresists, plastics, and brazing compounds.

A European patent with virtually identical claims was issued in 1994.[2] Continued research efforts aimed at improving lamination speed and machine design have resulted in a third patent, issued in late 1994.[3]

14.2 Principles of LOM

The patented LOM process is an automated fabrication method in which a 3-dimensional object is constructed from a solid CAD representation by sequen-

By Sun Pak, Staff Scientist; Gregory Nisnevich, Systems Analyst; Jennifer Maher, Marketing Manager; and Michael Feygin, President, Helisys, Inc., Torrance, California.

tially laminating the constituent cross-sections. The process consists of three essential phases: preprocessing, building, and postprocessing.

Preprocessing

The preprocessing phase encompasses several operations. The initial steps involve generating an image from a CAD-derived STL file of the part to be manufactured, sorting input data, and creating secondary data structures. These are fully automated by LOMSlice™, the LOM system software that controls the machine and performs the slicing operation as indicated in Figure 14-1.

LOMSlice, running under Microsoft Windows NT™, receives input from an STL file (binary or ASCII), renders the image, sorts the data and—based on the STL files—creates special data structures such as the list of neighboring triangles. This list is later used by LOMSlice's on-line slicing algorithm. Orienting and merging the part(s) require the LOM system operator to optimize part orientation and, as necessary, place multiple parts into the working envelope. The tasks are aided by LOMSlice, which provides a menu-driven interface to perform transformations (translation, scaling, and mirroring), as well as merging.

Building

During the building phase, thin layers of adhesive coated material are sequentially bonded to each other and individually cut by a CO_2 laser beam as shown in Figure 14-2. The building cycle follows these steps:

1. LOMSlice creates a cross-section of the 3-D model, establishing its height and slicing it in the horizontal plane. The software then images cross-sections that define the outer perimeter. It also converts surrounding material into a support structure.
2. The computer generates instructions that guide the focused beam so that it can cut the cross-sectional outline and, subsequently, the crosshatching. The laser beam power is designed to cut a depth equal to the thickness of one layer of material at a time. After the perimeter is burned, everything within the model's boundary is freed from the remaining sheet.
3. The platform descends with a stack of previously formed layers, and a new section of material is advanced. The platform then ascends. The heated roller laminates the material to the underlying stack with a single reciprocal motion, thereby bonding it to the previous layer.
4. The vertical encoder measures the height of the stack and relays this information to LOMSlice. The software then calculates the cross-section for the next layer as the laser cuts the model's current layer. This sequence continues until all the layers are built. The product emerges from the LOM machine as a completely enclosed rectangular block containing the part. This step completes the building phase.

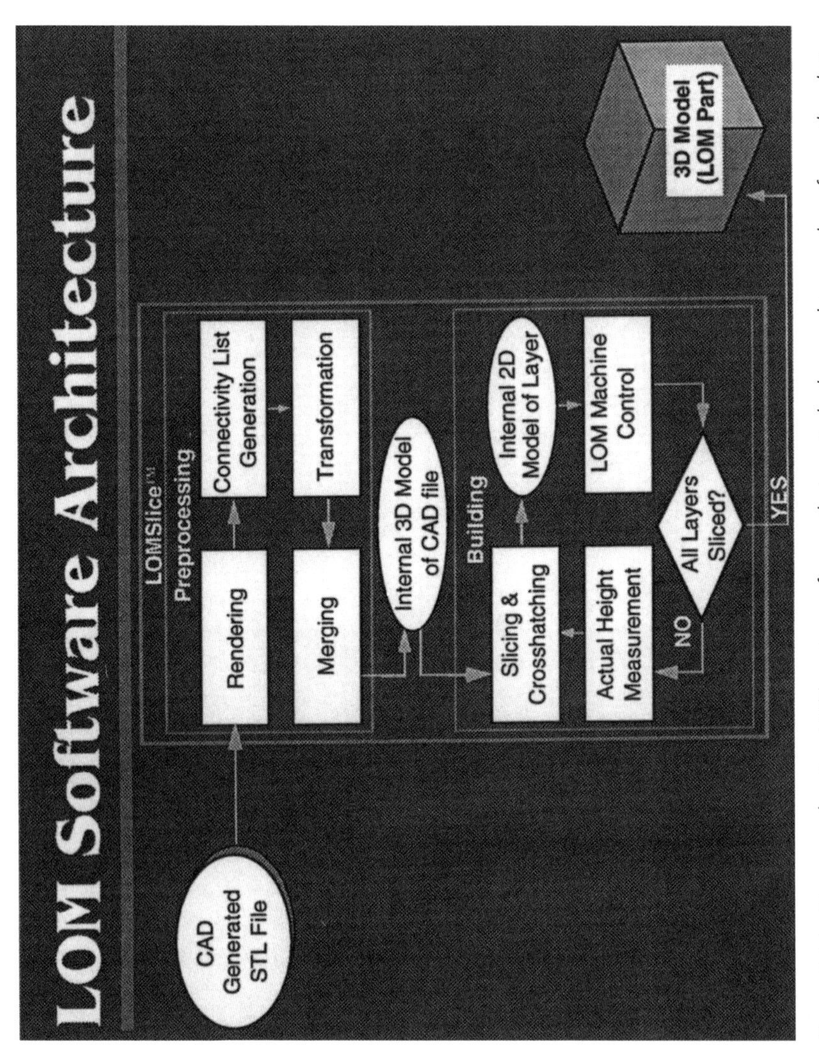

Figure 14-1. Diagram showing LOM system software that controls the machine and performs the slicing operation.

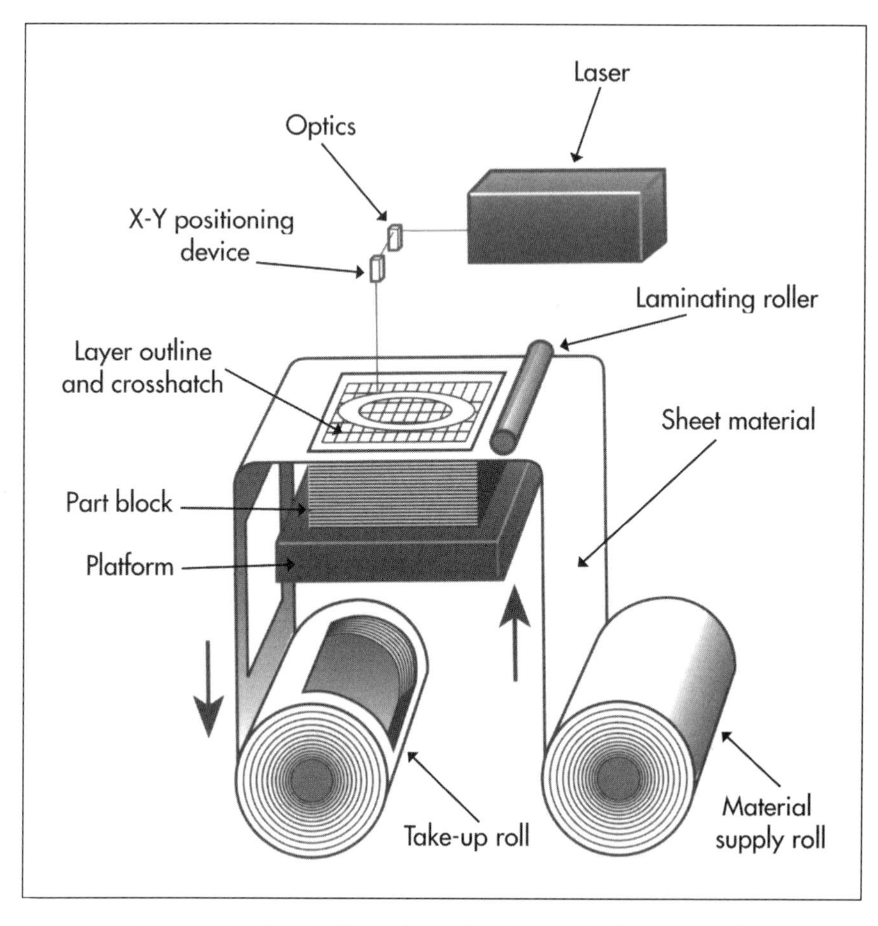

Figure 14-2. During the LOM building phase, thin layers of adhesive-coated material are sequentially bonded to each other and individually cut by a CO_2 laser beam.

Postprocessing

The final postprocessing phase of the LOM sequence includes separating the part from its support material and performing the finishing operation (Figure 14-3).

1. The metal platform and the newly created part are removed from the LOM machine. A forklift may be needed to remove the larger, heavier parts from the LOM-2030.
2. Normally, a chisel and a putty knife are the only tools required to separate the LOM block from the platform. However, a thin wire may also be used to slice through the double-sided foam tape that serves as the connecting point between the LOM stack and the platform.

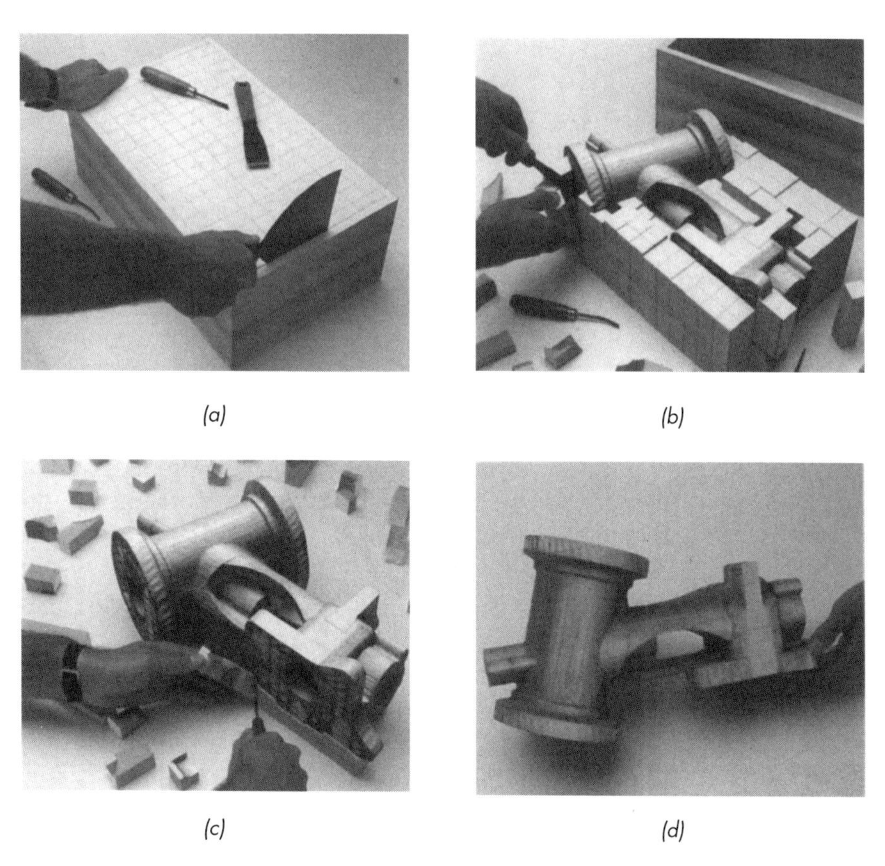

(a) (b) (c) (d)

Figure 14-3. Steps in separating the part from its support material and in performing the finishing operation: (a) laminated stack is removed from machine's elevator plate; (b) surrounding wall is lifted off the object to expose cubes of excess material; (c) cubes are separated from the object's surface; (d) object's surface can then be sanded, polished, or painted.

3. The surrounding wall frame is lifted off the block to expose the crosshatched pieces of the excess material. The excess crosshatched pieces may then be separated from the part using wood carving tools.

After the part is extracted from the surrounding crosshatched material, it should be finished. Traditional model-making techniques, such as sanding, polishing, and painting, can be used to finish the wood-like LOM part. It is strongly recommended that after the part has been separated, it be sealed immediately with urethane, epoxy, or silicone spray. Doing so reduces moisture absorption. LOM parts can be machined as though they were wood, by drilling, milling, turning, etc. If a part design is changed after building, the necessary modifications can be made by hand, using conventional woodworking methods.[4]

14.3 LOM System Structure

The LOM-1015 and LOM-2030 machines share the same basic structure, which can be broken down into several subsystems, each responsible for specific components of the manufacturing process. These are the computer hardware and software, the laser and optics, the X-Y positioning device, the platform and vertical elevator, the laminating system, and the material supply and take-up.

Computer Hardware and Software

The computer, an IBM-compatible 80486 system running Microsoft Windows NT, typically offers a 33 MHz CPU, 16 MB of RAM, and a 120 MB hard disk. The LOM software, LOMSlice, is a true 32-bit application, with a user-friendly interface, including menus, dialog boxes, progress indicators, etc., as shown in Figure 14-4.

LOMSlice is a fully integrated production tool, providing preprocessing, slicing, and machine control in a single program. The software allows a single PC to drive the entire LOM process without wasting either computing power or time. Accuracy in the Z direction is checked during the build through a closed-loop, real-time feedback mechanism, and is monitored upon the completion of each

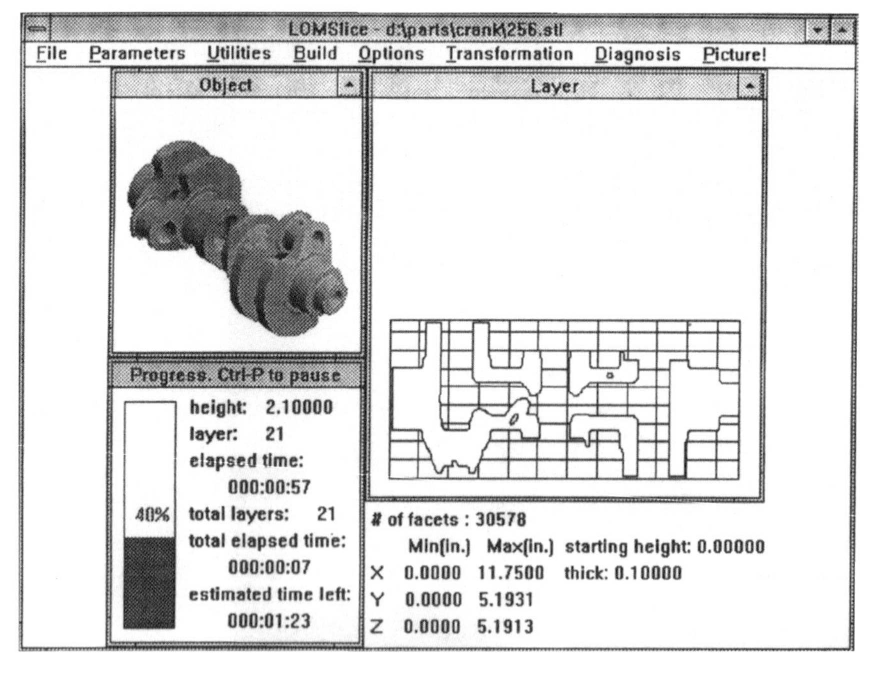

Figure 14-4. LOMSlice is a true 32-bit application with a user-friendly interface, including menus, dialog boxes, and progress indicators.

lamination. As the laser is cutting the model, the software is simultaneously planning the next layer's outline and crosshatching.

LOMSlice can also overcome STL file imperfections that violate normal vector orientation requirements, vertex-to-vertex rule specifications,[5] or even files that have incomplete surfaces (missing facets). In order to facilitate separation of the part from the excess material, LOMSlice automatically assigns reduced crosshatch size to intricate regions to enable burnout of the surrounding unwanted material.

Laser System

LOM machines use CO_2 lasers to cut the parts. The LOM-1015 uses a 25-watt laser and the LOM-2030 uses a 50-watt laser. The optical system, which delivers the laser beam to the top surface of the stack, consists of three mirrors that reflect the CO_2 beam and a lens that focuses the beam so as to cut only the uppermost layer of the stack. The diameter of the focused laser beam is about 0.010 inch (0.25 mm).

X-Y Device

The positioning table, which carries the optics, is a belt-driven X-Y plotting device allowing maximum speeds of 15 in/sec (38 cm/sec) in the LOM-1015 and 24 in/sec (61 cm/sec) in the LOM-2030, with a positioning accuracy of $\pm$0.001 inch (0.025 mm).

Z-axis Device

The laminated part is built on a platform capable of vertical incremental movements driven by a stepper motor. The platform features a precision ball screw (an Acme screw in the LOM-2030 to avoid backlash). This combination of LOM features delivers a reported platform position-accuracy of 0.001 inch (0.025 mm) per foot of part height.

Lamination System

Lamination is accomplished by applying both heat and pressure using a heated cylinder that rolls across a sheet of material coated with a thin layer of thermoplastic adhesive on one side. Recent studies have indicated that the interlaminate strength of LOM parts is a complex function of bonding speed, sheet deformation, roller temperature, and contact area between the paper and the roller.[6]

Specifically, by increasing the pressure of the heated roller, lamination is improved because air pockets are eliminated. Increased pressure also augments the contact area, thereby bolstering lamination speed. Pressure is controlled by a limit switch mounted on the heated roller. If compression is set too high, however, it may cause distortion in the part, and the system would then need to be adjusted.

In addition to ensuring appropriate pressure, maintaining a steady temperature within the LOM system is equally critical. To avoid temperature fluctuations, a thermocouple is built into the heated roller and monitored by a closed-loop feedback mechanism that maintains the temperature within ±35° F (±19° C), regardless of variations in the surrounding environment.

Material Supply and Take-up

The material supply and take-up system is comprised of two material roll supports (supply and rewind), several idle rollers to direct the material, and two rubber-coated nip-rollers (driving and idle). These rollers advance or rewind the sheet material during the preprocessing and building phases. The driving nip-roller is set into motion by a stepper motor, and the supply roll is powered by an AC motor that maintains a prescribed amount of tension on the material.

14.4 LOM Materials

Potentially, any sheet material with adhesive backing can be utilized in LOM. R&D at Helisys has demonstrated that plastics, metals, and even ceramic tapes can be used. However, the most popular material has been Kraft paper backed with a polyethylene-based heat seal adhesive system because it is widely available, cost effective, and environmentally benign.[7]

The engineering properties of parts fabricated with standard LOMPaper™ are more than adequate as models for numerous secondary fabrication processes. The low coefficient of thermal expansion of LOMPaper makes it attractive for investment casting.[8] Table 14-1 summarizes typical engineering properties of this material.

Sheet material and its adhesive can be readily tailored to suit specific applications. As an example, for increased lamination speed at low temperatures, a heat seal adhesive with a low softening point and viscosity is used. However, when fabricating molds for composite tape lay-up, where lamination temperatures and pressures exceed 250° F (121° C) and 100 psi (689 kPa) respectively, a special adhesive that resists deformation at elevated temperatures is used.

14.5 LOM Process Highlights

Helisys offers two models of its LOM systems. Models LOM-1015 and LOM-2030 have working envelopes of 10 inches W × 14.5 inches L × 14 inches H (25.4 cm × 37 cm × 35.5 cm) and 22 inches W × 32 inches L × 20 H (56 cm × 81 cm × 51 cm) respectively. LOM systems operate in both office and manufacturing environments. They are user-friendly, environmentally safe, and employ materials that require no special disposal facilities.

Materials

Paper, plastics, metals, and composites and ceramics (see Figure 14-5) can be used in sheet form in LOM systems. Commercial availability of different sheet materials allows users to vary the type and thickness of the manufacturing material for their specific applications. Paper is the simplest and least expensive material, producing rigid, durable parts that have characteristics similar to wood. Special

Table 14-1. Engineering Properties of Models Fabricated with LOM Paper

Property	Value
Density	1.449 g/cc
Specific heat, in-plane, 77° F (25° C)	1.455 J/g °C
Specific heat, transverse, 77° F (25° C)	1.410 J/g °C
Specific heat, in-plane, 122° F (50° C)	1.786 J/g °C
Specific heat, transverse, 122° F (50° C)	1.686 J/g °C
Thermal diffusivity, in-plane, 66° F (19° C)	0.00107 cm^2/sec
Thermal diffusivity, transverse, 66° F (19° C)	0.000344 cm^2/sec
Thermal diffusivity, in-plane, 212° F (100° C)	0.0023 cm^2/sec
Thermal diffusivity, transverse, 212° F (100° C)	0.00073 cm^2/sec
Thermal conductivity, in-plane, RT	0.002256 W/cm °K
Thermal conductivity, transverse, RT	0.000703 W/cm °K
Thermal conductivity, in-plane, 194° F (90° C)	0.005954 W/cm °K
Thermal conductivity, transverse, 194° F (90° C)	0.001784 W/cm °K
Coefficient of thermal expansion, in-plane, RT - 140° F (60° C)	13.1×10^{-6}/ °C
Coefficient of thermal expansion, transverse, RT- 151° F (66° C)	15.1×10^{-5}/ °C
Glass transition temperature	194° F (90° C)
Tensile strength, in-plane, RT	66 MPa (9.5 kpsi)
Tensile modulus, in-plane, RT	6.7 GPa (971 kpsi)
Tensile strain at failure, in-plane, RT	2%
Compressive strength, in-plane, RT	26 MPa (3.8 kpsi)
Compressive strength, transverse, RT	3.9 MPa (0.57 kpsi)
Compressive modulus, in-plane, RT	9.3 GPa (1350 kpsi)
Compressive modulus, transverse, RT	814 MPa (118 kpsi)
Compressive strain at failure, in-plane, RT	1%
Compressive strain at failure, transverse, RT	12.9 %
Ash content	2.7 to 3.1%

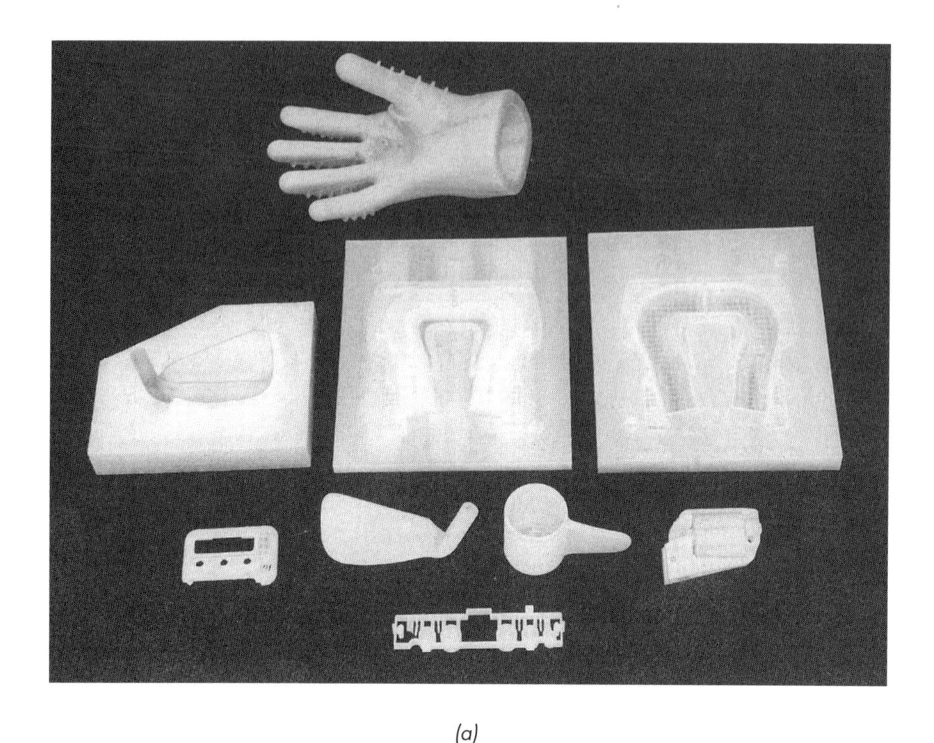

(a)

(b)

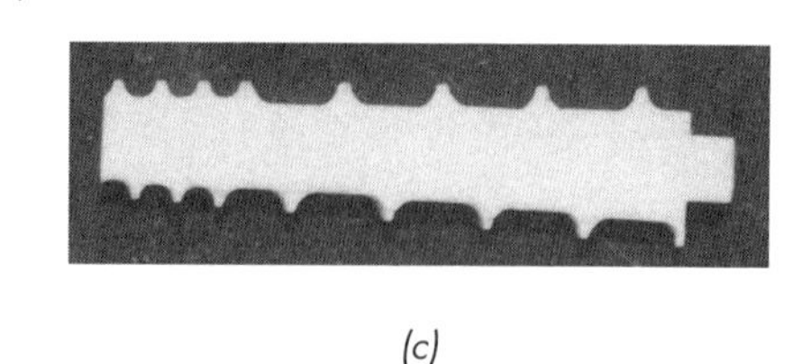

(c)

Figure 14-5. Paper and plastics (a); metals (b); and composites and ceramics (c), can be used in sheet form in LOM systems.

adhesive pre-impregnated composite materials, ceramic tapes, and sheet metals, as well as certain plastic films, are currently under development and are targeted to produce high strength, functional prototypes, and small batches of working components.

Several significant benefits are derived from using commercially available sheet materials. The LOM process spends no time converting expensive and, in some cases, toxic liquid polymers to solid plastic, or powders into sintered objects. Because sheet materials are not subjected to either physical or chemical phase changes, the finished LOM parts do not experience shrinkage, warpage, or internal stress from these causes.

Accuracy

The feature-to-feature accuracy that can be achieved with LOM machines is usually better than 0.010 inch (0.25 mm) across the entire working envelope. Through the design and selection of application specific parameters, LOM customers have achieved higher accuracy levels in the X-Y dimensions. Accuracy in the Z dimension is typically not as good as that in X and Y.

If, during lamination, the layer does shrink horizontally, there is no actual distortion because the contours are cut postlamination, and laser cutting itself does not cause shrinkage. Furthermore, if the layers shrink vertically, a closed-loop feedback system provides the true cumulative part height upon completion of each lamination, and the software then slices the 3-D model in the horizontal plane at the appropriate location. Other RP&M processes calculate the thickness of each layer based on empirical data prior to initiating fabrication. With these techniques, special leveling compensation routines are needed to correct the parameters during a run should there be any deviation between the calculated and the actual results.

Because LOM material has proven dependable in the LOM working environment, producing precise parts hinges upon the accuracy of the X-Y laser positioning, maintaining correct Z-axis measurements, and high resolution CAD data files. The LOM system uses an X-Y positioning table to guide the CO_2 laser beam, which is monitored throughout the build process by the closed-loop, real-time motion control system, resulting in an X-Y accuracy of $\pm$0.010 inch ($\pm$0.25 mm) regardless of part size.

The Z-axis is also controlled using a real-time, closed-loop feedback system. It measures the cumulative part height at every layer deposition and then slices the CAD geometry at the correct location. Finally, because the laser cuts only the perimeter of a slice, there is no need to translate vector data into raster form. Consequently, the accuracy of the cutting procedure depends on the resolution of the CAD model and the STL file tessellation characteristics, including the selected chordal deviation.

Speed

Unlike other RP&M technologies, the laser in the LOM process need not scan the entire surface area of each cross-section. Rather, it has only to outline the object periphery. Therefore, parts with thick walls are built just as quickly as those with thin walls, making the LOM process especially advantageous for producing large and bulky parts, as well as patterns and two-part negative tooling.

The speed of the LOM system is limited only by the availability of a faster X-Y positioning system, because the power of CO_2 lasers used in the LOM machines is sufficient (up to and beyond 200 watts) to achieve the maximum available X-Y positioning system translational speed.

Simplicity

The simplicity of the LOM process and systems allow them to be practical extensions of many manufacturing and design environments. LOM machines are like peripheral devices to CAD workstations allowing the designers to output directly to a LOM system. The need to create additional support structures is eliminated with the LOM process, an essential component for some RP&M systems. The LOM process is straightforward and the user needs no specific knowledge of chemistry, physics, mechanics, or electronics in order to operate the machine. Its inherent user friendliness results in a short learning curve, even for those new to LOM processing.[9] The systems are designed using standard electromechanical components, which makes them easy to maintain and troubleshoot.

Because no physical or chemical phase transformations are involved and no special manufacturing conditions are required by the LOM process, the build operation can be paused at any instant (e.g., to remove excess material) and then continued without sacrificing LOM part quality.

LOM systems are integrated with an IBM-compatible 80486-based computer and can be connected to an existing network of CAD stations or other computer driven equipment. The software that processes the data and controls LOM systems runs under the Microsoft Windows environment, making the system accessible to numerous personnel within an organization.

14.6 Advances in LOM

Helisys invests a major portion of its revenue into research and development. Always seeking improvements to continue offering customers a competitive edge, Helisys has improved performance of the LOM-1015 and LOM-2030 systems in several key areas: laser optics, motion control, lamination, and software.

Laser Optics

Among the second generation enhancements, Helisys has made it significantly easier, faster, and safer to align the laser beam. By implementing a visible HeNe laser, which projects a red beam of light collinear with the high power but invisible infrared CO_2 laser beam, the operator can switch on the red laser beam and watch as the mirrors are aligned, rather than using trial and error with the powerful CO_2 beam.

Motion Control

With a state-of-the-art closed-loop motion control system, LOM machines receive feedback on every machine access. The encoders track and record the motions of several elements: the laminated roller as it moves back and forth across the material, the Z-axis platform raising and lowering the stack, and the sheet material as it is being fed into the system. Because the motion control system has its own processor, the feedback is offered in real-time. This assures the operator that the system is running smoothly. If the sensor does detect any inconsistencies, the LOM system will pause or halt activities until the problem is resolved, ensuring an accurate build.

Lamination

In order to maintain uniform lamination across the entire working envelope, it is critical that the temperature remain constant. The new temperature control system, also utilizing closed-loop feedback, holds the system's temperature fixed within ±35° F (±19° C), regardless of variations in the surrounding environment. The introduction of mechanical nip rollers allows the material to flow through the LOM systems more smoothly. The friction resulting from compressing moving material between rubber-coated rollers on both the feed and wind mechanisms ensures a clean feed and avoids jamming. An embedded encoder alerts the operator when the paper supply has been depleted.

Software

The upgrade to the 32-bit Windows NT environment dramatically increases LOM processing speed and power, offering improved performance throughout the LOM process. Real-time calculations result in processing that is 100 to 120 times faster than in the original 16-bit environment. The NT software architecture allows multiple parts to be computed quickly despite complex geometries, and easily supports large databases from 50 to 100 Mbytes.

Grant Awards

In late 1994, Helisys won a 3-year ARPA (Advanced Research Project Agency) contract for upward of $3,500,000 to develop a solid-freeform fabrication method, based on the Helisys LOM process. The resulting technique would manufacture fully functional ceramic matrix composite parts with virtually no human intervention. The process, referred to as LOM/COM, is currently under development. The project team consists of research centers, material suppliers, federal agencies, and end users, including Wright-Patterson Air Force Base, The University of Dayton Research Institute, and Advanced Refractory Technologies, Inc. Helisys has also won an R&D contract from the U.S. Army to develop improved RP models that could be used for extensive functional testing.

R&D Activities

Helisys is committed to continued innovations and engineering advances. The focus of current research activities is in accuracy improvements through software modifications and laser spot size reduction. Establishing a thorough understanding of residual stress through mathematical modeling and simulation studies will aid in understanding associated issues in the LOM process. Methods to further increase the bonding and laser cutting speeds are also currently under examination. The latest laboratory results indicate that by modifying the chemical makeup of the adhesive, a factor of two or more improvement in interlaminate bonding strengths and almost a 6-fold increase in bonding speed are possible, as indicated in Table 14-2.

Table 14-2. Improvements Gained Through Adhesive/Sheet Chemical Modifications

Property	Standard LOM Paper/adhesive	New Paper/adhesives
Interlaminate Bond Strength	0.49 - 0.63 lb/in	1.26 lb/in
Minimum Bonding Temperature	150° F (66° C)	125° F (52° C)
Maximum Use Temperature	180° F (82° C)	> 315° F (157° C)
Maximum Bonding Speed	2 in/sec (5.1 cm/sec)	11 in/sec (27.9 cm/sec)

14.7 Applications of Laminated Object Manufacturing

Laminated Object Manufacturing's industrial applicability is virtually limitless, spanning a variety of fields including equipment for aerospace, automotive, consumer products, medical devices including prostheses, and reaching into the research and development arena for even greater product possibilities. LOM parts are ideal in design applications where it is important to visualize what the final piece will look like, or to test for form, fit, and function, as well as in a manufacturing environment to create prototypes, generate production tooling, or even produce low-volume finished goods.

Design Visualization

Many companies use LOM's ability to fabricate a model of a potential product purely for visualization. A LOM part's wood-like composition allows it to be painted or finished as a true replica of the product. Because the LOM procedure is inexpensive, several models can be created, giving sales and marketing executives an excellent opportunity to utilize these prototypes for consumer testing, product marketing introduction, packaging studies, samples for vendor quotations, and many others.

Form, Fit, and Function

In addition to viewing the prototype as a finished product, LOM parts also lend themselves to design verification or performance evaluation. In low-stress environments, LOM parts can withstand basic tests, giving manufacturers the chance to make changes, as well as to evaluate the aesthetic properties of the prototype in its total environment. Again, the wood-like qualities are especially useful because the LOM part can be modified easily to create a close fit, if the original design warrants changes.

Prototypes and Low-volume Manufacturing

Typically, a company must see a prototype before spending hundreds of thousands of dollars on full-scale production. As with other RP&M systems, LOM speeds the entire product development cycle by slashing days to weeks off the typical schedule. LOM part composition, based on the sealant or finishing products used, can be further machined for use as a pattern or mold for various secondary tooling techniques, including investment casting, sand and plaster casting, injection molding, silicone RTV molding, and vacuum forming. LOM parts offer several advantages important for the secondary tooling process, namely: predictable accuracy across the entire part; stability and resistance to shrinkage, warpage, and deformity; and the flexibility to create a master or a mold.

In many industries a master created through secondary tooling, or even when the LOM part serves as the master (e.g., vacuum forming), withstands enough injections, wax shootings, or vacuum pressure to enable a low production run up to 1,000 pieces. The surface finish of the LOM part allows molds to retain detailed aspects of a model and, depending on the production tooling used, pass it on to the finished product.

Production Tooling

The following are techniques to produce metal and plastic components using LOM as a pattern, mold, or master for tooling. In principle, any process using wooden or plastic patterns can be duplicated using LOM parts. The stability, accuracy, and cost effectiveness of LOM patterns result in substantial time and money saved in the production of metal components.

Investment Casting. There are two investment casting applications of LOM, direct and indirect. As RP&M is integrated into production, foundries are adopting new techniques. Among them is the "lost paper" shell investment casting process. Here, LOM parts are used for low quantity production of metal castings. This new process converts the paper master into a mold by burning the LOM part from within a ceramic shell. LOM's stability and lack of expansion during the burnout cycle helps prevent cracks in the ceramic shell. Metal is then poured into the ceramic shell, which is then broken away to expose a single metal component. This method of direct investment casting is applicable to complex parts. Using traditional methods, wax is injected into either aluminum or steel investment casting dies. Because wax is inexpensive, thousands of highly accurate components can be manufactured at a reasonable cost. Investment casting techniques are also used in the manufacture of two-part tooling, called indirect investment casting. Here, LOM systems create two-part negative molds to generate wax patterns. The mold is then usually sealed with a protective coating, and can be used to manufacture up to 100 parts. In the event that production requires several hundred components to be fabricated, an LOM master is used to create inexpensive metal-filled epoxy tooling for wax injection. This epoxy tooling lasts for up to 1,000 parts.

Sand Casting. Sand casting is used for production of metal components when accuracy is not critical. This is a high-volume production technology that requires stable foundry patterns, cores, and core boxes. The LOM process is well suited for creating the often large, bulky patterns and cores used in sand casting. When up to 100 components are needed, the LOM parts can also be finished, sealed, painted, and used directly to create impressions in the sand. If thousands of components are needed, the LOM patterns and cores can be converted to hard plastic permanent patterns via the rubber molding process. For high-temperature and high-pressure thermoplastic sand injection, LOM patterns and cores can be converted to aluminum or steel patterns via investment casting.

Rubber Plaster Casting. Similar to sand casting, rubber plaster casting is used to manufacture complex aluminum components that have better surface finish than sand cast components. This process is employed in applications that simulate die casting. LOM parts are used in this technique as patterns around which a rubber sheet is pulled. The plaster is then packed against the rubber sheet, which is subsequently removed. Metal is poured into the plaster cavity, later broken open to expose the metal part.

Silicone Rubber Molding. Silicone rubber molding is a quick process used to create duplicate plastic components using a LOM master. Room temperature vulcanizing (RTV) silicone is poured around a LOM master and after setting, the mold is cut along a parting line. The LOM master is removed and the plastic polyurethane or epoxy is poured into the mold. Several components per day can be created with this technique.

Vacuum Molding. Vacuum molding is similar to silicone RTV rubber molding, but differs in that the silicone is cast under both vacuum and pressure. This results in better mold quality, as well as improved surface and feature definition. Pressure injection under vacuum produces denser components. Through traditional vacuum forming methods, LOM patterns can be used to create prototype parts in a wide variety of plastic materials including polycarbonate, ABS, polypropylene, and others.

Applying vacuum and heat to a plastic sheet, the latter is formed to the shape of the LOM pattern. The strength of LOM objects makes them well suited to withstand exposure to high compression forces. In addition, the durability and rigidity of LOM models make them suitable for low-to-medium-volume prototype and production runs. Further, LOM parts serve well for this process because they do not react with silicone.

Spray Metal Tooling. With spray metal tooling, a low melting point metal is used to coat the surface of a LOM master. The resultant tooling is used for short run prototype injection molding. Capable of producing up to 1,000 injection molded parts, LOM is also well suited to this application. Almost all thermoplastic materials can be used with this type of prototype tooling.

Prototype and Low-production Injection Molding. Typically, prototype injection molding is accomplished using machined aluminum molds. The reasonably accurate nature of the investment casting process, combined with the ability to create two-part solid CAD negatives, allows the creation of aluminum molds using investment cast LOM patterns. These molds are subsequently polished and machined as necessary, and can then be used for injection molding. In some cases, these tools may accommodate the total production of plastic components.

Rapid Tooling. Two-part negative tooling can be created with LOM systems. Since the material is solid and inexpensive, bulky complicated tools are cost effective to produce. These wood-like molds can be used for injection of wax, epoxy, or other low-pressure and low-temperature materials. Also, the tooling can be converted to aluminum or steel via the investment casting process for use with other materials and high temperature molding processes.

14.8 Summary and Conclusions

As today's global market offers an increasingly wide range of products at competitive prices, companies offering these products are subjected to far greater competitive pressures. Companies that want to compete in the global marketplace must constantly seek ways to develop their products in an increasingly faster, more cost-effective manner.

Turning ideas into prototypes in a matter of hours, LOM accelerates product development cycles and reduces time to market. Bringing RP&M in-house, manufacturers gain tighter control over product costs, design schedules, proprietary information, and specific build parameters such as part resolution, orientation, and materials.

Helisys' LOM systems are well suited for today's most demanding RP&M applications. By using inexpensive, solid-sheet materials, LOM models offer the advantage of material safety, resistance to deformity, and extreme cost effectiveness regardless of geometry. Whether engineers require models for visualization, form, fit, and function, or secondary tooling purposes, LOM systems can rapidly produce complex 3-dimensional objects.

LOM systems can be successfully implemented to serve the rapid development and manufacturing needs of a company and its products. The sometimes intangible benefits of improved design, improved quality, and faster market introduction, coupled with quantifiable cost and time savings to produce plastic or metal components, enable a company to gain the most valuable asset: a competitive edge.

References

1. Feygin, M., "Apparatus and Method for Forming an Integral Object from Laminations," U.S. Patent No. 4,752,352, June 21, 1988.
2. Feygin, M., "Apparatus and Method for Forming an Integral Object from Laminations," European Patent No. 0 272 305, March 2, 1994.
3. Feygin, M., "Apparatus and Method for Forming an Integral Object from Laminations," U.S. Patent No. 5,354,414, October 11, 1994.
4. Rapid Prototyping Report, January 1994, Volume 4, Number 1, CAD/CAM Publishing, Inc., San Diego, CA.
5. Burns, M., *Automated Fabrication*, Prentice Hall, 1993, p. 228.
6. Pak, S. and Nisnevich, G., "Interlaminate Strength and Processing Efficiency Improvements in Laminated Object Manufacturing," Proceedings of the Fifth International Conference on Rapid Prototyping, University of Dayton, Dayton, OH, June 12-15, 1994, pp. 171-180.
7. Wood, L., *Rapid Automated Prototyping: An Introduction*, Industrial Press, Inc., New York, NY, 1993, p. 98.
8. Sarkis, B., "Rapid Prototype Casting (RPC): The Fundamentals of Functional Metal Parts from Rapid Prototype Models Using Investment Casting," Proceedings of the Fifth International Conference on Rapid Prototyping, University of Dayton, Dayton, OH, June 12-15, 1994, pp. 291-300.
9. Pak, S., "Laminated Object Manufacturing," *Medical Device & Diagnostics Industry*, November 1994, pp. 47-48.

Index

D

T